T0143173

Healing the Land and the Nation

Healing the Land and the Nation

Malaria and the Zionist Project in Palestine, 1920–1947

SANDRA M. SUFIAN

THE UNIVERSITY OF CHICAGO PRESS   CHICAGO AND LONDON

SANDRA M. SUFIAN is assistant professor of medical history and humanities at the
University of Illinois–Chicago College of Medicine.

The University of Chicago Press, Chicago 60637
The University of Chicago Press, Ltd., London
© 2007 by The University of Chicago
All rights reserved. Published 2007
Printed in the United States of America

16 15 14 13 12 11 10 09 08 07   1 2 3 4 5

ISBN-13: 978-0-226-77935-5 (cloth)
ISBN-10: 0-226-77935-1 (cloth)

Library of Congress Cataloging-in-Publication Data

Sufian, Sandra M. (Sandra Marlene)
    Healing the land and the nation : malaria and the Zionist project in Palestine,
1920–1947 / Sandra M. Sufian.
        p. ; cm.
    Includes bibliographical references and index.
    ISBN-13: 978-0-226-77935-5 (cloth : alk. paper)
    ISBN-10: 0-226-77935-1 (cloth : alk. paper)   1. Malaria–Palestine–History–20th
century.   2. Malaria–Israel–History–20th century.   3. Zionism–Palestine–History–
20th century.   4. Zionism–Israel–History–20th century.   I. Title.
    [DNLM: 1. Malaria–prevention & control–Middle East.   2. Colonialism–history–
Middle East.   3. Culture–Middle East.   4. History, 20th Century–Middle East.
5. Politics–Middle East.   6. Public Health–history–Middle East. WC 765 s946h 2007]
    RA644.M2S84   2007
    614.5′32095694–dc22   2007015180

TO MY PARENTS, DAVID AND BEVERLY SUFIAN, AND TO MY HUSBAND, ELIAS MURCIANO ATTIAS, FOR GETTING THROUGH STORMS AND SAVORING THE BREAK OF DAY.

Contents

Figures and Tables

Figures

Tables

Measures and Currency

Measures

1 Turkish dunam	=	919.3 square meters
1 metric dunam	=	1,000 square meters or 1/4 acre
1 acre	=	0.40 hectares or 4 metric dunams
1 square mile	=	640 acres or 2,560 metric dunams
1 square kilometer	=	1,000 dunams

Currency

Until 1927, the monetary unit in Palestine was the Egyptian pound (£E).
In 1927 the Palestine pound (£P) was introduced.

1 Egyptian pound	=	1,000 millimeters
	=	100 piastres
	=	£1.0s. 6d. sterling
1 Palestine pound	=	1,000 mils = £1 sterling
	=	97 ½ Egyptian piastres
	=	5 U.S. dollars

Sources: Warwick P. N. Tyler, *State Lands and Rural Development in Mandatory Palestine, 1920–1948* (Brighton and Portland: Sussex Academic Press, 2001), ix; Kenneth W. Stein, *The Land Question in Palestine, 1917–1939* (Chapel Hill: University of North Carolina Press, 1984), xxi.

Acknowledgments

Just as the land of Palestine has undergone significant ecological, environmental, and demographic changes, so too has this book undergone a long process of evolution subject to many challenges, reexaminations, and reconstructions. Throughout its development, this manuscript has been strengthened by the input of many supportive colleagues and friends.

The seeds of this project began ten years ago as a graduate paper for Talal Asad's course on the history of colonialism at the New School for Social Research. It then quickly blossomed into a full-fledged dissertation project at New York University (NYU). I owe a great deal of thanks and gratitude to my dissertation advisor, Zachary Lockman, for supporting what, at the time, seemed like an unlikely field of inquiry for the history of Palestine and a relatively untraditional methodology of mixing history of medicine with the history of the modern Middle East. I am also grateful to the rest of my dissertation committee, Lila Abu-Lughod, Michael Gilsenan, Tim Mitchell, and Salim Tamari, for demanding scholarly rigor and, in particular, for raising questions about the historical agency of nonhumans. Samira Haj deserves particular mention as well for she taught me the long-lasting skill of how to read and analyze primary and secondary historical texts.

A historian cannot do her work without texts; it is food for the project's soil and the author's soul. During the research period of this project, I received incalculable help from the staff at the Central Zionist Archives, Jerusalem; the Israel State Archives, Jerusalem; the Histadrut Archives, Tel Aviv; the Keren Keyemet L'Israel Photo Archives, Tel Aviv; the Jabotinsky Archives, Tel Aviv; the Rav Kook Archives, Jerusalem; the Yad Ben Tzvi Archives, Jerusalem; the Public Records Office, London; the St. Antony's Archives, Oxford; Hadassah Archives, New York; the

Joint Distribution Archives, New York; the U.S. National Archives, Washington, DC, and the Rockefeller Foundation Archives, Tarrytown. I was also aided by staff at the Rhodes House, Oxford; the Hebrew Union College Library, New York; the New York Academy of Medicine Library, New York; the New York Public Library, New York; the College of Physicians of Philadelphia Library, Philadelphia; the Ein Kerem Medical Library, Jerusalem; the Hebrew University Library at Mt. Scopus, Jerusalem; the Islamic Center at Abu Dis, the National Library of Israel, Jerusalem; the Truman Institute for Peace Library, Jerusalem; and the Orient House Library, Jerusalem. My deep thanks go to all of the professionals working at these locations for easing the arduous work of historians.

Hagit Lavsky and Shifra Shvarts provided incomparable advice during the research and writing phases of this project. Their knowledge of health services and the Zionist project in Palestine helped clarify and remind me of the wider context in which this study is situated. My decade-long correspondence and conversations with Dr. Aftim Acra and my interviews with several key physicians and malariologists of the Mandate period offered vivid recollections of Arab and Jewish health conditions in Palestine during British rule that many documents could not capture.

I am privileged to have many colleagues who are also friends. They have provided me with the balance so central to the ecology of academic life. This project would not have taken its unique shape without the scholarly advice and emotional support of my dearest friend Eugene Sheppard and my good friend Mark LeVine. Both are exceptional scholars in their own right, and I am lucky to have been able to have access to their wisdom and humor throughout these years. The esteemed U.S. history and disability scholar Paul Longmore has continually provided exceptional advice and help in making this project understandable to a larger audience and also opened my eyes to the vast uses and value of disability theory and history. I owe a special debt of gratitude to Sylvia Fuks-Fried who immediately saw the value of this project and gave me precious suggestions and guidance during the writing and submission phases. Tom Abowd, Michael Adas, Samer Alatout, Martin Bunton, Ellen Amster, Zvi Ben-Dor, Avner Ben-Zaken, Jennie Brier, Licia Carlson, Nadav Davidovitch, Geremy Forman, Tamara Giles-Vernick, Douglas Haynes, Salmaan Keshavjee, Julie Livingston, Neil Maher, Michael Osborne, Shobita Parthasarathy, Stephen Pemberton, Derek Penslar, Gayatri Reddy, Adam Rubin, Gershon Shafir, Bob Vitalis, Keith Wailoo, Amy Zalman, Yael Zerubavel, and others have all offered their scholarly expertise, critiques, and friendship throughout the various phases of this project. They have been smart enough not to question that there actually was malaria in Palestine in the interwar period (why else would I be writing about it?) and to understand that health issues do play a role in nationalist movements. My colleagues at the 2000 National Endowment for the Humanities (NEH) Summer Institute hosted by the San Francisco State University Institute on Disability and at Rutgers Center for Historical Analysis have all impacted my thinking in complex ways and have exemplified the fact that it is possible to maintain integrity, collegiality, and friendship while also striving for intellectual rigor. To all of you, I am extremely appreciative.

Sincere thanks go to Ted Brown for first taking an interest in this project and for always supporting my professional path and decisions. I credit Mitch Greenlick with providing me the tools in public health and health services research in order to rethink the dissertation and make it a more sophisticated book that speaks to audiences in both history and public health. Ken Stein first ignited my passion for the period of Mandate Palestine as an undergraduate and has been a confidante during the past twenty years. In my late graduate and more recent professional years, Sander Gilman has endlessly encouraged my perhaps idiosyncratic professional direction, has given me invaluable advice on academia and publishing and has truly understood my scholarly intentions and diverse intellectual passions.

The past five years at the University of Illinois–Chicago have introduced me to a whole new set of scholars and colleagues with diverse interests. Les Sandlow, Suzanne Poirier, and my other colleagues in the Department of Medical Education have afforded me their utmost encouragement and patience during the final writing stages of this book. My assistants and friends, Sara Vogt and Mike Gill, have tirelessly helped with the various drafts and bibliographic work while offering important insights into its content and style. My student assistants, Julia Geynisman, Asim Farooq, and Ahmed Khan, helped with bibliographic work and commented on revised drafts while learning about the importance of the history of medicine for their own careers in clinical medicine. Colleagues in the Department of History and the Department of Disability and Human Development have also been generous in recognizing the contributions this project makes to the history of medicine, history of the Middle East, Jewish Studies, and Disability Studies.

I am deeply indebted to Catherine Rice for acknowledging the worth of this project for publication, especially in such a competitive academic market. She was a hands-on editor with excellent insights and suggestions for framing the book. I thank her for being so consistently responsive to my questions and needs. Pete Beatty and new history of science editor Karen Darling at the University of Chicago Press, as well as manuscript editor Maia Rigas, assisted me tremendously in putting the final touches on the manuscript. Gaia Guirl-Stearley provided priceless recommendations for editing and strengthening the manuscript.

I must thank all my friends in Chicago, New York City, Boston, Berkeley, Philadelphia, Jerusalem, and Tel Aviv, as well as other places in the United States and around the world, for supporting and sustaining my mental health during what is a very long process of writing an academic book. Some have been generous enough to read drafts of my manuscript and have provided excellent recommendations for improving clarity and interest. My doctors, Drs. Eitan Kerem, Patricia Walker, Michael Jones, and Manu Jain, and my nurse and physical therapist have made sure that I retained my health to complete this project and hopefully other books in the future.

A special word to my parents, David and Beverly Sufian, and to my family, Beth Sufian, Aviva Sufian, James Passamano, Isabella Passamano, Leila, and my new family, the Murcianos, for their constant support, understanding, and patience.

One of my longtime dreams has been to be able to acknowledge a life partner in the first book that I wrote. I feel lucky that I can do that now. My deepest gratitude of all, therefore, goes to my husband, Elias Murciano Attias, who appreciates my love for my subject and my drive to achieve and who understands what it means to be a caring and devoted researcher and teacher from the example of his mother. Not only did he live with me during the writing stage of this project and survived, but he has added joy, laughter, music, dance, and happiness to my life.

This project has been funded during its various stages by many fellowships and grants: Fulbright-Hays Dissertation Fellowship, SSRC Dissertation Fellowship, NYU Dean's Dissertation Fellowship, Erna Wolff Memorial Scholarship of Jewish Endowment Foundation of New Orleans, Woodrow Wilson Women's Health Fellowship, Francis Clark Wood Institute for the History of Medicine of the College of Physicians of Philadelphia Fellowship, Memorial Foundation for Jewish Culture Fellowships, and the NEH Fellowship for University Professors. The Littauer Foundation has generously provided a subvention grant for publication.

Abbreviations

Archival and Bibliographical Citations

CZA   Central Zionist Archives, Jerusalem
ISA   Israel State Archives, Jerusalem
JDC   Joint Distribution Committee Archives, New York City
PRO   Public Records Office, London

Organizations

EMICA   ICA and the Palestine Emergency Fund
HMO   Hadassah Medical Organization
HMG   His Majesty's Government
ICA   Jewish Colonization Agency
JA   Jewish Agency for Palestine
JNF   Jewish National Fund
KKL   Keren Keyemet L'Israel (Jewish National Fund)
MRU   Malaria Rescarch Unit
MSS   Malaria Survey Section
PICA   Palestine Jewish Colonization Agency
PLDC   Palestine Land Development Company
PZE   Palestine Zionist Executive
RBM   Roll Back Malaria
SPNI   Society for the Protection of Nature in Israel
WHO   World Health Organization
ZO   Zionist Organization

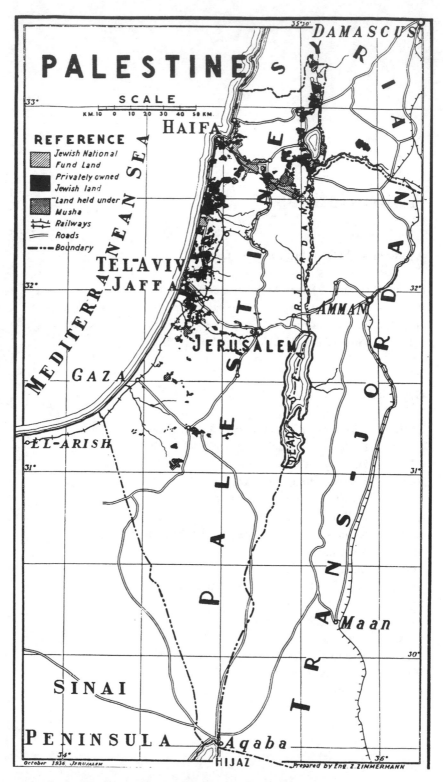

Map of Palestine, October 1936. Prepared by Engineer Z. Zimmermann.

A History of Malaria and Zionist Nationalism in Mandatory Palestine

Drainage work is the most important project of our time in Palestine. —Central Executive of Keren Keyemet L'Israel[1]

In the 1990s, fertilizer nitrates began to pollute Lake Tiberias, Israel's largest water reservoir. Deprived of water, the peat soil in the Huleh Valley, an area in northern Israel adjacent to Lake Tiberias, began to spontaneously combust, and the ground began to sink. Dust storms eroded valuable agricultural lands.[2] These environmental problems prompted the Israeli government between 1992–1994 to invest and reflood part of the old Huleh Lake environment in the hope that the original flora and fauna would return and restore the area's ecological balance. Indeed, the Huleh region was and still is one of the world's most traveled migration routes for birds. Ducks, storks, herons, cranes, cormorants, pelicans, other waterfowl, and fish have begun to return, and papyrus has begun to grow in the area.[3]

The drainage of the Huleh swamp, completed in 1958, set in motion the ecological crisis of the 1990s. This is not surprising. Despite the warnings

1. "Metoch din veheshbon shel hanhala mirkazit le KKL" [Out of the minutes of the Central Executive of the Keren Keyemet L'Israel (KKL)] (1921–1923), 2, Central Zionist Archives (hereafter CZA), Jerusalem, CZA A246/441.

2. Haim Shapiro, "Reclaimed Hula Now Being Recreated," *Jerusalem Post*, October 20, 1995, 11.

3. Liat Collins, "After the Flood," *Jerusalem Post*, July 19, 1996, 14.

FIGURE O.1. Reeds with mountains in background. Huleh Nature Reserve, December 2005. Photograph by the author.

of scientists about potential environmental damage at the time of drainage, the Keren Keyemet L'Israel (Jewish National Fund, the Israeli government's land agency) decided to drain the Huleh swamp. Swamps were by then recognized as the breeding place of malaria-transmitting mosquitoes, and malaria was a scourge that had interfered with human habitation projects throughout history. But malaria was largely under control in Israel by the late 1950s.[4] The malaria risk posed by the Huleh was minimal, the ecological damage extensive. In addition, according to Rabinowitz and

---

4. Ch. Dimentman, H. J. Bromley, and F. D. Por, *Lake Hula: Reconstruction of the Fauna and Hydrobiology of a Lost Lake* (Jerusalem: Israel Academy of Sciences and Humanities, 1992), 1.

Khawalde, most of the settled bedouin had been forcibly removed in the 1948 war.[5]

So why did the Keren Keyemet L'Israel proceed with the project? Some explain the drainage of the Huleh swamp in terms of political conflict: the vast project was an Israeli response to Jordanian attempts to divert waters from the Jordan River during that decade. Others claim the opposite, that the Huleh drainage was part of a larger Israeli diversion project that would allow the new state to exploit the headwaters of the Jordan River.[6] Rabinowitz and Khawalde note that the Israelis drained the Huleh in order to strengthen their position in the Huleh Demilitarized Zone.[7] Still others insist on a public health explanation: that drainage was necessary to reduce the risk of malaria and make the land suitable for much-needed agricultural use.[8] Yet the project to drain the Huleh also points to much deeper issues in the historical relationship between Zionism, health, and the environment.

This book documents the history of swamp drainage projects as part of a broader history of malaria control in mandatory Palestine, the period of British rule between the years 1920 and 1947. In the process, it raises fundamental questions about the relationship between technology, disease, and nationalism. Ideology is central to this history but so are powerful biological forces and material conditions. Through a focus on malaria, the most thorny and complex of public health problems, I explore the at-times troubled relationship between ideology and materiality. My central argument is that Zionist swamp drainage projects, like the drying of the Huleh, effected a triple transformation of Palestine's landscape—technological, ecological, and perceptual—in an effort to promote Zionist ideological,

5. Dan Rabinowitz and Sliman Khawalde, "Demilitarized, then Dispossessed: The Kirad Bedouins of the Hula Valley in the Context of Syrian-Israeli Relations," *International Journal of Middle East Studies* 32, no. 4 (Nov. 2000); and Sliman Khawalde and Dan Rabinowitz, "Race from the Bottom of the Tribe That Never Was: Segmentary Narrative amongst the Ghawarna of Galillee," *Journal of Anthropological Research* 58, no. 2 (Summer 2002): 229, 234.

6. Allison Berland, "The Water Component of the Peace Process between the Israelis and the Palestinians" (M.A. thesis, Tufts University, 2000), chapter 1, "Politics of the Jordan Watershed," available online at the Oregon State University Department of Geosciences Transboundary Freshwater Dispute Database, http://www.transboundarywaters.orst.edu/publications/related_research/berland/berland_part1.html#_ftn19. Also Stephen Kiser, *Water: The Hydraulic Parameter of Conflict in the Middle East and North Africa*, INSS Occasional Paper, 35 (USAF Academy, CO: USAF Institute for National Security Studies, 2000), 4–5.

7. That is the reason, they claim, that Israel first started its work on the disputed land. This led to the dispossession of the Kirad. Rabinowitz and Khawalde, "Demilitarized," 522–24.

8. State of Israel, Ministry of Environment, *Conservation of Wetlands in Israel: Israel National Report on the Implementation of the Ramsar Convention* (Tel Aviv: Ministry of Environment, 1999), 3.

practical, and political agendas. The history of efforts to combat malaria offers a unique perspective on Zionist history itself, with its simultaneous focus on people, place, science, and politics.

To understand the centrality of such a seemingly technical subject as malaria control, we might start with the symbolics of the land and the Huleh Lake in the contemporary Israeli social imagination. The modern drive in Israeli society to rule over nature is commonly conveyed in stories about malaria and swamp drainage in Israeli/Zionist history. Commenting in 1998 on the allegorical role of swamp drainage (*yibush habitzot*) in Zionist mythology, Israeli author Meir Shalev described the drainage of the Huleh Lake and its contemporary, environmental consequences:

The Hula [*sic*] Lake has been dried out, killed. It is gone but, like many other great victims, it refuses to be forgotten.

Ecologists, hydrologists, and agronomists have discussed and will go on discussing the professional aspects of this drying. For us laymen, the Hula will forever be an allegory about brutality, blindness, arrogance, and the abuse of ideals. If this sounds familiar, let us not forget that we have acted the same way in other areas—hence the power of the allegory.

A review of the history of those days, just like a walk through the scarred Hula landscape, tells an irritating tale. Fifty years after the pioneers of the Second Immigration (1904–1914), the State of Israel found itself a new swamp to fight, as if to relive its youth. Those who looked for signs of the old enemy in the Hula Lake—water, reeds, buffaloes, and mosquitoes—found them all over the place. Those who wished to find land in it found it at the bottom of the lake. And it was not just any land, either. It was virgin land that had never been possessed by man. Without Turkish footprints, Roman relics, Arab leftovers, or Canaanite remains. It was land, just like our ethos and pathos love it—pure, anticipating, waiting for redemption.

The newspapers of the time describe the drying of the Hula in the all-too-familiar and enthusiastic tone: "a great miracle," "a giant enterprise," "a pioneering act." It seems that in those days, everyone thought and acted according to the same formula. What's so surprising? We have been brought up that way. As soon as we are given some still waters, a few mosquitoes, and some reeds, we take off on a tour of our conditioned and well-trodden thought patterns: You say drying, you mean swamp; by swamp, you mean malaria; malaria means pioneers; and they mean settlement, plowing the first furrow, dancing an energetic hora, values, vision, land, Zionism. And so, by the power of words that are too lofty and professionalism that is too low, the lake has been dried, killed, and turned into an allegory.

Several years ago, I heard a very edifying little tale: Before the Hula was dried, a Dutch expert was brought here to study the issue and give some advice. He was not too enthusiastic about the entire affair and said, among other things, that our expectations regarding the quality of the turf supposedly waiting to be rescued at the bottom of the lake should not be too high. One of our experts stood up, banged on the table with his fist, and thundered: "Our turf is good turf', it is Zionist turf!"

Had this story been told about any other fool, we would laugh. Because he is our fool, we can only weep and mourn for him, for us, for the dying lake, and worst of all, for the kind of thought processes that dried it out. We find the very same approach today, everywhere. It is the attitude of 'I know best,' and 'It will work out fine in the end,' of arrogance and machismo.

It has been forty years since the lake was dried and, I am sorry to say, things did not work out so fine. It seems that the dying lake is fighting back with a vengeance. The water we had expected to collect is contaminated and contaminates Lake Kinneret. Wild animals and plants are disappearing, mice are multiplying, and the land- -that virgin land that awaited us so—is not good Zionist turf, but simply barren wasteland. We could almost say that the Hula waters cry to us from the ground. However, because we have already attributed human qualities to the turf, let us leave the water alone.

The lake is dying. Part of it has been re-flooded, but not revived. The allegory, becoming clearer before our very eyes, teaches us the hard way (do we know of any other way?) that brute force is a double-edged saber—it swings right back and wounds those who rattled it. The Hula is dead, and this is its last will and testament.[9]

In the pages that follow, I explore this powerful ethos that Shalev points to and examine its historical roots. Why is malaria control symbolic in the creation of a uniquely Zionist "turf," of modern, technological expertise in engineering and medical science, and in the attainment of the status of a "nation among nations"? This book traces the foundations of the ethos of Jewish redemption and the social and physical transformation of the Jewish body in the contested land of Palestine. It examines the land's ecological transformation in its connection to nationalist ideology, the power of science and technology coupled with nationalist doctrine, and the consequences

9. Meir Shalev, "Drying the Hula: The Dying Lake," in *Those Were the Years,* ed. Nissim Mishal (Tel Aviv: Miskal, 1998), 37. Shalev was born in Nahalal in 1948, and his first novel, *The Blue Mountain (Roman Russi* in Hebrew), was published in 1988. He writes fiction, nonfiction, and children's books.

FIGURE O.2. Stream, Huleh Nature Reserve, December 2005. Photograph by the author.

of that doctrine upon the topographical, demographic, and epidemiological landscapes of Mandate Palestine.

Inquiry into this era is not only useful for understanding current conditions in the Middle East but also sheds light on the relationship between health and nationalism or statehood. In the example of Palestine, we can see how malaria-eradication measures legitimized Zionist claims to the land of Israel in a period of intense nationalism and colonial development, how medical knowledge affected transformation of the landscape and demography, and how Zionist medical discourse reflected Zionist images of the land, the Jewish people,[10] and the indigenous Arab inhabitants. Zionist medical discourse promoted the nation-building processes of settlement, labor, and land purchase as well as the creation and consolidation of national scientific institutions.

The legacy of malaria and swamp drainage in Israel continues today. But to understand its powerful role in contemporary ecology and ideology, we must start by first understanding the history of Palestine and Zionism.

10. Images of the Jewish people would include their physical bodies and their mental states of being.

The Biology of Nationalism

With the fall of the Ottoman Empire after World War I, the French and British colonized regions in the Middle East according to a secret wartime pact the Sykes-Picot Agreement of 1916, then renegotiated and finalized its conditions in the Treaty of Sèvres (1920). The final agreements charged the British with ruling over the territories of Palestine, Transjordan, and Iraq. The French took over what would later become Lebanon and Syria. In addition to their negotiations with the French, the British had promised the Zionist movement that it would facilitate "the establishment of a Jewish national home in Palestine" if it eventually gained Palestine after the war, as long as Zionists did not "prejudice the civil and religious rights of existing non-Jewish communities in Palestine or the rights and political status enjoyed by Jews in any other country" under the Balfour Declaration of 1917. The Zionist movement was a Jewish nationalist movement that sought the return of European Jews to the Holy Land in order to establish political and social autonomy. The Zionist Organization, the official umbrella organization of the Zionist movement, was founded at the first Zionist Congress is Basel, Switzerland, in 1897. The main reason the British decided to support the Zionist cause was their perception of the need for international Jewish support, especially in Russia and the United States, during the war.[11] However, the British had also promised the Arabs "independence" in certain parts of the Middle East if they rose up against the Turks during World War I. These conflicting promises were made as part of British attempts to forward their own territorial interests in the Middle East against those of the French, Germans, and Russians. The contestation of the Balfour Declaration by the Palestinian Arab population, along with the resentment of Arab nationalists against Britain for faltering on its (intentionally vague) promises grew throughout the period of British rule in Palestine. It remains a point of bitter contention in debates about political-territorial rights over the land even today.

The League of Nations gave Palestine the status of a mandate territory after the war, which meant that, in theory at least, the British would have a presence there to help the "native" population set up their own independent government, at which point the British would leave. In reality, however, the British treated Palestine like their other colonies, particularly with regard to health services.

11. Benny Morris, *Righteous Victims: A History of the Zionist-Arab Conflict, 1881–1999* (Knopf: New York, 1999), 74–75.

Because of the lack of comprehensive and coordinated medical services under Ottoman rule and the exacerbation of disease during World War I, British officials who arrived in Palestine found a poverty-stricken population of approximately 600,000 Arabs and 85,000 Jews.[12] Both communities experienced high rates of disease and famine that raised mortality rates among all segments of the population: urban and rural; Muslim, Christian, and Jewish. At the beginning of the Mandate period, malaria, tuberculosis, gastrointestinal diseases, and trachoma were among the most prominent diseases in Palestine. Other diseases included typhus, smallpox, and epidemic cholera, typhoid, leprosy, and dysentery.[13] Infant mortality rates were high and life expectancy was low.[14] Scarce and infected water also affected health conditions, depending upon residents' social class and location.

Within this context, the British Mandatory government focused first and foremost on ridding the country of infectious disease by installing new sewage and drainage systems, draining swamps, initiating widespread vaccination, and instituting hygiene education campaigns, including a school hygiene service. The government also inaugurated the registration of all cases of infectious diseases in 1918; launched the first population census in 1922; legislated food, pharmaceutical, and licensing ordinances; and strengthened quarantine measures.[15] All of these measures, however, were only partially implemented or limited in scope, and British investment in them was restricted.[16]

Despite its limitations, the institution of public health and medical controls in Palestine under British rule brought demographic and epidemiological shifts. By 1931, the main causes of death were pneumonia (23 percent of all causes of death), diarrhea and enteritis (15.6 percent of all causes of death before two years of age, and 6.5 percent after two years

12. Medical services, both private and governmental, existed in Palestine during the late Ottoman period.

13. Personal correspondence with Dr. Aftim Acra, professor emeritus of environmental science, letter dated 26 September 1998.

14. Infant mortality was 205 per 1,000 births in towns and 134 per 1,000 in rural districts during the first year of life in 1922. British Health Department of Palestine, annual report for 1922, p. 21, and annual report for 1923, p. 6, both in Israel State Archives, Jerusalem (hereafter ISA), M4475/06/1.

15. Some of these measures remained more theoretical than practical. British Health Department of Palestine, annual report for 1922, p. 20. For a critique of British health ordinances, see Abd al-Aziz Al-Labadi, *Al-Ihwal al-sahiya wa al- ijtima'ia lil-sha'ab al-Filastin, 1922–1972* [The health and social conditions of the Palestinian people, 1922–1972] (Amman: Dar-al Karmel, 1986), 42.

16. Gideon Biger, *An Empire in the Holy Land: Historical Geography of the British Administration in Palestine, 1917–1929* (New York: St. Martin's Press, 1994), 83, 85.

of age), while infectious and parasitic diseases (9.5 percent) like malaria became the third most common cause of death.[17] Infant mortality rates also declined during the Mandate period.[18]

The British authorities were largely responsible for malaria eradication and other public health issues for the Arab population in Palestine, and therefore there was no autonomous Arab agency for malaria control during this time. In contrast, malaria eradication was a central goal of the Zionist movement. For various reasons, a separate Arab health system was not set up during the Mandate period: a lack of funds and Palestinian Arab physicians, the relative fragmentation of the Palestinian community and its rural concentration, a lack of encouragement from the British, and a strong political emphasis on mobilizing Palestinian nationalist resistance against the Zionists and the British.[19] While the Mandatory government did initiate many public health measures, the general health of the Palestinian Arab population was clearly not a fiscal priority. The government understood, however, that the health of the "natives" was necessary for the economic development of the country and, more importantly, for securing the health of the English officials and soldiers serving in Palestine. This attitude was consistent with a colonial policy for which health measures were "not only parceled out selectively but were robed ideologically, geared more toward protecting imperial armies and settlers than indigenous populations."[20] Between 1920 and 1923, the Mandatory government allocated between 6.2 and 9.6 percent of its annual budget to health: this fell to an average of 4 percent during the first half of the Mandate period and approximately 3 percent in the second half.[21]

---

17. British Health Department of Palestine, annual report, 1931, 14, ISA M4475/06/1.

18. Aziza Khalidi, "Indicators of Social Transformation and Infant Survival: A Conceptual Framework and an Application to the Populations of Palestine from 1927–1944" (Ph.D. diss., Johns Hopkins University, 1996), 443. By the end of the Mandate period, life expectancy for Muslims reached about fifty years of age.

19. Mohammed Karakara, "Ma'arechet habriut hamandatorit vehavoluntarit ve'aravi Eretz Israel: 1918–1948" [Development of public health services to the Palestinians under the British Mandate, 1918–1948 (English title given)] (master's thesis, University of Haifa, 1992), 37–38.

20. Nancy Kreiger and Anne Emmanuelle Birn, "A Vision of Social Justice as the Foundation of Public Health: Commemorating 150 Years of the Spirit of 1848," *American Journal of Public Health* 88, no. 11 (November 1998): 1604.

21. British Health Department of Palestine, annual report for 1940, 2, Joint Distribution Committee Archives, New York City (hereafter JDC), AR21/32/291, microfilm 40; Institute of Palestine Studies, *Survey of Palestine: Prepared in December 1945 and January 1946 for the Information of the Anglo-American Commission of Inquiry* (Washington, DC: Institute of Palestine Studies, 1991), 2: 630.

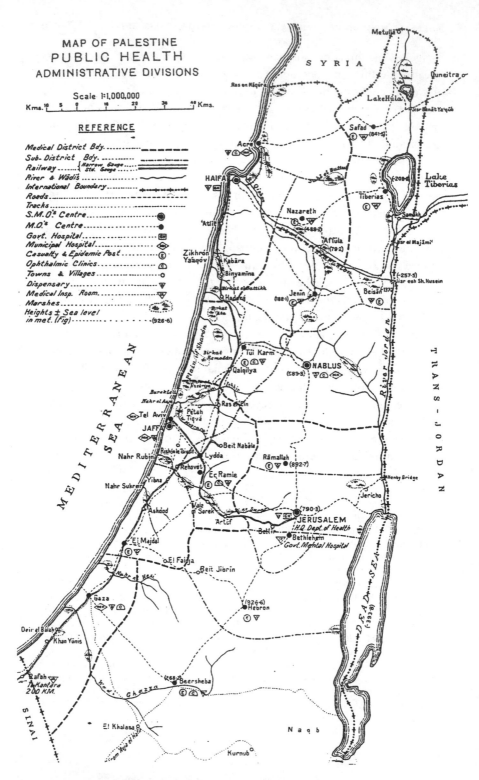

FIGURE 0.3. Map of Public Health Districts of Palestine. From front matter to Palestine Department of Health annual report for 1933, Ein Kerem Medical Library.

Where access to government hospitals was difficult or the range of treatment was insufficient, the Arab population was largely serviced by Christian missionary groups who had established medical facilities in most cities and holy places (Jerusalem, Nazareth, Nablus, Jaffa, Hebron, Haifa, Bethlehem, Gaza, Tiberias, and Safed) under Ottoman rule.[22] In striking contrast, the *Yishuv* (Jewish community in Palestine) possessed a largely autonomous health system built predominantly upon financial and technical assistance from abroad and only minimally subsidized by the British Mandatory government.[23] The Zionist movement took care of public health and health care efforts for the *Yishuv* (Zionist and non-Zionist alike). The two main Zionist health organizations servicing the Jewish community in Palestine were the U.S.-sponsored Hadassah Medical Organization and Kupat Holim Clalit, the health insurance provider of the Zionist labor union. The Hadassah Medical Organization was officially established in 1920, but its direct predecessor was the American Zionist Medical Unit, founded in 1918. Kupat Holim (Sick Fund) was founded in 1911 by a group of Jewish workers in Palestine as a prepaid health care organization.[24] The National Council (Vaad Leumi), the representative of the Jewish population in Palestine (Knesset Israel), created the Health Council (Vaad Briut), which served as the central institution charged with coordinating all health institutions in addition to overseeing and allocating resources for all health activities in the *Yishuv*. For the most part, these agencies worked to alleviate Jewish health problems without ignoring broader British colonial health and land policies.

These Jewish (mainly Zionist) organizations cooperated with the land reclamation and public health efforts of the Keren Keyemet L'Israel (the main Zionist land-purchasing agency during the Mandate period) and the Jewish Agency. Activities included, among other things, establishing

22. Care was also provided by Muslim health institutions on *waqf* property, although these were minimal and included clinics rather than hospitals. *Survey of Palestine*, 2: 617–618.

23. It is worth noting that not all Jews, even in Palestine, were Zionists. Many Jews in America or Europe, for instance, did not support the Zionist movement's goals because they thought it would threaten their standing in their own countries. Although some of these Jews—Zionist and non-Zionist—did not desire to immigrate to Palestine, many of them did contribute monies for health activities there.

24. Shifra Shvarts, "Kupat Holim and Jewish Health Services during the Mandate," in *Health and Disease in the Holy Land: Studies in the History and Sociology of Medicine from Ancient Times to the Present,* ed. Manfred Wasserman and Samuel Kottek (Lewiston: Edwin Mellon Press, 1996), 329–360. For more on the sick funds, see Shifra Shvarts, *Kupat Holim Clalit: itzuva vehitpatchuta kegorem hamirkazi beshirutei habriut be Eretz Israel, 1911–1937* [Kupat Holim of the Histadrut: Its formation and development as the central actor in health services in Palestine, 1911–1937] (Beersheva: Ben Gurion University Press, 1997).

hospitals and dispensaries, providing health insurance to its respective members, instituting settlement arrangements relating to health and hygiene, and implementing campaigns against infectious disease.[25] Jewish health organizations also collaborated with the Mandatory government in producing and disseminating health education propaganda and conducting scientific research, particularly with regard to malaria.

The resulting situation—an autonomous, well-organized, and relatively well-financed health system run by the Zionists, and an Arab population dependent upon a limited British health system and missionary efforts—maintained and perhaps even deepened the social gap between the two communities. This disparity and relative separation in health services reflected the broader political framework of the British Mandate.[26]

Zionism, Nationalism, and Health

Zionism was, in essence, a Jewish nationalist movement that emerged in the mid- nineteenth century in response to the oppressive socioeconomic conditions and persecution under which Jews lived and the strong anti-Semitism they faced, especially in Eastern Europe. As a modern Jewish nationalist movement, Zionism attempted to radically transform the political status of the Jews, their socioeconomic profile, and their physical and psychological disposition and self-perception.[27]

Zionism evolved alongside the rise of European national movements and colonialism, drawing upon the former in its search for political and national autonomy and the latter for its methods of implementation. As Zachary Lockman has pointed out, this mixture of influences was unique to Jewish territorial dispersion and different from other European nationalist movements in which the majority population in a contiguous territory sought to achieve national sovereignty. Jewish masses in Europe lived among non-Jewish populations and therefore had to find a way to distinguish themselves from their non-Jewish counterparts. Zionism's dis-

25. *Kupat Holim Amamit*, the Hadassah-sponsored health fund, and *Kupat Holim Klalit*, the Kupat Holim–sponsored health fund.

26. Barbara Smith, *The Roots of Separatism in Palestine: British Economic Policy, 1920–1929* (Syracuse: Syracuse University Press, 1993), 59–60.

27. These ideas were shared by Hadassah, a Jewish Zionist women's organization, and the General Zionists, another strand of Zionist thought that promoted the primacy of nationalist interests over class issues and the principle of self-help. G. Shimoni, *The Zionist Ideology* (Hanover: Brandeis University Press, 1995), 117–126.

tinction, then, lies in it being out of necessity "not simply a conventional national movement but a colonizing and settlement movement as well": a movement that gathered Jews in a territory outside of Europe in order to establish Jewish sovereignty.[28]

That territory would be Palestine. Even before the emergence of the Zionist movement, *Eretz Israel* (the Holy Land, Zion, the land of Israel) held a sacred place in the religious map of Judaism. For most Jews living in Europe, Israel was a source of longing, an idealized, sacred terrain that was connected with the coming of the messiah rather than a site for resettlement. For those Jews already living in Palestine, largely detached from the experiences of European Jewry, the ideas of a Jewish national revival and the pursuit of political independence were conceptually remote. By rejecting the belief that Jewish redemption lay in divine intervention, yet appropriating the importance of *Eretz Israel* in Jewish history and in the Jewish imagination, the Zionists quickly made Palestine into a secularized, territorial object upon which they focused their political and cultural aspirations. In this way, they drew upon the ideas of secularism and rationalism espoused in the Haskala movement (Jewish Enlightenment) but reformulated them in order to promote a specifically Jewish identity in another land.[29]

Although the movement was in no way monolithic, most Zionists agreed that attaining autonomy in their own land would ultimately solve the "Jewish problem," rendering Jews a "normal" nation among nations instead of the "abnormal," pariah status they possessed in Europe. The most desirable path for achieving that end, however, was a matter of debate. Differences of opinion were accompanied by divergent emphases and ideas about the nature of Jewish identity and the elements needed to achieve the Zionist goal.

Zionist opinion converged, however, when it came to thinking specifically about how to attain and maintain health. Three themes held particular relevance with regard to Zionist health concerns (including malaria): the transformation of the Jewish people, images of the land of Palestine, and the perceived political and developmental status of the indigenous Arab population. These themes reflected particular strands of Zionist thinking.

The paradigms and institutions of labor Zionism were the most influential in medical and antimalaria projects during British rule (1920–1947), the period on which this book focuses, including the creation of sustainable

---

28. Zachary Lockman, *Comrades and Enemies: Arab and Jewish Workers in Palestine, 1906–1948* (Berkeley: University of California Press, 1996), 27.

29. Michael Mayer, *Jewish Identity in the Modern World* (Seattle: University of Washington Press, 1990), 60–61.

rural settlement and the mission of shifting Jewish work from one based in industry in Europe to one based in agriculture in Palestine. Optimal health and hygiene were vital to achieve these goals. Although there were certainly other contexts in which Zionist visions manifested themselves—much of it examined fully in the scholarly literature—only those related to malaria and swamp drainage are addressed here.[30]

Labor Zionists believed that delivering the Palestinian environment and Jewish immigrants from their perceived pathological condition in the Diaspora was a necessary step for redirecting Jewish labor and establishing a tangible (rather than merely spiritual) attachment to the Holy Land. The Hebrew phrase *havra'at hakarka vehayishuv* (healing the land and the nation) captures this sentiment, the ambition behind it, and the actions that followed from it.

This nationalist phrase connected disease and medicine with statehood, politics, and geography. It is this paradigm within which malaria-eradication programs operated. The phrase frequently appeared in public health documents. Although *havra'at hakarka vehayishuv* literally means "healing the land and the Jewish community settled in Palestine," I translate it here as "healing the land and the nation" because the Zionist community in Palestine (the *Yishuv*) was the group that came from Europe to Palestine to be healed as a nation through their settlement in the Holy Land. In Zionist ideology, immigration and settlement and all the health efforts that were part of these activities were meant to create a new Jewish nation that was considered proud, self-sufficient, and healthy. As an ideal, the *Yishuv* was to become the "Jewish nation" at its best.

Precisely because malaria involves the interaction among land, water, an insect vector, and a human host, it is a useful vehicle for exploring the effort to "heal the land and the nation." Indeed, disease is a problem that all individuals and societies face, but it is also something that can simultaneously alter the landscapes of politics, demographics, topography, ideology, and ecology, as will be evident in the story of swamp drainage in Palestine. The treatment of malaria on the human and physical landscape provides a microcosmic view of the ideological and practical efforts of Zionists to achieve their nationalist goals. In its linkage with Zionist national transformation, making the land healthy through swamp drainage was intended to facilitate an attachment between the land of Palestine and the Jewish people through clear, practical measures. Swamp drainage demanded

---

30. See Morris, *Righteous Victims*, for a relatively recent, general history of the conflict.

hard labor on the land, a type of work that Jewish immigrants were not accustomed to in Europe but set out to undertake in Palestine.

Health education efforts that sought to reform and transform the habits and bodies of members of the *Yishuv* tried to further consolidate this attachment. Malaria-education efforts tried to inculcate hygienic principles among the *Yishuv* in order to create a new type of Jew—one completely different from the one in the Diaspora—and to establish a new relationship between the Jew and his or her environment. Malaria propaganda was part of the proliferation of Zionist health education activities that produced a nationalized culture of health stressing individual responsibility and discipline for the benefit of the nation. Health behavior was considered to have life-or-death ramifications for the national project. In this way, malaria and its eradication tapped into the ideological dreams of the Zionist movement while forwarding its practical agendas. Furthermore, unlike other colonial contexts where "natives" were the object of "sanitary gestures,"[31] under Zionist health activities, the Jewish community was the unique, primary recipient of the enlightened medical message of Jewish doctors and medical bureaucrats.

Nationalism, Land, and Colonial Medicine

In Mandatory Palestine, the *Anopheles* mosquito, the vector of malaria, commonly bred in natural and man-made swamps, pools, and cisterns. Thus like their colleagues in colonies around the world, Zionist scientists and physicians studying endemic malaria directed particular attention to the issue of swamps. The swamp was the place where mosquitoes developed and therefore where they could be destroyed, thus breaking the chain of malaria transmission.

Yet the swamp was more than a site of disease. As William Cronon observes,

> An ecological history begins by assuming a dynamic and changing relationship between environment and culture ... Environment may initially shape the range of choices available to a people at a given moment, but then culture re-

31. John Comaroff and Jean Comaroff, *Of Revelation and Revolution: The Dialectics of Modernity on a South African Frontier,* vol. 2 (Chicago: University of Chicago Press, 1997), 336.

shapes environment responding to those choices...Changes in the way people create and re-create their livelihood must be analyzed in terms of changes not only in their social relations, but in their ecological ones as well.[32]

Scholarly discussions of "landscape" have centered on the imaginative and the functional, the integrated whole of the perceptual and the practical. Each chapter of this book uses the word "landscape" in its title in order to capture both the imaginative and practical aspects of the Zionist malaria project and to reemphasize the simultaneous changes taking place. Like those scholars who have inaugurated a new field of critical medical geography (health geography), I see the term "landscape" as embodying not only the physical features of the land, but also its engagement with societal activities, its ability to represent a certain relationship with nature, and its reflection of shared and contested meanings among different social groups.[33]

Throughout this book I show that landscape and people are closely intertwined. In Zionist discourse, this was true both practically and symbolically. The history of the Jewish people was seen as intertwined with the history and changes of Palestine's terrain. Zionists believed that Palestine's terrain reflected Jewish history and that Zionist nationalism promised civilizing progress. In tracing these relationships and in showing the dynamics between them, the book investigates the social, cultural, political, and institutional context in which swamp drainage took place. This environmental change was not straightforward, linear, or consistently successful. Like other forms of change, swamp drainage engineers experienced shifts in plans, delays, mistakes, and unexpected trajectories. As we see today, contemporary Israel has experienced the long-term outcomes of the envi-

32. William Cronon, *Changes in the Land: Indians, Colonists and the Ecology of New England* (New York: Hill and Wang, 1983), 13–14, as quoted in Ian Scoones, "The Dynamics of Soil Fertility Change: Historical Perspectives on Environmental Transformation from Zimbabwe," *Geographical Journal* 163, no. 2 (July 1997): 163. For the American case, see Conevery Bolton Valencius, *The Health of the Country: How American Settlers Understood Themselves and Their Land* (New York: Basic Books, 2002), 3–4, 10–11.

33. Denis Cosgrove, *Social Formation and Symbolic Landscape* (Totowa: Barnes and Noble Books, 1984), 13, 15; Robin A. Kearns and Wilbert M. Gesler, eds., *Putting Health into Place: Landscape, Identity and Well-Being* (Syracuse: Syracuse University Press, 1998), ix–x, 11; Tom Griffiths and Libby Robin, eds., *Ecology and Empire: Environmental History of Settler Societies* (Seattle: University of Washington Press, 1997); Roza El-Eini, "British Agricultural-Educational Institutions in Mandate Palestine and Their Impress on the Rural Landscape," *Middle Eastern Studies (England)* (January 1999): 98–114; Ian Scoones, "The Dynamics of Soil Fertility Change: Historical Perspectives on Environmental Transformation from Zimbabwe," *Geographical Journal* 163, no. 2 (July 1997): 161–169; Alon Tal, *Pollution in a Promised Land: An Environmental History of Israel* (Berkeley: University of California Press, 2002).

ronmental changes that began during the Mandate period. Like the Zionist engineers of the past, Israeli scientists today are constantly responding to unexpected and many times unwelcome or unintended consequences.

Zionist medical and engineering professionals during the Mandate period utilized the same general methods and medicines to overcome malaria as other colonial scientists did, but their scientific goals united with a sociopolitical mission of national transformation. In this way, Zionist swamp drainage projects fit within the larger trends of colonial medicine in the early twentieth century, which tried to rid the colonies of infectious diseases in order to facilitate European settlement. Zionist projects, however, were fairly unique within the colonial medical milieu in their clear linkage to nationalist agendas. Zionist malariologists' employment of the term *yedi'at Ha'Aretz vehaOiyev* (knowing the land and the enemy, i.e., the mosquito) in malaria experiments and surveys was indicative of the way Zionist malariologists used science, medicine, and rationality as the basis of building a Zionist homeland. Science and technology was seen as contributing to the moral regeneration of the Jewish people by increasing their productivity.[34] Palestine became a territory that was molded and furnished as a national landscape, infused with meaning through medical strategies, settlement agendas, and Zionist ideological claims.

Representations and Experiences

Achieving the dual, nationalist transformation of land and people was no simple task. Ideological obligations were based in and subject to material realities in Palestine. Malaria in late Ottoman and Mandatory Palestine was both an epidemic and endemic disease. Although the Ottoman Empire engaged in some swamp drainage measures to alleviate the situation in Palestine, malaria remained the most prevalent infectious disease in the country at the beginning of the Mandate period and throughout much of its tenure. Malaria posed a serious challenge to Zionist leaders and settlers. Anxieties especially rose when residents and experts expected or experienced a malaria epidemic, but the apprehension about malaria (and infectious disease at large) that Zionist leaders and settlers felt throughout this period reflected the larger implications endemic malaria had for the success of Zionist colonization. Indeed, the rugged and marshy terrain of

---

34. Derek Penslar, *Zionism and Technocracy: The Engineering of Jewish Settlement in Palestine, 1870–1918* (Bloomington: Indiana University Press, 1991), 17, 30.

Palestine, and the malarial fevers that came with it, shaped the possibilities for Jewish settlement. At the same time, Zionist culture and its reliance upon technology and science shaped the environment within the constraints of Palestine's ecological conditions. This book concentrates on the endemic (constantly present) picture of malaria in Palestine rather than specific epidemic moments in order to illustrate the ongoing relationship between disease (and its epidemiology), cultural values, social institutions, political and demographic realities, and the nationalist goals of the Zionists.[35] An endemic analysis also seems fitting for a disease like malaria, because its symptoms lie dormant in individuals but then erupt in febrile episodes.

Endemic malaria not only consistently shifted the ecological and social relations among Jews living in Palestine but also changed ongoing relations, both representational and material, between Zionist settlers and their Arab counterparts. Because of the complexity of environmental change—the new social relationships it produced and meanings it evoked—ridding the topography of Palestine of swamps affected not only the number of mosquitoes and swamps and the rates of malaria, but it severely impacted the lifestyles of the bedouins and Arab peasants who lived off the swamps for their livelihood.

Although the story of Jewish/Arab interaction and of the larger context of communal politics plays a part in this book, the work is not intended to be a comparative history of malaria between the Zionist and Palestinian Arab communities. The paucity of Arabic sources on this subject poses a major difficulty to framing and writing it in this way. Rather, the book describes the interface of Zionist antimalaria efforts with the Palestinian Arab inhabitants, showing not only how that interaction influenced the Zionist reformative mission, but also how it revealed the communities' differing cultures of agricultural development. I use existing Arabic sources, where relevant, to examine this dynamic. The swamp became a location of contestation; a place that manifested a clash of visions about the landscape, about productivity, and about land use.

Conclusion

In my focus on malaria as a case for exploring issues of health in the Zionist nationalist movement, I do not claim that health was the most impor-

---

35. Charles Rosenberg, *Explaining Epidemics and Other Studies in the History of Medicine* (Cambridge: Cambridge University Press, 1992), 4.

tant component in Zionist national ideology or the movement's highest practical priority, either in an organizational or financial sense. Yet an analysis of malaria in Palestine reveals how Zionist public health measures were not only an important part of a comprehensive health project in Palestine, but that they were constituted by and constitutive of larger Zionist discourses and practices that promoted the establishment of the Jewish national home in Palestine.[36] As this book shows, and as the contemporary Israeli discussion over the Huleh continues to demonstrate, these visions and practices are culturally and physically encoded in the landscape of Palestine through swamp drainage measures (or in the case of Israel today, through reflooding/natural reserve measures). Intersections between issues of health and immigration, settlement, labor, land purchase, agriculture, and other areas of the Zionist project make malaria and issues of health at large a ripe field for studying the complexities of Palestine/Israel's social history.[37] Perhaps this case can help serve as a model for studying the place of health/disease in the complex mélange of the social histories of other countries as well. My specific examination of malaria in Mandate Palestine here shows that culture matters as much as place and that the discourses of medicine and health matter as much as the epidemiology of the disease itself. These elements of analysis must be woven together in order to capture the workings of the Zionist project in Palestine and its long-term outcomes.

36. Shula Marks, "What Is Colonial about Colonial Medicine? And What Has Happened to Imperialism and Health?" *Social History of Medicine* 10 (1997): 206–207; Uriel Kitron, "Malaria, Agriculture and Development: Lessons from Past Campaigns," *International Journal of Health Sciences* 17, no. 2 (1987): 303.

37. D. J. Bradley has pointed out that malaria in particular has had "so great an effect on the discourse of health that we must ask both why and how." D. J. Bradley, "The Particular and the General: Issues of Specificity and Verticality in the History of Malaria Control," *Parassitologia* 40, nos. 1–2 (June 1998): 5.

CHAPTER ONE

Archetypal Landscape

*Healing the Land and the People in the Zionist
Imagination*

And when, O Man, you will return to Nature—on that day your eyes will open, you will gaze straight into the eyes of Nature, and in its mirror you will see your own image. —A.D. Gordon, "Logic for the Future" (1910)

In his opening speech for the scientific convention of the Hebrew Medical Organization of Palestine in 1940, Dr. Hermann Zondek (1887–1979), a professor of medicine at the University of Berlin before immigrating to Palestine in 1934, asked the rhetorical question, "Is Judaism a disease?" Zondek's answer to his query was based on health conditions in Palestine, specifically, the prevalence of malaria. His answer explicitly evoked the image of malaria: "If it is a disease, then it is a very acute Jewish disease. This disease appears intermittently after long or short periods of inhibitions, followed ... by acute explosions during which the world is shaken by feverish religious, political, or economic crises."[1] We know Zondek is referring to malaria here because of its characteristic symptom of intermittent, explosive fevers. In this way, Zondek likened the symptoms of Judaism as a social disease to the physical symptoms of malaria.

Zondek's use of the metaphor of malaria could seem somewhat surprising when considering the state of malaria in Palestine in the 1940s.

1. Hermann Zondek, "Contemporary Medicine and Jewish Physicians," *Medical Leaves* 3, no. 1 (1940): 10. Zondek pioneered the study of the physiopathology of endocrine disorders.

Unlike in Europe, where malaria was largely under control by the end of the nineteenth century and the earliest years of the twentieth century, malaria had been, and continued to be, an endemic disease in Palestine in the first half of the twentieth century. Malaria was first brought under control in Palestine during the British Mandate period (1920–1947). By the time of Zondek's speech, however, malaria had significantly declined in Palestine in comparison to the beginning of the Mandate. In addition, the Jews had established the infrastructure of a relatively autonomous Zionist society in Palestine by the 1940s.

At the same time, Zondek's connection between malaria, disease, and Judaism is not so surprising. Zionist doctors and ideologues had already tried to answer Zondek's question about the "pathological" nature of Judaism in the early years of the Zionist movement as a response to ongoing anti-Semitic claims about the "abnormal" state of Jews in Europe.[2] The issue of Jewish stereotypes and discrimination became especially acute with the rise of the Third Reich. But even before the rise of Hitler, anti-Semites had long looked upon Judaism and Jews as contagious and deviant. Two thousand years in the Jewish ghetto had supposedly helped produce the sick state of Judaism and the Jew. Although religious (largely Christian) views in earlier periods of history upheld notions of Jewish difference(s), a new, scientific, and allegedly neutral vocabulary emerged in the modern era to describe the dissimilarities between the Jew and the Christian. Anti-Semitic scientists and politicians represented the Jewish body, for instance, in scientific terms as both physically and mentally ill. Because of its broad symbolic and practical power in the Western world, these medical and scientific descriptions of the Jewish body possessed social authority.[3]

The problematic condition of Jews in Europe has its roots decades earlier in modern anti-Semitism. Anti-Semitic rhetoric of the late nineteenth century was informed and substantiated by eugenics and theories of degeneration and racial hygiene. Anti-Semites asserted that Jews were

2. S. Winter and N. Levy, "Medicine in Palestine following the Flight of Jewish Physicians from Nazi Germany," *Adler Museum Bulletin* 12, no. 19 (1986): 21; Shmuel Almog, "Hayahadut kemachala: stereotyp antisemi vedimui 'atzmi" [Judaism as a disease: Anti-Semitic stereotype and self-image], *Leumiut, Zionut, antishemiut: masot vemechkarim* [National, Zionism and Anti-Semitism: Essays and studies] (Jerusalem: Hasifri'a hatziyonit, 1992), 250–252.

3. Sander Gilman, *The Jew's Body* (New York: Routledge, 1991), 233–236; see also chapter 9, "The Jewish Disease: Plague in Germany, 1939/1989."

a morally and physically debased race, rootless, shamed, and ashamed.[4] This belief, in its basic form, actually went back at least as far as the Middle Ages, when Jews were blamed for the plague—and before that, in the early Christian era, when Jews were persecuted by Romans who held similar beliefs. What made this anti-Semitic racial discourse uniquely modern was its representation of individual Jews as possessing physically and mentally afflicted bodies as based upon modern scientific theories and experiments. Believers of this theory purported that Jews evidenced their diseased status in their physical appearance and in their psyche. Their deviant physicality was allegedly symbolic of a national, moral disorder. Anti-Semites believed that it was impossible to reform the physical and moral characteristics of the Jewish race; such traits were inescapable and immutable.

Late nineteenth-century European medical ideas of racial hygiene deeply influenced debates among Jewish intellectuals about the physical and mental quality of the Jews in the Diaspora as well as its transformative Zionist counterpart. Jewish racial science developed within the academic study of Judaism, the *Wissenschaft des Judentums*. Jewish doctors and scientists in the Diaspora used this research for both internal (Jewish) and external (Gentile) consumption to address and refute common anti-Semitic claims.[5] Mitchell Hart has described Jewish social scientists' responses to this racial, anti-Semitic discourse as complex: "Resistance and rejection were accompanied by acceptance and appropriation." Although they defended European Jewry against claims of racial pathology,

4. Daniel Pick, *Faces of Degeneration: European Disorder, 1848–1918* (Cambridge University Press: Cambridge, 1989); John Efron, *Defenders of Race: Jewish Doctors and Race Science in Fin-de-Siècle Europe* (New Haven: Yale University Press, 1994), 131, 179; Michael Berkowitz, *Zionist Culture and West European Jewry before the First World War* (Chapel Hill: University of North Carolina Press, 1993), and *Western Jewry and the Zionist Project, 1914–1933* (Chapel Hill: University of North Carolina Press, 1993); Sander Gilman, *Difference and Pathology: Stereotypes of Sexuality, Race, and Madness* (Ithaca: Cornell University Press, 1985), and *The Jew's Body*; George Mosse, *Nationalism and Sexuality: Respectability and Abnormal Sexuality in Modern Europe* (New York: H. Fertig, 1985), idem, *The Image of Man: The Creation of Modern Masculinity* (New York: Oxford University Press, 1996), and idem, *Confronting the Nation: Jewish and Western Nationalism* (Hanover: Published for Brandeis University Press by University Press of New England, 1993).

5. Efron, *Defenders of Race*, 127; Mitchell Hart, *Social Science and the Politics of Jewish Identity* (Stanford: Stanford University Press, 2000); Comaroff and Comaroff, *Of Revelation and Revolution*, 324; Sander Gilman, "The Jewish Nose; or, Are Jews White?" in *The Other in Jewish Thought and History: Constructions of Jewish Culture and Identity,* ed. Laurence J. Silberstein and Robert L. Cohn (New York: New York University Press, 1994), 375.

they simultaneously reiterated many of those same ideas in their proposals for self-improvement. Many Zionist social scientists (and physicians) acceded that modern Jewry in the Diaspora was undergoing a period of physical and moral decline, evident in lower fertility rates, intermarriage, and increased rates of nervous disorders.[6] But instead of explaining causality through racial determinism, these social scientists argued that such traits were environmentally and historically contingent; they could be (and should be) transformed. Certainly, according to them, Jews could engage in national improvement. Accordingly, Zionist ideologues and social scientists proposed that reformation in a new place would restore Jewish dignity and Jewish physicality. It is important to note that by locating the etiology of Jewish degeneration in environmental conditions rather than in race, Jewish social scientists raised the task of environmental transformation to parity with physical improvement. As we will see, this linkage plays a key role in the malaria project in Palestine.

It is this linkage between malaria and national improvement that Zondek implicitly refers to in his speech. Zondek's answer to the age-old question of the diseased nature of Judaism likely referred not only to the declining position of Jews in the "feverish crisis" of Nazi-controlled Europe but also to the enduring potency of the image of disease, and the metaphor of malaria in particular, in the Zionist imagination. Because of the etiology of malaria, which involved the intricate interaction of people and their environmental conditions, the disease's use as a metaphor could capture both the long-standing anti-Semitic idea about the pathological nature of Judaism and the importance of place and environment in producing or alleviating that characteristic. In essence, the ethos surrounding the metaphor of malaria and its eradication maintained that without healing *both* the Jewish body and the Jews' new habitat and attacking the "acute" pathological characteristics of person and place, both Judaism and the Jew would continue to be characterized as diseased. This ethos, which will be explored fully in this chapter, signifies the inseparability of malaria, physical health, and social health in the Zionist nationalist project.

Within the framework of that ethos, Zionist health officials both before and during the Mandate period sought to destroy the mosquito and malarial parasite through antimalaria campaigns and tried to eradicate what they had internalized as the parasitic nature of the Jewish immigrant through malaria and hygiene education efforts. Indeed, many Zionists who came

6. Hart, *Social Science,* 12–14, 21, 100.

to Palestine in the early twentieth century considered it to be stagnant and malaria-infested, like the Jewish ghetto. Zionist political, ideological, and medical leaders explicitly used the metaphor of parasitism—usually meaning *economic* parasitism in Zionist discourse, but referring to mosquitoes, malaria, and the malarial parasite in Zondek's speech—to communicate the Zionist program for sociocultural reform. As Zionist ideologue A. D. Gordon urged, "We must wage war against parasitism of every kind, parasitism that is also rooted among us."[7]

Zondek's speech used malaria as a way to evoke two key nationalist aims of the Zionist project: building a nation-state in Palestine and reimagining the character of the Jewish people and the body of the individual Jew. The idea of recreating the Jew and the Jewish people in order to achieve redemption had its roots in the *Haskala* (Jewish Enlightenment) movement, starting in the late eighteenth century in Europe and continuing into the nineteenth century. Most followers of the Jewish Enlightenment (*maskilim*) tried to fortify secular education, encourage entry into other occupations, and emphasize Bible study.[8] Before the Jewish Enlightenment, Jewish redemption was conceptualized as collective. During and after the *Haskala,* individual incarnations of redemption were accepted and paired with collective notions. The innovation Zionism brought to this transformative tendency was less the idea of reimagining the Jew than the particular territory of Palestine and emphasizing improving both the individual and the nation by settling Jews there. Zionist thinkers essentially believed that Jews could change their tenuous diasporic condition and gain the capacity for nation building through self-transformation.[9]

The Zionist impulse for national transformation was European in its origin. Many nationalist movements in the late nineteenth and early twentieth centuries held fast to visions of self-transformation. This impulse

7. A. D. Gordon, "A Definition of Our Attitude," 226, as quoted in Zeev Sternhell, *The Founding Myths of Israel: Nationalism, Socialism and the Making of the Jewish State* (Princeton: Princeton University Press, 1998), 63. German, Austrian, French, U.S., and English anti-Semites also used the language of parasitism to describe the Jewish race. See Hart, *Social Science*, 98.

8. Richard Cohen, "Urban Visibility and Biblical Visions: Jewish Culture in Western and Central Europe in the Modern Age," in *Cultures of the Jews: A New History*, ed. David Biale (New York: Schocken Books, 2002), 733, 765–766; David Biale, "A Journey between Worlds: East European Jewish Culture from the Partitions of Poland to the Holocaust," *Cultures of the Jews: A New History*, ed. David Biale (New York: Schocken Books, 2002), 820–834.

9. Mayer, *Jewish Identity in the Modern World*, 67.

reflected a desire to fulfill the Western Enlightenment promise to remake or re-form each individual and society. The goal of mental and physical transformation incorporated the ultimate ideal: racial perfection.[10] In response to anti-Semitic claims and consistent with this general European nationalist inclination, Zionists attempted to "cure" the Jewish people by furnishing a representational and physical conversion whereby a new Hebrew man would be born, healthy in body and mind. The new Jewish body (primarily conceived of as male) would be muscular, strong, virile, proud, and productive. It is interesting to note that the use of the word "Hebrew" in the phrase "Hebrew man" refers, in part, to the revival of Hebrew as a national language in Zionism. The revival of Hebrew as a language is another transformative aspect of the Zionist project not examined here. It should be noted, however, that in Zionist health documents during the Mandate period, acquiring Hebrew language was considered another way to become physically and mentally "healthy."[11]

Just as the alleged physical degeneration of Jews symbolized a deviant Jewish society for anti-Semites (and even for some Jewish social scientists), creating a new Jewish body and psyche became symbolic of creating a new Jewish nation for Zionists. Individual Jewish health became a metonym for national health.[12] This Zionist conception of individual health as integral to nationalist success stood in stark contrast to the treatment of Palestinian Arab health. As mentioned in the introduction, while the Mandatory government held primary jurisdiction over Arab health matters during the Mandate period, some Zionist health organizations also extended public health services to the Arab population. Both Zionist and British health officials addressed Arab health in the aggregate as compared to the treatment of Jewish health, a colonial view in which "native" communities were seen as part of the natural environment. Palestinian Arab health was therefore largely managed and modeled in broad, composite terms, conflating important differences in class, religion, and residence.[13]

10. Nancy Stepan, *Picturing Tropical Nature* (Ithaca: Cornell University Press, 2001), 139.

11. Georges Canguilhem, *The Normal and the Pathological* (New York: Zone Books, 1991), 43, 91, 100–101; Ludwik Fleck, *Genesis and Development of a Scientific Fact* (Chicago: University of Chicago Press, 1979), 37; Efron, *Defenders of Race,* 138; Lennard Davis, "Constructing Normalcy," in *The Disability Studies Reader,* ed. Lennard Davis (Routledge: New York, 1997), 9–28.

12. David Biale, *Eros and the Jews: From Biblical Israel to Contemporary America* (New York: Basic Books, 1992), 178; Gilman, "The Jewish Nose," 367; Hart, *Social Science,* 12–13.

13. Palestinian Arab health officials worked for the Mandatory government. For aggregation in colonial medicine, see Megan Vaughn, *Curing Their Ills: Colonial Power and*

In contrast, the vision of a new, Hebrew man—at once individual and national—is most widely attributed to Max Nordau (1849–1923), a physician-neurologist, who was one of the closest associates of Theodor Herzl (1860–1904), the founder of the international Zionist movement. Nordau was a critic of modernism and of the decadence of fin de siècle European culture. Most noted for his book entitled *Degeneration* (1892), Nordau attacked conservatives who viewed Jews as degenerate, a view "so ubiquitous in the nineteenth century that even Jews themselves learned to believe it." Zionist leader Israel Zangwill also spoke about the future of the Jewish race at the Universal Races Congress in London (1911), a conference where the white race stood as the standard to which others should aspire.[14] Six years after the publication of his book, at the Second Zionist Congress (1898) and a year after Ronald Ross's discovery of the malaria parasite, Nordau proposed a solution that called for the reformation of Jewish bodies through the creation of a nation of "muscle Jews." Reflecting the prevailing notion that physiological and psychological conditions were interconnected, Nordau asserted that diasporic Jews could overcome their hereditary nervous state by becoming physically and athletically fit. Nordau recommended that Jews move away from the nervous stimuli of the city to more "healthy" rural surroundings. He saw the acquisition of physical fitness, like he saw Zionist settlement in Palestine, as a return to the bodily status of the Jews before exile.[15] Nordau, a neurologist by profession, addressed this point when he wrote that "[the Jews'] terrible posture does not come from any natural trait. It is but a result of a lack of physical education. In this way, there is not

---

African Illness (Stanford: Stanford University Press, 1991), 11; and David Gilmartin, "Models of the Hydraulic Environment: Colonial Irrigation, State Power and Community in the Indus Basin," in *Nature, Culture, Imperialism: Essays on the Environmental History of South Asia,* ed. David Arnold and Ramachandra Guha (Delhi: Oxford University Press, 1995), 214.

14. Stepan, *Picturing Tropical Nature,* 141–142; Hans-Peter Soder, "Disease and Health as Contexts of Modernity: Max Nordau as a Critic of Fin-de-Siècle Modernism," *German Studies Review* 14, no. 3 (1991): 473–487, especially 478–79, and *Degeneration* (1968; Lincoln: University of Nebraska Press, 1993); M. Foster, "The Reception of Max Nordau's *'Entartung'* in England and America" (Ph.D. diss., University of Michigan, 1954), 317, cited in Soder, "Disease and Health as Contexts of Modernity," 474, 474n6, 474n9. For a critique of Nordau, see *Regeneration: A Reply to Max Nordau* (Westminster: Archibald Constable and Co., 1895).

15. Max Nordau, "Yehadut hashririm" (1900–1902), *Max Nordau el 'amo: ketavim mediniim* (Tel Aviv: Medinit Press, 1936), 171–172; and Biale, *Eros and the Jews,* 178–179.

really a difference between the Jew and Aryan." To Nordau and other Zionists, the Jews had the capacity to improve their bodies, particularly their muscles.[16]

In fact, the idea of a "new Jew" was not a Zionist invention. Like the idea of transformative redemption, the notion of the "new Jew" resulted from the formulation of a secularized identity constructed by the Jewish Enlightenment movement. The idea of a "muscle Jew" in Zionist discourse is a reformulation of the *Haskala*'s "new Jew." Still, Nordau's views had a great impact on subsequent labor Zionist efforts to reform the Jewish body. In their basic form, Zionists of other orientations replicated these ideas, such as Vladimir Jabotinsky, the founder of Revisionist Zionism, who in 1925 espoused Jewish athleticism and strength as a transformative measure to "rebuild the generation" and also emphasized industrial development as the basis of a Jewish state.[17] Abraham Isaac Kook, chief Ashkenazi (i.e., originating from Central and Eastern Europe) rabbi in Palestine from 1921 to 1935, also supported Jewish self-transformation through Torah study and physicality but supported the creation of a Jewish dominion in the British commonwealth.

In his call for Jews to move out of the city into rural surroundings, Nordau underscored one of the most critical factors in inaugurating a transformation: location. Creating a "territorialization of identity," a tie to place, was a key strategy in the Zionists' attempt to turn the *Galut* (Diaspora) Jew into a new Zionist Jew.[18] Zionists rejected the disembodied, exilic existence of the Diaspora Jews and firmly believed that the Zionist transformation could not be achieved within the diasporic context. Because the Jewish diasporic condition was caused by the diseased nature of modern European culture, Zionists argued that the desired transformation could be achieved only when the Jews were transplanted to a fixed place such as Palestine. Other Zionist ideologues believed that such transplantation would protect the Jews of Europe not only from anti-Semitism but also from the adverse effects of the Jewish Enlightenment. It would safeguard Jews from the demographic decline to which assimilation and

16. Nordau, "Yehadut hashririm," 174, 176; Mosse, *Nationalism and Sexuality,* 143.

17. Ze'ev Jabotinsky, "Silk and Steel" (1924), quoted in Meira Weiss, *The Chosen Body: The Politics of the Body in Israeli Society* (Stanford: Stanford University Press, 2002), 2.

18. Jonathan Boyarin, "A Response from New York: Return of the Repressed?" *Grasping Land: Space and Place in Contemporary Israeli Discourse and Experience,* ed. Eyal Ben-Ari and Yoram Bilu (Albany: SUNY Press, 1997), 218; and Mayer, *Jewish Identity in the Modern World,* 10–32. I thank Hagit Lavsky for pointing out this distinction.

FIGURE I.I. Ayanot during exercise, February 1940. Courtesy of the Keren Keyemet L'Israel
Photo Archive.

other modern social ills—conversion, intermarriage, mental illness, migration, low fertility rates, and venereal disease—could lead. It would
save them from the depravity of Europe itself.[19]

In 1903, the leadership of the Zionist Organization, the official international organization of the Zionist movement, decided that Jewish
regeneration could only take place in Palestine. Although the Zionist Organization initially considered other locales, the majority of Zionists concentrated on Palestine. This focus was due to the significant role of Palestine in Jewish history. *Eretz Israel* (the Land of Israel), the Holy Land,
was the site of Jewish statehood in the biblical period and is invoked in
Jewish liturgy.[20]

19. Efron, *Defenders of Race,* 144, 163; Biale, *Eros and the Jews,* 178; Mayer, *Jewish Identity in the Modern World,* 61; Hart, *Social Science,* 3, 137.
20. Lockman, *Comrades and Enemies,* 27.

Transforming People, Conquering Labor and Land

The Zionist Organization's focus on Palestine reflected a core component of Zionist ideology about national transformation: the notion that
the Jewish people were inextricably linked to the land of Israel. Historian
Yael Zerubavel writes,

> Influenced by European romantic nationalism on the one hand and drawing
> upon a long, distinctively Jewish tradition of longing to return to the ancient
> homeland on the other, Zionism assumed that an inherent bond between the
> Jewish people and their ancient land was a necessary condition for the develop-
> ment of Jewish nationhood. Indeed, the movement's name, Zionism, was based
> on the Hebrew name of the ancient homeland, Zion, articulating the centrality
> of this bond between the people and the land.[21]

Such Zionist notions of an organic bond between the land and the Jewish
people were rooted in earlier expressions of German *volkish* nationalism
in which man is bound up with the soil. In this paradigm, the landscape
was thought to reflect the soul of the nation.[22] Shaul Tzernichovsky, a
famous Zionist poet and doctor whose work forms part of the Zionist
canon, conveyed this union in "Ha'adam eino ela" (1930): "Man is nothing but piece of soil [*karka*] of a small land/Man is nothing but the body
[*tavnit,* also structure] of the landscape of his native land."[23] Here Tzernichovsky, uses the metaphor of the landscape as a natural way of talking
about the body in the world. He taps into the labor Zionist ethos of returning to the land in order to become one with it, the return to oneself
and one's nation.

21. Yael Zerubavel, *Recovered Roots: Collective Memory and the Making of Israeli National Tradition* (Chicago: University of Chicago Press, 1995), 15–17, 28–29, 90; Tsili Doleve-
Gandelman, "The Symbolic Inscription of Zionist Ideology in the Space of Eretz Israel: Why
the Native Israeli is Called Tsabar," *Judaism Viewed from Within and Without: Anthropological Studies,* ed. Harvey E. Goldberg (Albany: State University of New York Press, 1987), 259;
Shmuel Almog, "People and Land in Modern Jewish Nationalism," in *Essential Papers on
Zionism,* ed. Yehuda Reinharz and Anita Shapira (New York: New York University Press,
1996), 15–16, 29, 54–60.
22. Berkowitz, *Western Jewry and the Zionist Project,* 93; Sternhell, *Founding Myths of
Israel,* 6–8, 10–16, 25, 31–32, 36, 47, 53–73; Shimoni, *Zionist Ideology,* 208–216.
23. Shaul Tzernichovsky, "Ha'adam eino ela," *Shirim* (Jerusalem: Shocken Publishing,
1943), 466–469; Anita Shapira, *Land and Power: The Zionist Resort to Force, 1881–1948*
(Oxford: Oxford University Press, 1992), 29–32, 142.

The conceptual wedding of land and man constituted the ideological underpinning for antimalaria and other public health projects in the *Yishuv* (Jewish community in Palestine).[24] According to this ideology, swamp drainage measures helped create the bond between the land and the Jewish people. The importance of these projects in creating that bond operated according to the following logic: if the Jewish people and Jewish individuals are connected to the soil in complex ways and can only become whole nationally, spiritually, and physically through that relationship, then a precondition for the fulfillment of this equation is that the land itself had to be whole and healthy. But what if this equilibrium could not be struck in reality—on the ground, so to speak? If a "healthy" state of being was not a given, which it was not in swamp-infested Palestine, then the land had to be *made* healthy. European ideas about the degeneration of the Jewish body relative to the Jews' environment were therefore imported and translated into actions to make Palestine—the Jews' new environment—hygienic.[25] Perfecting both the land and the Jewish people through malaria control therefore became a vital part, if not a precondition, for the Zionist project and its promise of national redemption. One could not be done without the other; both had to be healed synchronically for redemption.

Hebrew phrases that described the interdependence of healing man and land in Zionist public health discourse included *havra'at hakarka ve hayishuv* (healing the land and the *Yishuv*) or *havra'at ha'am ve haaretz* (healing the [Jewish] nation and the land of Israel). Although the basic, literal English translation of *havra'at hakarka vehayishuv* is "healing the land and the Jewish community in Palestine," the term *yishuv* here carries with it, in Zionist parlance, the connotation of the new Zionist, Jewish nation. Throughout the book and in its title, I therefore use the broader English translation of this phrase: healing the land and the nation.

Zionist physicians like Hermann Zondek articulated these phrases before and during the Mandate period with regard to clinical and preventive medicine. For example, at the ninth meeting of the Hebrew Medical Society of Jaffa in 1912, Dr. Yunis-Gutman (a founding member) argued for a string of Jewish hospitals in Palestine to be established in order to work toward *havra'at haaretz* (healing the land). These phrases were also

24. Penslar, *Zionism and Technocracy*, 1.

25. These ideas were based in European social and medical sciences. Hart's book deals with social science in Europe, and does not show how these notions were applied in the context of Mandatory Palestine. Hart, *Social Science*, 18.

used in many popular contexts.[26] A blessing in a New Year's card in 1938 distributed by the Histadrut Nashim 'Ivriot (Hebrew Women's Union), an organization active in social welfare activities in Palestine, read, "May the coming year be a year of peace and building, a year of creation and development and of our realization of all the aspirations of 'Am Israel (nation of Israel)—the healing of the nation and the land. havra'at ha'am ve ha'aretz."[27]

The phrase "healing the land and the nation" was closely related to the well-known phrase Geulat ha'aretz, "redemption of the land." In the Mandate period, "redeeming the land" typically referred to the purchase of land when it was transferred to Jewish hands. As material conditions changed in Palestine, the concept of "redeeming the land" went through many changes as well. In the second Aliya (Zionist wave of immigration from 1904 to 1914), "redeeming the land" meant the full redemption of the Jewish people and the land. But from the fourth Aliya (1924–1931) onward, other terms denoted the full redemption of the land and the people. "Redeeming the land" referred only to the purchase of land.[28] The phrase "healing the land and the nation" eventually replaced "redemption of the land" in evoking the full redemption of the people and the land. In addition to affirming the end result of a healthy land and people, "healing the land and the nation" described the actual land reclamation that would bring the improvement of the soil and the people after land purchase. "Healing the land and the nation" reflected the notion that the "reclamation of the land and of the body" was intimately connected in Zionism.[29]

The chief Zionist organization in Palestine that "put into concrete form the connection between this ideology and Eretz Israel" was the Keren Keyemet L'Israel (KKL), the main land-purchasing agency of the Zionist Organization. The KKL was charged with producing the redemption of

26. Minutes of the ninth meeting of the Hebrew Medical Society, May 12, 1912, Zichronot HaDavarim 2–3 (May 1913): 47, CZA Library. The Hebrew Medical Society for Jaffa and the Jaffa district was founded by nine physicians (out of a total of thirty-two Jewish physicians in Palestine at the time) on January 11, 1912.

27. Memo from Histadrut Nashim Ivriot, September 23, 1938, CZA J17/669.

28. Henry Near, "Geulat hakarka," in Geulat hakarka beEretz Israel: re'ayon vema'ase [Redeeming the land in Eretz Israel: Idea and practice], ed. Ruth Kark (Jerusalem: Yad Ben Zvi Press, 1990), 41, 44; Shmuel Almog, "Hageula berhetorica bezionit," in Geulat hakarka beEretz Israel, 16.

29. I borrow this term from Biale, Eros, 177.

the land and its dialectical counterpart, the Jewish people.[30] It is no surprise, then, that in the field of malaria eradication and land reclamation, largely under the direction of the KKL, the phrases *havrá'at hakarka vehayishuv* (healing the land and the nation), *havrá'at HaAretz* (healing of the land), or *geulat há'aretz* (redemption of the land) frequently surfaced in land purchase correspondence and drainage reports.[31]

Healing the land and the nation through land reclamation was inseparable from another Zionist ideological paradigm, *kibush há'avoda* (conquest of labor).[32] The idea of conquest of labor and the related concept, *'Avoda 'ivrit* (Hebrew labor), referred to an effort by Jews to be productive in Palestine, to overcome their bourgeois past, undertake agricultural, manual labor and physically transform themselves into an "authentic Jewish proletariat" in the process.[33] Zionist labor parties upheld these ideas as they espoused the synthesis of socialism and Jewish nationalism in distinct ways.[34] The leitmotif of redemption drew upon Western physiocratic ideas, which stressed that a nation's wealth derived from the land and that a healthy economy was dependent upon scientifically proven agricultural methods, and was strengthened by agrarian reforms in Russia, England, and France.[35]

The Zionists' emphasis on manual labor in *Kibush há'avoda* was a response to the accusation that Diaspora Jews were economic parasites because their work was limited to commerce and trade. The idea of "normalizing" Jewish life through the restratification of Jewish labor and the idea of gaining spiritual inspiration by cultivating the soil were strongly linked. This idea is commonly referred to as *Dat há'avoda* (Religion of Labor) and is attributed to A. D. Gordon (1856–1922), a leader of the workers' movement of the second wave of Zionist immigration. Gordon added a spiritual dimension to Hebrew labor with the idea that agricultural

30. Doleve-Gandelman, "The Symbolic Inscription of Zionist Ideology in the Space of Eretz Israel," 258–261; Near, "Geulat hakarka," 44; Zvi Shilony, *Ideology and Settlement: The Jewish National Fund, 1897–1914* (Jerusalem: Magnes Press, 1998), 15–18.

31. Letter from Ruppin and Ettinger of the KKL to N. De Lieme, "Rechishat hakarkaot be'Emek Yizra'el" [Purchase/acquisition of the lands in Jezreel Valley] (March 27, 1921): 2, CZA KKL3/53.

32. Gershon Shafir, *Land, Labor and the Origins of the Israeli-Palestinian Conflict, 1882–1914* (Cambridge: Cambridge University Press, 1989), 188.

33. Lockman, *Comrades and Enemies,* 48–49.

34. Shimoni, *Zionist Ideology,* 166; Lockman, *Comrades and Enemies,* 17.

35. Ruth Kark, "Land-God-Man: Concepts of Land Ownership in Traditional Cultures in Eretz-Israel," in *Ideology and Landscape in Historical Perspective: Essays on the Meanings of Some Places in the Past,* ed. Alan Baker and Gideon Biger (Cambridge: Cambridge University Press, 1992), 71.

work would heal the diasporic condition; in his words, cure the "disease that attacked us."[36] Here Gordon appropriated the religious concept of *geula* (redemption) for Zionist purposes and drew upon messianic notions of redemption. These notions included the rejection of Jewish dispersion in the Diaspora, the (re)settlement of Jews in the Holy Land, and for many Zionists, active and constructive human collaborations to hasten the coming of the messiah.[37] In Zionism, the creation of a new spiritual bond with *Eretz Israel* would replace the land's traditionally religious sacredness. Man would instead regenerate the land of Israel with a secular act.[38] This spiritual element in nationalism, Gordon believed, resulted in the "blending of the natural landscape of the homeland with the spirit of the people inhabiting it."[39] Gordon's ideas strongly influenced labor Zionism, the prevailing ideology of the Mandate period.[40]

Another meaning of *kibush ha'avoda* (conquest of labor) developed as a result of labor conditions in Palestine, where Arab workers provided cheap wage labor for Jewish capitalist farmers in the first and second waves of Jewish immigration (1882–1914). Palestine's labor market presented considerable employment competition for the newly arrived Jewish immigrants. As a result, the conquest shifted from an individual and collective drive to proletarianize the Jewish worker to a tactical and defensive struggle against the hiring of Arab laborers in order to protect employment for Jews and to secure a European standard of living by obviating competition with Arab labor. Workers did not generally think of this policy as exclusive or discriminatory but primarily as a struggle to defend the rights of organized Jewish labor.[41] This labor strategy was largely ineffectual in practice because many Jewish farmers refused to stop hiring Arab laborers out of self-protective economic interest.

As a result of this tension, another ideological paradigm, *kibush hakarka* (conquest of the land), became the vehicle through which *kibush ha'avoda* would be achieved. Conquest of the land stressed Hebrew labor

---

36. A. D. Gordon quoted in Mitchell Cohen, *Zion and State: Nation, Class and the Shaping of Modern Israel* (Oxford: Basil Blackwell Inc, 1987), 97; Doleve-Gandelman, "The Symbolic Inscription of Zionist Ideology in the Space of Eretz Israel," 261; Shafir, *Land, Labor and the Origins of the Israeli-Palestinian Conflict,* 191; Shapira, *Land and Power,* 65.

37. Shimoni, *Zionist Ideology,* 54–55.

38. Penslar, *Zionism and Technocracy,* 1.

39. Gordon, "Our Tasks Ahead," 379.

40. Near also notes that the religious aspect of geula becomes less relevant in the Mandate period. Near, "Geulat hakarka," 41, 45.

41. Shafir, *Land, Labor and the Origins of the Israeli-Palestinian Conflict,* 60–90; Lockman, *Comrades and Enemies,* 47–57.

but bypassed the labor market by shifting focus to the nationalization of land.[42] Land bought through a national agency such as the KKL guaranteed and required Hebrew labor.[43] In this way, the strategies of conquest of labor and conquest of land became interdependent. By the Mandate period, these strategies were firmly in place, emphasizing self-sufficiency among Zionist settlers for the development of a Jewish economy in contrast to an earlier dependence upon Arab labor.[44] Health bureaucrats, who composed part of the non-agricultural Hebrew workforce, reinforced the aim of self-sufficiency through their sanitation measures for workers in Jewish settlements. Towards that end, Jewish medical agencies operating in Palestine welcomed financial support from a variety of sources, mostly from the United States.[45]

Malaria, Swamps, and National Transformation

Within the dual conquests of labor and land (and the related religion of labor), however, there was an inherent tension. On one hand, Jews would return to the land to become closer to it. On the other hand, Jews would achieve this closeness and redemption by controlling nature and molding it to their needs. Though a primary reason for antimalaria projects in most colonial contexts was to enable European settlement in tropical locales, the Zionist project of taming an environment and changing immigrants' habits to prevent physical and mental degeneration revealed the settlers' great vulnerability and anxiety about diseases that could seriously threaten their nation-building endeavor.[46] Given the high incidence of malaria in the first decade of the Mandate and its endemic status throughout the entire period, these fears were not unfounded.

Efforts to rid Palestine of malaria through swamp drainage before and during the Mandate period became emblematic of the self-transformation

---

42. Shafir, *Land, Labor and the Origins of the Israeli-Palestinian Conflict*, 162, 189.
43. Shafir, *Land, Labor and the Origins of the Israeli-Palestinian Conflict*, 135–145, 162.
44. Lockman, *Comrades and Enemies*, 55.
45. Letter from Szold to Judge Mack, December 26, 1920, CZA J113/554.
46. On anxiety about disease, colonialism, and settling in the "tropics," see David Arnold, *Colonizing the Body: State Medicine and Epidemic Disease in Nineteenth-Century India* (Berkeley: University of California Press, 1993), 42; and Mark Harrison, *Climates and Constitutions: Health, Race, Environment and British Imperialism in India, 1600–1850* (Oxford: Oxford University Press, 1999), 224.

process inherent in conquest of labor. Eliminating malaria specifically advanced the goals of labor Zionism by facilitating the transformation of Jewish society from one alienated from the land in Europe to one based on the land in Palestine. The role of malaria eradication in transforming Jewish society was critical: the more Jews worked the land, the fewer places existed for the mosquito to reproduce and develop. Reduced mosquito presence would produce better conditions in the colony, fewer cases of malaria, and lower mortality rates among the colonists, allowing further beautification and transformation of the settlement areas.[47] Thus, according to Dr. Israel Kligler (1889–1944), the chief Zionist malariologist in Palestine during the Mandate period, eradicating the mosquito and malaria was vital to the fortitude of the Jewish people in Palestine. It was an essential move toward "success in rebuilding our home in Palestine." Although he produced this report for the Rockefeller Institute, Dr. Israel Kligler used the language of the Balfour Declaration—the concept of the Jewish National Home—which showed his Zionist sentiments.[48] However, as in other colonial narratives, settlers, physicians, and scientists could not foresee the future cumulative side effects of their conquest of nature; their primary focus at this time was the redemption of the land and the Jewish nation.[49]

Even though the majority of Jewish immigrants preferred to continue their urban existence, labor Zionism emphasized agricultural settlement in rural areas as the most effective way to establish "the territorial basis of the new Jewish society" and "to strengthen" the emotional and cultural ties between land and people.[50] Malaria was prevalent in both the rural and urban areas of Palestine, albeit caused by different conditions. In the urban areas, malaria was mostly caused by mosquito breeding in cisterns. In the rural areas, Jewish settlers contracted "pioneer malaria," as medical

47. *Conversation on Malaria,* no. 3 (Jerusalem: Hadassah National Library, 1921), 12, CZA Library.

48. Israel Kligler, *Sanitary Survey of Palestine* (New York: Rockefeller Institute, 1917), 43. Rockefeller Foundation Archives, Tarrytown, New York (hereafter RF), RF5/2/61/399; and J. P. Goubert, *Conquest of Water: The Advent of Health in the Industrial Age* (Princeton: Princeton University Press, 1986), 6.

49. David Lowenthal, "Empires and Ecologies: Reflections on Environmental History," in *Ecology and Empire: Environmental History of Settler Societies* (Seattle: University of Washington Press, 1997), 233.

50. Nachum T. Gross and Jacob Metzer, "Public Finance in the Jewish Economy in Interwar Palestine," *Research in Economic History* 3 (1978): 121, as quoted in Smith, *Roots of Separatism in Palestine,* 87, and Shafir, *Land, Labor and the Origins of the Israeli-Palestinian Conflict,* 147.

historian Erwin Ackerknecht described it in his study on the Upper Mississippi Valley, which resulted when a continual influx of susceptible immigrants came into an agricultural setting and therefore increased the rates of malaria occurrence.[51]

Real vulnerability to pioneer malaria was countered by the invocation of heroic infection. Jewish immigrants of the second and third waves of immigration (1904–1914 and 1919–1923, respectively) believed malaria infection was a form of self-sacrifice that exemplified Zionist patriotism. Leaders extolled the virtues of martyrdom to mobilize Jews before and during the Mandate period and deemed individual sacrifice as essential for the realization of Zionism.[52]

Choosing the hardest forms of labor, 'Avoda shechora (literally "black work" and meaning very hard, manual labor), signified the highest form of commitment to the national project, even if it meant a risk to life. During the second Aliya, draining the swamps—engaging in "black work"—was most strongly associated with heroism in Zionist mythology. Yet even after this ideological wave of immigration, devotion to this principle continued. One striking example is the case of the Kabbara swamps.[53] During the beginning of the drainage process in 1924, the Palestine Jewish Colonization Association (PICA), a Jewish land-purchasing agency involved in reclamation works, hired neighboring bedouin workers while limiting Jewish employees. This situation caused much controversy because it went against the spirit of conquest of labor and conquest of land. By that time, the KKL had adopted a Jewish-only labor policy (at least in theory) for their reclamation schemes. These workers believed that self-sacrifice authenticated the relationship between the land and the Jewish people and wrote to HaPoel HaTzair (The Young Worker), the publication of the labor party under the same name, that

> Zionist workers want to stand in the Kabbara swamps neck high in water and feel the harms of creation. There is no hard work in front of us and there is no

51. Erwin Ackerknecht, *Malaria in the Upper Mississippi Valley, 1760–1900* (New York: Arno Press, 1977), 130, 132; League of Nations Health Organization Malaria Commission, *Principles and Methods of Antimalarial Measures in Europe* (Geneva, 1927), 45.

52. Lockman, *Comrades and Enemies*, 55.

53. Martin Bunton, "Land, Law and the 'Development' of Palestine: The Case of the Kabbara/Caesarea Concession" (unpublished paper, 1998); Geremy Forman and Alexandre Kedar, "Colonialism, Colonization and Land Law in Mandate Palestine: The Zor al-Zarqa and Barrat Qisarya Land Disputes in Historical Perspective," *Theoretical Inquiries in Law* 4, no. 2 (July 2003): article 11.

fear of death [from malaria] because the work for building and creating is our work. Because it makes us closer to the end and quickens redemption. And this is our goal...We overcame the Nahalal swamps, and the Nuris swamps and so for the swamps of Kabbara it is our duty to stand at the front line and win. And if they will demand victims from us, [we will] give them. And then we will have a more healthy feeling...It is our obligation and our right to die in this place so as to leave behind our right to live in this place.[54]

Acknowledging this ethos, Dr. Isaac M. Rubinow (1875–1936), a proponent of social insurance in the United States and first director of the American Zionist Medical Unit in Palestine from 1918 to 1922, wrote to the Zionist Commission in 1921 that insistence upon the use of mosquito veils and gloves "shows an utter lack of familiarity with the habits and psychology of the *Halutzim* [pioneers]." Here he implied that the Hebrew pioneers would not comply with these prescriptions, given their belief that contraction of malaria reflected their devotion to Zion.

During the early years of the Mandate period, Dr. Israel Kligler argued against the settlers' view of heroism by teaching them how to protect themselves against the scourge. He stressed that a Jewish settler did not have to die to show patriotism.[55] In the eventual shift of the settlers' attitude toward Kligler's view, the meaning of self-sacrifice and heroism changed but it was not discarded. Where settlers previously attained heroic status by contracting malaria, they now achieved heroic status by avoiding infection. Kligler commented on this important shift in health awareness after the first experimental malaria campaign in Migdal, Kinneret, and Yavniel in 1921. As he wrote in 1928,

The first, and perhaps the most important, fruit of this experiment was the change in the attitude of the population toward malaria...the attitude of the population had changed; they had come to realize that malaria was a preventable disease...Most important is the fact that the old indifference is gone... The settlers themselves now recognize that it is a foolhardy, quixotic kind of martyrdom to succumb to malaria...In other words, malaria has been robbed

---

54. *HaPoel HaTzair,* November 1924 as quoted in Y. Ayalon, "Yibush habitzot Kabbara" [The drainage of the Kabbara swamps], *Zev Vilnay Jubilee Volume* (1987): 235–237; Don Handelman and Lea Shamgar-Handelman, "Presence of Absence," in *Grasping Land: Space and Place in Contemporary Israeli Discourse and Experience*, ed. Eyal Ben-Ari and Yoram Bilu (Albany: SUNY Press, 1997), 95.

55. Also Kitron, "Malaria, Agriculture and Development," 300.

of its mystic attributes, its inevitability; it has been revealed as a preventable—
and often eradicable—disease.[56]

The attitude of Jewish officials also changed. Kligler noted, "The colonization agencies now realize that it is neither good economics nor sound
practice to settle people in malarious areas unless provision is made to
render these areas habitable. *The land, they now realize, has to be rendered fit for colonization.*"[57] This change in the meaning of heroism was
evident later in the 1949 annual report of the Israeli Antimalaria Division, which stated: "The only reliable source is the pioneering spirit of
our workers...This pioneering spirit should be maintained under the assumption that the antimalarial front is part of the nation's struggle for
survival."[58]

Pathological Topography: Connections to Colonialism

Zionism and Colonial Medical Landscapes

Before healing the land and making it suitable for European settlement,
Zionist ideologues, settlement engineers, health professionals, administrators, and settlers—like their political counterparts—first conceptualized Palestine's swamp areas as barren and unhealthy.[59] They considered swamps
as deviations in the topography of Palestine and therefore as one of the
main targets in reordering the landscape. In this way, antimalaria programs helped create a new topography of Palestine.

Zionist health officials sought to develop their work and medical institutions "to adjust to ethnic, scientific foundations on the one hand, and
on the other hand, to think deeply about the special pathology of *Eretz
Israel* [Land of Israel], to take a part in the clarification of the medical geography of *Eretz Israel.*"[60] Kupat Holim, the medical insurance provider
of the labor movement founded in 1911, was no exception. Dr. Cohen of

56. Israel Kligler, *The Epidemiology and Control of Malaria in Palestine* (Chicago: University of Chicago Press, 1930), viii–x.

57. Kligler, *Epidemiology and Control of Malaria,* ix, emphasis added.

58. Anti-Malaria Division annual reports (Jerusalem, 1949–1960), as quoted in Kitron,
"Malaria, Agriculture and Development," 303.

59. Mark Levine, "The Discourse of Development in Mandate Palestine," *Arab Studies
Quarterly* 17, nos. 1–2 (Winter and Spring 1995): 101; and Arnold, *Colonizing the Body,* 29.

60. Dr. M. Cohen, "Tafkidei Kupat Holim," *Briut Ha'Oved* (Kupat Holim, 1924), 23.

Kupat Holim illustrated this sentiment when he said "*leyashev* (to settle) means *lehavri* (to heal)."[61]

Although Zionist settlement in Palestine contained some unique elements in its self-transformative prescriptions, it was founded on colonial notions regarding ecology and the landscape. Many Zionists were enamored with the idea of an exotic Palestine, one with a glorious Biblical past, a view that fit into broader nineteenth-century European conceptions of the "the Orient" and the Holy Land.[62] In addition, Zionist images of Palestine, including its watered places, were consistent with colonial views about the tropical environment.[63] The Zionists largely envisioned rural landscapes in Palestine as intrinsically untamed wildernesses and romantic paradises, at once centers of pastoral calm and places of inhospitable climes. This combination of influences of romanticism, exoticism, degeneracy, and rationalism projected onto Palestine's landscape was repeated in accounts of other places deemed "tropical" at the time.[64] Although the Zionists did not generally consider Palestine a Garden of Eden, they did consider some areas as healthful.[65] Mostly, however, immigrants expected to find an uncultivated, desolate land that needed intensive work before it would truly be a "land flowing with milk and honey" (in Hebrew *eretz za-vat cha-lav oo–dvash*), a Biblical phrase referring to God's description of the land of Canaan as fecund and essentially female.[66]

To create rich, agricultural land, the Zionists emulated European capitalists' relationship to their physical environment. Land management was

61. Furthermore, in health affairs the term "shvut"—usually meaning "return"—meant settling the land through human reproduction. Men in this instance were obliged to reproduce, not women. Cohen, "Tafkidei Kupat Holim," 22.

62. Edward W. Said, *Orientalism* (New York: Vintage Books, 1979); Lockman, *Comrades and Enemies,* 29; Yehoshua Ben-Arieh, "Perceptions and Images of the Holy Land," in *The Land That Became Israel: Studies in Historical Geography,* ed. Ruth Kark (New Haven: Yale University Press, 1989), 37–56.

63. Richard Grove, *Green Imperialism: Colonial Expansion, Tropical Island Edens and the Origins of Environmentalism, 1600–1860* (Cambridge: Cambridge University Press, 1995); Vaughn, *Curing Their Ills,* 2, 158; Mary Dobson, *Contours of Death and Disease in Early Modern England* (Cambridge: Cambridge University Press, 1998), 2; and David Arnold, introduction to *Imperial Medicine and Indigenous Societies,* ed. D. Arnold (Manchester: Manchester University Press, 1988), 6.

64. For Brazil, see Stepan, *Picturing Tropical Nature*; Tal, *Pollution in a Promised Land,* 24; Richard Grove, "Conserving Eden: The (European) East India Companies and Their Environmental Policies on St. Helena, Mauritius and in Western India, 1660 to 1854," *Comparative Studies in Society and History* 35, no. 2 (April 1993): 326.

65. Near, "Geulat hakarka," 39.

66. The phrase is found in Solomon's Song of Songs.

based primarily on a scientific definition of the environment that, according to scholar David Gilmartin, separated Western colonists from the natural world that they tried to control. It also separated them from the indigenous society that traditionally resided there, a point to which we shall later return.[67]

Two of the land transformation approaches used by Zionist engineers and scientists included the drainage of marshland and the reclamation of other "wasteland." They did not, however, undertake the traditionally colonial clearance of woodland as a transformative mode. Experiences by the imperial powers in the colonies had shown that deforestation was a risky proposition. Vast deforestation efforts throughout the British Empire had resulted in salinization, abandonment of wells, water table elevation, and the spread of malaria. In addition, the Ottoman deforestation of Palestine during World War I had reduced its fuel resources, devastating to Palestine's landscape and population. The Zionists therefore engaged in afforestation campaigns throughout the Civil Administration and Mandate periods between 1918 and 1947.[68]

In fact, forests and swamps hold a symbolic place in Zionist history. As much as they saw draining swamps as healing act, settlers saw planting trees as a way to connect to the land, to redeem the landscape by literally securing roots in the soil.[69] Tree planting, according to Zerubavel, was a way to reintroduce both nature and the exiled Hebrew nation into the "native" landscape and to overcome the suffering of Jewish exile. It also announced ownership of a particular space, one that became increasingly contested. As had occurred in other settlement movements around the world, an alleged, preconceived chaos was replaced with a new, Zionist-infused natural order.

For practical purposes, Otto Warburg, third president of the Zionist Organization and a botanist and expert in Asian and African flora, argued that planting trees would improve the country's climate and help its

---

67. On the Indus Basin, see Gilmartin, "Models of the Hydraulic Environment," in Arnold and Guha, *Nature, Culture, Imperialism,* 226.

68. H. C. Darby, "The Changing English Landscape," *Geographical Journal* 117 (1951): 377–398; Tal, *Pollution in a Promised Land,* 76–86; Michael Williams, "Ecology, Imperialism and Deforestation," in *Ecology and Empire: Environmental History of Settler Societies,* ed. Tom Griffiths and Libby Robin (Seattle: University of Washington Press, 1997), 176; Grove, "Conserving Eden," 320, 350, and "Climatic Fears: Colonialism and the History of Environmentalism," *Harvard International Review* 23, no. 4 (Winter 2002): 50–55.

69. Yael Zerubavel, "The Forest as a National Icon: Literature, Politics and the Archeology of Memory," *Israel Studies* 1, no. 1 (Spring 1996): 62, 63; and Roza El-Eini, "British Forestry Policy in Mandate Palestine, 1929–48: Aims and Realities," *Middle Eastern Studies* 35, no. 3 (July 1999): 72–155.

economy. He urged the importation of Asian, Australian, and African fruit trees into Palestine,[70] and before and during the early Mandate period, Australian eucalyptus trees were imported to dry swamps. In this way, the eucalyptus brought together the symbolically charged activities of tree planting and swamp draining to advance Zionist reclamation goals. The eucalyptus became such a common feature of Jewish settlements that the Arab inhabitants called it *shajarat al-Yahud* (the Jews' tree). By the 1930s, as the incidence of malaria decreased and antimalaria measures loosened, the Zionists favored the eucalyptus less as new findings emerged about the tree's' adverse effects upon wildlife, soil, biodiversity, vegetation, and the water cycle. Zionist scientists soon replaced the eucalyptus with a variety of other trees, including pine, acacia, and cypress. Tree planting not only fulfilled Zionist afforestation efforts but also prevented the reversion of Jewish land into Arab hands (a goal of conquest of land) because of a British policy related to the protection of trees.[71]

As Warburg suggested in his call for tree planting, harsh climate and "primitive conditions" were two factors that caused human degeneration in the colonial imagination. Palestine was deemed, like other tropical environments, to be "no place for a white man, and yet just the place for white dominion over man and nature."[72] Colonial medical debates about human and plant acclimatization to conditions in the colony also shaped the Zionists' particular relationship to the land. The debate regarding adaptation to new, tropical surroundings was based on the belief that the tropical environment and its diseases were distinct from those in the temperate world.[73]

70. Hillel Yofe, a key figure in malaria control in Palestine, also promoted the importation and use of eucalyptus trees to dry swamps. For Warburg, see Derek Penslar, "Zionism, Colonialism and Technocracy: Otto Warburg and the Commission for the Exploration of Palestine 1903–07," *Journal of Contemporary History* 25 (1990): 143–160.

71. Tal, *Pollution in a Promised Land*, 77–79; Zerubavel, "The Forest as a National Icon," 60.

72. Grove, *Green Imperialism*, 14; Warwick Anderson, "Disease, Race and Empire," *Bulletin of the History of Medicine* 70, no. 1 (1996): 63; Philip Curtin, *Death by Migration: Europe's Encounter with the Tropical World in the Nineteenth Century* (Cambridge: Cambridge University Press, 1989); Daniel Boyarin, *Unheroic Conduct: The Rise of Heterosexuality and the Invention of the Jewish Man* (Berkeley: University of California Press, 1997), 274, 276–277, 302–305; Gilman, *Difference and Pathology*, 31–35; and Smith, *Roots of Separatism in Palestine,* 147, on the demand of the Zionist Commission in 1921 for conditions "suitable for white men" for the Jewish Labor Company.

73. Mark Harrison, "'The Tender Frame of Man': Disease Climate and Racial Difference in India and the West Indies, 1760–1860," *Bulletin of the History of Medicine* 70, no. 1 (1996): 68–93; Anderson, "Disease, Race and Empire," 62–67; Curtin, *Death by Migration*; Grove,

As European immigrants, the Zionist settlers recognized that they would have to adapt to and endure the harsh conditions in Palestine. Though in other colonial contexts support of acclimatization dissipated in the twentieth century, a continued confidence in Zionist acclimatization was necessary to fulfill the practical and political desire of building Palestine as a Jewish national home.[74] Despite this resolve, Jewish emigration from Palestine occurred, as many settlers found they were unable to bear the country's difficult health conditions.

Human acclimatization methods in the *Yishuv* closely followed medical historian Warwick Anderson's characterization of twentieth-century colonization, in which an earlier belief in "seasoning" through long-term settlement eventually shifted to a growing confidence in sanitary barriers as sufficient to protect from tropical pathogens.[75] Sanitary engineers in European colonies expressed a growing pessimism about climatic determinism at the same time the development of sanitary barriers occurred. Colonial scientists strongly believed in the distinction between European and "native" bodies and the notion that immunity to tropical diseases would not occur with long-term residence. As Mark Harrison explains, in the case of India, these changing approaches resulted from the solidification of racial categories of colonizer and colonized. Despite such developments in the larger British Empire, Zionist medical professionals persisted in their belief in human acclimatization alongside a fastening of racial difference between Arab and Jewish populations. Notwithstanding this difference, Zionist swamp drainage methods followed the typical malaria vector control program (drainage and prophylactic measures) characteristic of twentieth-century acclimatization techniques.

In the early years of the Mandate period, Zionist imaginings about the swamp and antimalaria measures to "correct" its wasted state involved a hybrid formulation of miasmatic and germ theories.[76] From the medical perspective of the late nineteenth and early twentieth centuries, swamps contained two crucial characteristics of a disease-ridden place in miasma

---

Green Imperialism, 14; Stepan, *Picturing Tropical Nature;* and Richard Grove, "Indigenous Knowledge and the Significance of South West India for Portuguese and Dutch Constructions of Tropical Nature," *Modern Asian Studies* 30, no. 1 (February 1996): 140.

74. Harrison, "'The Tender Frame of Man,'" 90–91.

75. Anderson, "Disease, Race and Empire," 62–67; Harrison, "'The Tender Frame of Man,'" 68–93; and Arnold, *Colonizing the Body,* 40.

76. Grove, *Green Imperialism,* 14; Dobson, *Contours of Death and Disease,* 15.

theory: stagnant water and bad air. Zionist scientists integrated vestiges of such miasmatic views into ideas about the environment, even though their public health interventions utilized classic bacteriological approaches intended to eradicate the malarial parasite and mosquito vector.

Upon their arrival in Palestine, most Zionist leaders, settlers, and health professionals grafted these ideas and concerns about geography and topography onto Palestine's landscape.[77] This is not to say that a strict imposition occurred, but rather a dialectical relationship emerged between Zionist imaginings of the landscape tempered by constraints on the ground. Such constraints included ecological conditions such as rainfall, climate, and disease along with political circumstances such as land tenure practices and indigenous resistance.

While Jewish settlement and medical professionals may have seen a "patchwork of different ecological and economic communities" when arriving in Palestine, they soon began to impose a consistent pattern onto the landscape.[78] Their attempts to invest in, inhabit, and transform their own society and the natural world were situated firmly within the general environmental ethos of settler movements and societies of this era.[79] Imposing a systematic arrangement onto the landscape involved the modern, scientific process of destroying what was considered a neglected or pathological existence in order to construct a more "healthful" one. This dual process of destruction-construction was meant to not only rid the Jewish physical and mental *nature* of its contemptible diasporic traits but was also intended to cure *nature* of its diseased state by erasing undesired physiographic characteristics and replacing them with new "fertile and productive" land.[80] Once in Palestine, Zionist scientists engaged the knowledge of the applied "settling" sciences (tropical medicine, entomology, and agricultural and veterinary science), imported European and

77. The same process occurs in British India and in Brazil. Arnold, *Colonizing the Body,* 9; Stepan, *Picturing Tropical Nature,* 158–170.

78. I take this phrase from Cronon, *Changes in the Land,* 33. Such a vision applies to the British as well. Gideon Biger, "Ideology and the Landscape of British Palestine, 1918–1929," *Ideology and Landscape in Historical Perspective,* 173–196, Biger, *Empire in the Holy Land*; Levine, "Discourse of Development in Mandate Palestine," 95–104; Arnold, *Imperial Medicine,* 17; and Tom Griffiths. "Ecology and Empire: Towards an Australian History of the World," in *Ecology and Empire: Environmental History of Settler Societies,* ed. Tom Griffiths and Libby Robin (Seattle: University of Washington Press, 1997), 3.

79. Griffiths, "Ecology and Empire," 10.

80. Thomas Kuhn, *The Structure of Scientific Revolutions,* third edition (Chicago: University of Chicago Press, 1996), 66; Fleck, *Genesis,* 38–45, 79, 83, 108, 154; and David Harvey, *The Condition of Postmodernity* (Oxford: Oxford University Press, 1989), 258.

U.S. medical technologies, and recruited foreign capital to "restore" the land to its "original" state.[81]

Altneuland as an Expression of the Zionist Imagination

Elements of the Zionist imagination just reviewed are evident in the works of Zionist thinkers, political leaders, and health professionals. Herzl integrated the topic of malaria and swamps in his famous utopian novel *Altneuland: Old New Land* (1902). In this fictional account of the creation of a Jewish homeland in Palestine, Herzl used a great deal of medical imagery, including images pertaining to malaria, such as mentions of swamps being drained and eucalyptus trees being planted to "cure" the marshy soil in his discussion of a barren wasteland and a diseased people.[82] Indeed, as Davidovitch and Seidelman argue, Herzl bridges his thinking about utopianism and pragmatism in *Altneuland* through a discussion of health issues. Questions of health point to the abstract mission of Jewish rejuvenation while they also emphasize the practical, scientific practices required for achieving physical, mental, symbolic, and national transformation.[83] A main character in the novel, the Jewish European ophthalmologist Frederich, for example, insisted that Palestine was a "primitive and neglected" territory after viewing the marshy landscape. He compared Palestine to the state of Diaspora Jewry: "If this is our homeland, then it has been brought just as low as we are."[84]

Herzl's novel, set twenty years in the future when written, describes the Jews of the old *Yishuv* (the Jewish community in Palestine before Zionist settlement) and of the Diaspora as a population in desperate need of reformation. At the beginning of the novel, Frederich points out to

81. The term "sciences of settling" is used by Libby Robin, "Ecology: A Science of Empire?" in *Ecology and Empire: Environmental History of Settler Societies,* ed. Tom Griffiths and Libby Robin (Seattle: University of Washington Press, 1997), 65; Gilmartin, "Models of the Hydraulic Environment," 226–227; Penslar, *Zionism and Technocracy,* 3; Alan Baker, "On Ideology and Landscape," in *Ideology and Landscape in Historical Perspective: Essays on the Meanings of Some Places in the Past,* ed. Alan Baker and Gideon Biger (Cambridge: Cambridge University Press, 1992), 3, 3n14; memorandum to Mrs. Schoolman from Denise Tourover, dated August 6, 1942, re: Atabrine, Hadassah Archives, New York, NY, RG1/100/2.

82. Nadav Davidovitch and Rhona Seidelman, "Herzl's Altneuland: Zionist Utopia, Medical Science and Public Health," *Korot* 17 (2004): 2–3; Theodor Herzl, *Altneuland/Old-New-Land,* trans. Lotta Levensohn (Princeton: Markus Wiener Publishers, 1997), 122 and also 42, 238.

83. Davidovitch and Seidelman, "Herzl's Altneuland," 5.

84. J. Kornberg, introduction to Theodor Herzl, *Altneuland: Old New Land,* trans. Lotta Levensohn (New York: Markus Wiener Publishing, 1987), viii.

Kingscourt, his traveling companion, that "no one can be deader than the Jewish people." Whereas Frederich is described as having a "flat-chested" body, the protagonist David Litwak is described as a "tall, well-built, sunburnt young man...brave and free, healthy and cultured, a man who could stand up for himself."[85] Litwak is the Zionist pioneer, a character modeled after David Wolffsohn, Herzl's associate and successor as head of the World Zionist Organization.

In *Altneuland,* physical changes of the new Jew, illustrated by the description of David Litwak, parallel the benefits of tending the soil: "Just as plants can be saved, if they are transplanted to the right soil in time, so can human beings. That is what we have done."[86] Once transplanted, man struggles with the "stubborn soil" in order to master it and make it his "friend."[87] Part of the encounter with nature involves a struggle with water. Litwak revels in the Jewish triumph over the flow of water when he acknowledges the work of the hydraulic engineers. He describes them as the "true creators of our Old New Land...they drained the swamps, irrigated the steppe...used water for electric power—[which] was the foundation of everything." A character who works for the British Health Department of Palestine adds, "We do a lot here for public health." The state of health achieved in the New Society, however, does not equal societal immortality as we see in the death of David's mother at the end of the novel.[88]

Altneuland also includes an Arab character, Rashid Bey. Bey is a European-educated Arab and a member of the utopian New Society in Palestine established by the Jews.[89] Through Rashid, Herzl tries to show that the Palestinian Arab community would benefit from Zionist immigration, agricultural technology, and public health interventions.[90] But in so doing, Herzl recapitulates Zionist images of Palestinian marginality and the European colonial racial typology of a hierarchical relationship between European peoples and the "natives" of the colonies. These are two issues to which we shall shortly turn. The inclusion of Rashid, however, differs from Herzl's earlier disregard for the Palestinian Arab population in *The*

85. Herzl, *Altneuland,* xix, 42, 47, 55.

86. Herzl, *Altneuland,* 62.

87. Herzl, *Altneuland,* 167.

88. Herzl, *Altneuland,* 175, 203.

89. Theodor Herzl, *The Jewish State* (New York: Dover Publications, 1988), 115, for a discussion of hygiene; Shapira's discussion of Herzl, *Land and Power,* 42; Elias Zureik, *The Palestinians in Israel: A Study in Internal Colonialism* (London: Routledge, 1979), 37.

90. Herzl, *Altneuland,* 94–95.

Jewish State (1896), a book commonly cited for its espousal of the *terra nullius,* or "empty land," doctrine. Under the *terra nullius* principle, applied particularly in colonial contexts, if the land was not being cultivated, then by Western standards it was considered as not being properly used. Those who could, therefore, cultivate the land had the right, if not an obligation, to do so.[91]

Combating disease is a central mission of the New Society in Herzl's novel. Herzl creates the character of Dr. Eichenstamm, an eye doctor and head of the New Society. His character is modeled after Max Mandelstamm, a Russian Zionist and Herzl supporter. Mandelstamm, along with Israel Zangwill, cofounded the Jewish Territorial Organization, whose mission was to find a territory for the Jewish state. Mandelstamm was an ophthalmologist who propounded the degeneration of Eastern Jewry and believed in a strong link between a healthy body and a healthy environment.[92] Herzl's inclusion of two ophthalmologists in *Altneuland,* Frederich and Dr. Eichenstamm, alludes to the prevalence of trachoma in Palestine at the time, and also perhaps to the new, hygienic habits needed for adoption in order for Jewish national transformation to take place. Indeed, trachoma was considered a "social disease" as it related to poor hygienic conditions. It was prevalent not only in Palestine but also in Eastern Europe at the time.[93] For Eichenstamm, although welfare work is said to be less necessary "than in many other communities, because social conditions are far healthier," hospitals, infant welfare, physical education, and other public health and clinical measures are considered central elements in the New Society.[94]

Herzl also portrays Zionism's place within the colonial world, especially in its relation to science, medicine, and malaria, using the character of Steineck as his vehicle. The architect, engineer, and laboratory doctor in the novel, Steineck praises the eucalyptus trees in their ability to absorb the water of the swamps. He explains the importance of drainage measures for settlement and the severity of the malaria problem in the

91. The *terra nullius* doctrine refers to a seventeenth-century legal principle that allowed European colonial powers to take control over land that was considered unclaimed. In the eighteenth century, the principle expanded to include taking control over land occupied or uncultivated by "native" peoples or people considered uncivilized. The most well-known case of the *terra nullius* doctrine in practice was the United Kingdom's claim over Australia and its final invalidation in 1992.

92. Hart, Social *Science,* 105.

93. Thanks to Nadav Davidovitch for pointing this out.

94. Herzl, *Altneuland,* 60.

entire colonial world.[95] By studying microbes in his laboratory, Steineck hopes to find the drug to cure malaria in Palestine and then export his knowledge to Africa.

Comparing Palestine to colonial Africa, Steineck situates Zionist antimalaria efforts within a larger colonial context. Steineck asserts that malaria has largely been eradicated in Palestine because of "our drainage, canalization, and eucalyptus groves." He compares these results to the situation in Africa, where he notes that "costly amelioration projects are not possible there, because there is no mass immigration."[96] In stark contrast to Palestine, the white man in Africa gets sick and dies, causing Steineck to remark that Africa will only be civilized when malaria has been eradicated. In his plan to export Jewish redemption, Steineck intends to transport both the white and black masses to Africa. His obligation to help in this process is based on the belief that "there is an unsolved national problem, a great tragedy of human suffering that only we Jews can fully comprehend ... Now that I have lived to see the return of the Jews, I wish I could help to prepare the way for the return of the Negroes."[97]

Steineck's commentary shows how malaria eradication and swamp drainage efforts closely overlap with nationalist aims in the Zionist imagination. As we have seen, the meanings of these efforts for the Zionist project are particular to the movement's larger ideologies. Yet as we observe in Herzl's novel, they also draw from and interact with colonial notions of (and connections between) the environment, disease, and race. In practice, these Zionist nationalist and colonial visions were physically and culturally projected onto the landscape of Palestine through swamp drainage measures. The story of malaria and its control in Palestine illustrates how nationalist means are often justified by scientific projects.

Redemptive Topography

Land and Water as Giver and Taker of Health

Despite the Zionists' negative view of swamps as stagnant and contaminating, the Zionists considered other (watered) areas of Palestine healthful,

---

95. Herzl, *Altneuland*, 121, 125.

96. Herzl, *Altneuland*, 129.

97. Herzl, *Altneuland*, 129. The analogy of the Jews and Africans and/or African American reflects ideas of racial hygiene prevalent during the time that equated Jews with Blacks. Gilman, *Difference and Pathology*, 31–35, and idem, "By the Nose: On the Construction of 'Foreign Bodies,'" *Social Epistemology* 13, no. 1 (1999): 49–58.

purifying, and exploitable for healing purposes.[98] Their effort to distinguish between healthy and unhealthy zones was consistent with the European colonial tendency to view landscapes as either therapeutic or pathological spaces.[99] The land of Palestine and water, in Zionist theory and practice, was at once both the giver and taker of health.

Springs suitable for bathing were part of the therapeutic landscape of Zionist Palestine. Bathing was historically practiced in all parts of the world to cure or heal skin diseases, chronic diseases, and other ailments. Many Zionist doctors believed in this practice, known as hydropathy, which became a popular method for improving health in Europe in the nineteenth and twentieth centuries, not only as a medical therapy but also for individual redemption and social uplift.[100] The spa "held the torch" for the bourgeoisie relationship to water and hygiene of the body during this period.[101] Zionist doctors were interested in replicating this practice in Palestine for the European Jewish population as one of the means of social, cultural, and physical regeneration. They intended to make healthful areas into resort spots for tourists. Instead of being an alternative therapy to a physician's allopathic cures, hydropathy was integrated into many Zionist doctors' therapeutic prescriptions.

The Tiberias Hot Springs project exemplifies the Zionist medical debate about hydropathy and its potential role in industrial expansion. Discussion about the Tiberias Hot Springs during the Mandate period was part of a larger effort to distinguish some areas in Palestine as "curing places" in order to facilitate the visitation to Palestine of sick people, particularly Diaspora Jews, to be cured of various ailments. Friedman performed scientific testing of the Tiberias waters before the Mandate period, in 1913. The water was found to possess all the needed qualities (temperature, salt content, etc.) in far greater number than "any medical spring in Europe." Three buildings for bathing were already in place by

98. Wilbert Gesler, *The Cultural Geography of Health Care* (Pittsburgh: University of Pittsburgh Press, 1991), 172, 176–181.

99. Grove, *Green Imperialism*, 13. For India, see Harrison, "'The Tender Frame of Man,'" 73; Kearns and Gesler, *Putting Health into Place,* 8; Wilbert M. Gesler, "Bath's Reputation as a Healing Place," in *Putting Health into Place: Landscape, Identity and Well-Being* ed. Robin A. Kearns and Wilbert M. Gesler (Syracuse: Syracuse University Press, 1998), 21–23, 27–28, 34–35; Martha E. Geores, "Surviving on Metaphor: How 'Health = Hot Springs' Created and Sustained a Town," in *Putting Health into Place*, 36–37; and Tawfiq Canaan, *Mohammedan Saints and Sanctuaries in Palestine* (Jerusalem: Ariel Publishing House, 1927).

100. Gesler, *Cultural Geography*, 180–181; and Grove, *Green Imperialism*, 13.

101. Goubert, *Conquest of Water*, 122–123.

1926—an old one, a new one, and one repaired by the Germans during World War I. European Jews only used the pools in the new and repaired buildings. Still, the conditions of these buildings and pools therein were described as dark and atrocious. Zionist doctors considered whether or not to send their patients to these unhygienic pools. On the one hand, they recognized the power of bathing practices in the public's mind and in religious traditions. On the other hand, they feared that the crowded conditions could transmit infections.[102]

The Hebrew Medical Committee strongly encouraged the hot springs project because they viewed the "question of curing places as one of the most pressing questions in the life of our country."[103] Even before additional scientific analysis of the waters and prior to the Jews attaining the concession to open the resort, an average of 3,000 visitors came to Tiberias each season, taking a total of 100,000 baths.[104] After the Zionists attained the concession for the Tiberias Hot Springs, an Arabic newspaper article noted that the Zionists rebuilt them in the "European style," making the area into a health resort and area for convalescence meant for people (Europeans) of all classes.[105] These new, modern facilities would hopefully attract people from surrounding countries and from the United States and Europe.[106]

The opening of the hot springs as a modern, Jewish health resort took on national value equal to swamp draining and tree planting. Invoking a healing ethos, the springs were thought to be not "just medical springs but *our* medical springs, a place where Jews from all over the world can feel themselves [a place that would] pave the way for the health of the nation and the land (*briut ha'am ve ha'aretz*)."[107]

Doctors conducted experiments on patient therapeutics before the hot springs buildings were remodeled (in 1924 and 1926). At the end of 1932, the reconstructed Tiberias Hot Springs opened and doctors began again to treat patients who had, among other things, rheumatoid arthritis;

102. Decisions of the Hebrew Medical Committee regarding Curing Places on its meeting of (December 25–26, 1928), ISA M1631/75/7.

103. Ibid., 2.

104. Memorandum on the Concession of the Baths at Tiberias, no date, 1, JDC Archives, file 275.

105. "Watan bil-mazad: imtiyaz al-Huleh" [Nation at auction: The Huleh concession], *Filastin* (November 7, 1934): front page.

106. Memorandum on the Concession of the Baths at Tiberias, 2.

107. Dr. M. Matmon, "Haderekh lebriut" [The way to health], *Briut* 1, no. 9 (December 15, 1932): 2.

neuritis; oophoritis and other reproductive problems; muscle, bone, or joint pain; old scars; skin diseases; neuralgia; ulcerated conditions; and trachoma. Other illnesses treated were ischias (caused by an adverse reaction to quinine shots used to prevent malaria), nervous disorders, and temporary partial paralysis.[108] The baths were contraindicated for heart diseases, tuberculosis (latent and active), nervous diseases like neurasthenia, and acute diseases.[109] Doctors commonly prescribed about twenty to thirty baths for each patient.[110]

The hot springs were considered a place free from malaria, a place where nature and science were allies.[111] In their pursuit of medical knowledge, doctors used bathing prescriptions as a way to establish the correlation between the quantity of baths and a cure for illness, the influence of climate on bathing for health, and the comparative effects of bathing on chronic and subacute rheumatism. However, no regime of water temperature or standard therapeutic number of baths or length of soaking was ever set. The price for the private (and cleaner) pools was high—twelve Egyptian liras. If one could not afford this fee, they would bathe in the public pools for a price that depended upon the number of bathers, for two, five, or eight Egyptian liras.[112]

Despite a mission of healing—or perhaps because of it—doctors involved in the Tiberias Hot Springs project referred to notions of racial distance. When the springs reopened, a particular recommendation was made to set hours for Jewish patients. Separation of European and Arab patients in the healing process was deemed preferable, especially in the "second season, when a great number of Arab bathers stream in to Tiberias." It is unclear whether or not this recommendation was actually implemented.[113]

108. Dr. Wilenska, "Harechitza beHamaei-Tiberias beshnat Tar"g" [Bathing in the Tiberias Hot Springs in 1923], *Briut Ha'Oved* 1 (1924): 1, binder 33:2, National Library of Israel at Givat Ram, Jerusalem; Dr. Mandelberg, "On Bathing in Tiberias Hot Springs" ["Al Harechitza behamei Tiberias"], *Briut Ha'Oved* 2 (1924): 6–8, National Library of Israel, binder 33.

109. Dr. M. Buchman, "Hamei Tiberias," *Briut ha'Am* 1, no. 3 (October 1926): 28–29, CZA J1/2689; Middleburg, "On Bathing in Tiberias Hot Springs."

110. See Dr. M. Buchman, "Hamei Tiberias," *Briut ha'Am* 1, no. 3 (October 1926): 28–29, CZA J1/2689; Wilenska, "Harechitza beHamaei-Tiberias beshnat Tar"g," binder 33:20–21; Memorandum on the Concession of the Baths at Tiberias, 4. A scientific analysis of the water is also attached to this memorandum.

111. Geores, "Surviving on Metaphor," 45–49.

112. Memorandum on the Concession of the Baths at Tiberias; Dr. Mandelberg, "On Bathing in the Tiberias Hot Springs," 3–5, binder 33.

113. Wilenska, "Harechitza behamai Tiberas beshnat tarp"d."

Pathological "Natives": Contested Visions and Claims to the Land

References by Zionist doctors to Arab patients at the Tiberias Hot Springs indicate the Palestinian Arab practice of bathing and washing in springs as a medical therapy as well. Palestinian *fellah* (peasant) medicine advocated therapeutic bathing, especially in watered places associated with holy places, to cure fever, eye afflictions, fear, or sterility. Dr. Tawfiq Canaan, a Palestinian Arab doctor before and during the Mandate period, performed extensive research on Palestinian folk practices and shrines. He noted that people with fevers (most likely due to malaria) usually took baths in Ein Silwan, Ein al-Samiya in Kolonia, Ein al-Nebi Ayyub, or the well of al-Sheikh Ibrahim in Beit Djibrin or they drank from the cistern of al-Suhada in Hebron. These baths were often bound to prayer time. Such rituals were quite complicated and variable.[114] Animals with afflictions were also bathed in these allegedly healing waters.

In certain cases, the Arab *fellahin* (peasant farmers) considered specific swamps as places of healing and holiness. This view sharply contrasted with the Zionist description of swamps as always dangerous to one's health, "an illness of the ground," and a formidable obstacle to Jewish redemption. For the *fellahin*, these swampy, sacred spaces formed part of an imagined therapeutic landscape of Palestine.[115] For the Zionists, the swamps were secular, pathological, instrumental, and economic objects.[116]

Al-Matba'a, a swamp in the Plain of Esdraelon at Tel al-Sammam, for instance, was associated with the *weli* (saint) al-Sheikh Ibrek and therefore considered holy and powerful by local peasants. According to Dr. Canaan, this marsh (like the Tiberias Hot Springs) had a widespread reputation for healing rheumatism and nervous pains and for curing sterility.[117] Canaan observed that the healing ritual at al-Matba'a for a barren woman involved washing herself at Ein Ishaq and then offering a present to al-Sheikh Ibrek.

In general, whether running or stagnant, water was a common feature of healing for the *fellahin*, as well as a central feature of many of their

---

114. Canaan, *Mohammedan Saints,* 110–112.
115. Y. Treidel, "Technical Problems of Soil Drainage: Address Delivered before the Jewish Engineers Association on the Occasion of the Convention of the Zionist Organization of Germany" (no date), 2, CZA J113/1430; Karakara, "Ma'arechet habriut hamandatorit vehavoluntarit ve'aravei Eretz Israel (1918–1948); Kearns and Gesler, *Putting Health into Place,* 8.
116. Kark, "Land-God-Man," 64.
117. Canaan, *Mohammedan Saints,* 42, 111.

holy places.[118] In its association with a *weli,* water was considered sacred and supernatural. According to the location of the shrine (hills, plain, or mountaintop), cisterns storing rainwater (*bir* or *hrabat,* an unplastered cistern-like hole), wells and running water from springs were the most common forms of water near holy places. Canaan recognized that the cisterns or wells were frequently in poor condition. Some of them did not hold water at all. Sometimes a *sabil* (reservoir) was attached to the sanctuary so that thirsty passersby could partake of water. Zionist doctors dealing with malaria described these reservoirs, cisterns, and wells as some of the main sources of mosquito proliferation.[119]

Broadly speaking, Palestinian Arab peasants' visions of certain watered places as therapeutic landscapes concurred with Zionist views, but with two main differences. These differences indicate the varied, contested meanings of watered places and the respective uses of those places by each community. The first difference lies in the distinction among the Zionists between stagnant water and running water with regard to healing properties. This dichotomous view contrasted with the Palestinian *fellahin*'s understanding of water (as healing or corrupting) correspondent with local religious practice and agricultural production. The urban Palestinian Arab population, especially the elite class, likely had still another view of springs, swamps, and bathing, probably based on modern scientific, medical principles. Yet it is the rural landscape of Palestine—a landscape covered with swamps—where the *fellah* population lived and upon which this study focuses. Another difference lay in the connections, or lack thereof, between healing and industry. Zionist visions included making medical bathing into a tourist industry, an approach that the *fellah* or Arab elite communities did not consider. On the other hand, *fellahin* integrated papyrus production from swamps into the capital economy. In general, each community treated watered places in Palestine in their relation to larger, respective processes of healing.

The Zionist settlers and the Mandatory government by and large failed to fully appreciate indigenous understandings of and interactions with the landscape such as saint worship in swamp areas. As many scholars of Palestinian life and of environmental history have noted, the same spaces seen as desolate by a Western eye were almost always filled with local history and meaning where animals, plants, and topographical features

118. Canaan, *Mohammedan Saints,* 111–112.
119. Canaan, *Mohammedan Saints,* 38–39.

had uses and functions not necessarily visible to or acknowledged by a foreign viewer. In contrast, the foreign observer often conflated the habitats and the inhabitants as unimproved and in need of scientific management.[120]

Although environmental historians today recognize that human relationships to nature have been consistently dynamic, variable, dialectic, and adaptive, regardless of place or culture, and that ecosystems and agricultural methods are neither continually stable nor stagnant, European settlers across the globe in the late nineteenth and early twentieth centuries were generally "hardly aware of indigenous impacts, blind to signs of non-European occupation." As David Lowenthal has explained, "They assumed that they saw virtually untouched virgin lands." They overlooked indigenous impact on land, especially those that made "improvements":

> Any impacts that settlers did note seemed to them trivial, wasteful or unproductive. Indigenes unable or unwilling to abandon "primitive" practices for permanent settlement were thus held doomed to give way to superior races with advanced technologie...Just as indigenes changed their environments more radically than settlers realized, so too they not infrequently wrought environmental havoc. No culture has a monopoly on ecological sanctity.[121]

Arab cultivation necessarily caused ecological changes both destructive and beneficial, yet the British and Zionists during the Mandate period consistently described Palestinian Arab cultivation as static and backward. They did not recognize it as responsive to regular environmental and social changes.[122] Such views persisted despite the acknowledgement that rainfall, soil quality, livestock populations, and other environmental

120. Mary Louis Pratt, *Imperial Eyes: Travel Writing and Transculturation* (London: Routledge, 1992), 61; Cronon, *Changes in the Land*, 161; and Gilmartin, "Models of the Hydraulic Environment," in Arnold and Guha, *Nature, Culture, Imperialism*, 227.

121. Lowenthal warns environmental historians against casting indigenes as ecological gurus. Lowenthal, "Empires and Ecologies," 234. See also Gilmartin, "Models of the Hydraulic Environment," in Arnold and Guha, *Nature, Culture, Imperialism*, 210; and Ian Scoones, "The Dynamics of Soil Fertility Change: Historical Perspectives on Environmental Transformation from Zimbabwe," *Geographical Journal* 163, no. 2 (July 1997): 162–163.

122. Richard Grove describes instances in other colonial contexts where officials were more critical of Western methods of cultivation. I did not find this in the Palestine case. Richard Grove, *Ecology, Climate and Empire: Colonialism and Global Environmental History, 1400–1940* (Cambridge: White Horse Press, 1997), 3.

factors varied temporally or spatially. These judgments persisted in the face of occasional imperial and Zionist efforts to solicit and utilize local knowledge in technical analyses and mapping of the land.

Although attitudes about and actions in agriculture in British colonies and among Zionist leaders were neither monolithic nor conclusively consistent, they were generally infused with racial overtones or nationalist and imperialist agendas. Scientists and other technical advisers at this time generally understood "native" lifestyles as unchanging and environmental transformations as essentially linear. These "apprehensive" technocrats customarily correlated indigenous agency with ecological decline and their own expertise with ecological progress.[123]

Indeed, the seemingly antithetical approach of a "romantic attachment to the land and a technologically driven, exploitative attitude toward it" espoused by the Zionists was likely alien to the Arab farmers' cultural experience.[124] The Zionist perception of an Arab indifference to and misuse and neglect of nature was verified to them by the existence of swamps.[125] Even in 1931, Jewish malariologist Dr. Israel Kligler noted that by allowing swamps to form, by letting nature assert *itself* rather than subduing it, the Arab population was "nature's unwitting ally" rather than its master.[126] For Kligler, failing to control nature or to directly subjugate it was completely undesirable, illustrating no strength and no initiative. Kligler added, "From the standpoint of malaria, Palestine was indeed a house in disorder, and the housewife in charge [was] ignorant and ineffective."[127] Kligler, as well as other Zionist leaders and scientists, projected the image

123. I borrow this term from John MacKenzie, "Empire and the Ecological Apocalypse: The Historiography of the Imperial Environment," in *Ecology and Empire: Environmental History of Settler Societies,* ed. Tom Griffiths and Libby Robin (Seattle: University of Washington Press, 1997), 216, 218, 226. For British-sponsored agricultural education efforts for the Arab population in Palestine, see Roza El-Eini, "British Agricultural-Educational Institutions in Mandate Palestine and Their Impress on the Rural Landscape," *Middle Eastern Studies* [England] (January 1999): 98–114; also Nancy Stepan, *The Idea of Race in Science: Great Britain, 1800–1960* (London: Macmillan Press, 1982), 124–126.

124. Tamar Katriel, "Remaking Place: Cultural Production in Israeli Pioneer Settlement Museums," in *Grasping Land: Space and Place in Contemporary Israeli Discourse and Experience,* ed. Eyal Ben-Ari and Yoram Bilu (Albany: SUNY Press, 1997), 163.

125. This also implied having no feelings for it. Yael Zerubavel, "The Forest as a National Icon: Literature, Politics and the Archeology of Memory," *Israel Studies* 1, no. 1 (Spring 1996): 72.

126. Kligler, *Epidemiology and Control of Malaria,* 12–13.

127. Israel Kligler, "Fighting Malaria in Palestine," April 16, 1931, Hadassah Archives RG1/100/2.

of weakness onto Arabs when Jews themselves were struggling with the same, oppressive view of weakness by the European population.

Racial stereotyping by Europeans of the indigenous population in tropical areas contributed to the image of malaria as a "disease of rural decay, spawn of an ill-kempt, savage and uncultivated land, yielding to good husbandry and civilization."[128] In the eyes of Zionist doctors and land experts in Palestine, the indigenous peasant population had allowed the land to become filled with swamps through its "improper" utilization. According to them, the land became desolate due to human neglect over many centuries. According to Zionism historian Anita Shapira, viewing "native" Arabs as backward supported the desirability and feasibility of Jewish settlement without the need to reconcile the implication of either for the local population.[129]

The concept of Arabs as primitive and infective was not just figurative. As they did in the recommendation to segregate patients at the hot springs, racial notions informed Zionist medical and settlement experts' ideas about the location of Jewish settlements. European colonies were generally situated far away from "native" areas because of the "natives'" high rates of infection with a variety of diseases (including malaria). This selective location, referred to as racial distancing, was thought to create a sanitary barrier for European acclimatization. Situating Zionist colonies at a distance from the indigenous Palestinian population in the rural areas of Palestine was more difficult than in other colonial contexts because of the small size of the country and the politics of land purchase. Nonetheless, Zionist sanitary engineers and doctors explicitly recommended establishing settlements far from swampy areas on Arab land, especially as the Mandate period

128. This description mostly referred to malignant malaria. Gordon Harrison, *Mosquitoes, Malaria and Man: A History of the Hostilities since 1880* (London: John Murray Press, 1978), 27; Arnold, *Imperial Medicine,* 8; Mark Harrison, "'Hot Beds of Disease': Malaria and Civilization in Nineteenth-Century British India," *Parrasitologia* 40, nos. 1–2 (June 1998): 11; and Elizabeth Whitcombe, "The Environmental Costs of Irrigation in British India: Waterlogging, Salinity, Malaria," in *Nature, Culture, Imperialism: Essays on Environmental History of South Asia,* ed. David Arnold and Ramachandra Guha (Delhi: Oxford University Press, 1995), 250.

129. Shapira, *Land and Power,* 43. Also see Sarah Graham-Brown, "The Political Economy of Jabal Nablus, 1920–1948," in *Studies in the Economic and Social History of Palestine in the Nineteenth and Twentieth Centuries,* ed. Roger Owen (Oxford: Oxford University Press, 1982), 98; Alexander Scholch, "European Penetration and the Economic Development of Palestine, 1856–1982," *Studies in the Economic and Social History,* ed. Roger Owen (Oxford: Oxford University Press, 1982), 55; and Bishara Doumani, *Rediscovering Palestine Merchants and Peasants in Jabal Nablus: 1700–1900* (Berkeley: University of California Press, 1995).

proceeded, as a way of avoiding not only the swamps themselves but also the Arab residents, who were believed to be the primary carriers of malaria.

Such racial assumptions were typical of late nineteenth- and early twentieth-century colonial thought. Race and environment were, according to Nancy Stepan, *the* signifiers of this period. The connections between race, the tropical environment, and national identity, she says, are what historians must turn to in order to make sense of this era.[130] Other scholars concur, noting that framing disease, environment, and "race" were all part of the same maneuver; a maneuver that, according to Warwick Anderson, had political and social consequences perhaps as significant as any military deployment.[131]

Specific agricultural methods formed one issue in this "framing" matrix in Palestine. The Zionist use of land stood in stark contrast to that of indigenous economies.[132] The Zionists viewed intensive cultivation as a means for production and agricultural development. Zionist malariologists believed that the bedouin population was "still in the pastoral stage," and that their constant migration assisted in spreading malaria.[133] Kligler and his colleagues deemed the bedouin's pastoral nomadism and use of swamps as a "wasteful system of nomadic grazing."[134] This attitude surfaced in the Kabbara swamp drainage concession documents drawn up between the Mandatory government and PICA. Arab inhabitants, if they were mentioned at all, were noted as those who would be employed in the drainage works or were referred to as bedouins who "pitched their tents" rather than people who resided in the area.[135] Ironically, David Ben-Gurion, the future first prime minister of the State of Israel wrote in one of his essays that reclamation should be made by "Jewish Bedouins," who would know how to support themselves like the Arabs but who "along with possessing primitive bedouin skills, will also be familiar with modern cultural, scientific and technical knowledge."[136]

130. Stepan, *Picturing Tropical Nature,* 124.

131. Anderson, "Disease, Race and Empire," 63.

132. For comparison with colonial New England, see Cronon, *Changes in the Land,* 169.

133. Kligler, *Epidemiology and Control of Malaria,* 12–13.

134. General Federation of Jewish Labour in Israel, *Documents and Essays on Jewish Labour Policy in Palestine* (Westport, CT: Greenwood, 1975), 231, as quoted in Ghazi Falah, "Pre-State Jewish Colonization in Northern Palestine and Its Impact on Local Bedouin Sedentarization, 1914–1948," *Journal of Historical Geography* 17, no. 3 (1991): 293.

135. Bunton, "Land, Law and the 'Development' of Palestine," 6.

136. David Ben-Gurion, "The Imperatives of the Jewish Revolution" (1944), in Arthur Hertzberg, *The Zionist Idea* (New York: Atheneum Press, 1959), 617.

Theoretical and Political Consequences of Zionist Visions

In Zionist discourse, attitudes about cultivation influenced views about the legitimacy of Palestinian Arab claims to the land.[137] Images of roaming bedouin or *fellahin* strengthened some Zionists' claim that Palestinian Arabs had no attachment to the land. These groups' supposed lack of real ownership or attachment to the land by the Palestinian Arab inhabitants, was mostly attributed to the Western determination that *musha'* land (collective village ownership system of land that was periodically distributed to qualified residents in shares) was totally economic ineffective.[138]

The Arabs' presumed lack of attachment to the land based on their agricultural relationship to it was reiterated in Zionist notions of nationhood. The imagined bond between the Jewish people and the land of Israel, their colonial visions and agricultural goals, had significant implications for their ideas about the status of the Palestinian Arab population. The idea of rooting Jewish identity in Palestine and returning Jews to their ancient homeland revealed a Zionist view of Palestine as essentially and eternally Jewish.[139] Palestinian Arabs, their national identities and histories, as well as their perceptions and use of land and swamps, were frequently overlooked or denied by Zionist concepts of *kibush ha'avoda* (conquest of labor), *kibush hakarka* (conquest of land), and *havra'at hakarka vehayishuv* (healing the land and the nation), and Palestinian Arab claims to the land were devalued or dismissed.

The Zionists believed that the indigenous Arab population did not possess the characteristics of a distinct, coherent nation. As such, they did not have national rights to the land. Disregarding the cultural, geographical, and social diversity within their own community, the Zionists perceived the Jews, in contrast, as one nation.[140] Ben-Gurion expressed this view

137. Ted Swedenburg, *Memories of Revolt: The 1936–1939 Rebellion and the Palestinian National Past* (Minneapolis: University of Minnesota Press, 1995), 59–60.

138. Zureik, *Palestinians in Israel,* 41; Yaakov Firestone, "Land Equalization and Factor Scarcities: Holding Size and the Burden of Impositions in Imperial Central Russia and the Late Ottoman Levant," *Journal of Economic History* 41, no. 4 (1981): 813–33, and "Crop-Sharing Economics in Palestine (Parts I and II)," *Middle East Studies* 11, nos. 1–2 (1975): 3–23 and 175–194; Sayigh in Zureik, *Palestinians in Israel,* 40; Jacob Metzer, *The Divided Economy of Mandatory Palestine* (Cambridge: Cambridge University Press, 1998), 95–96.

139. Lockman, *Comrades and Enemies,* 31.

140. Lockman, *Comrades and Enemies,* 31. The Zionist effort to homogenize and unify the nation had dire consequences for non-European Jews, especially after the State of Israel was established.

when he contrasted Jewish and Arab futures in the land according to their standing as a "nation." His evaluation ultimately obscured the national rights of the Palestinian Arab population by omitting their status in his considerations: "Palestine is destined for the Jewish nation and for the Arabs domiciled there."[141]

Given the Zionist linking of the nation and the land, Palestinian Arabs could not have a true attachment to the land if they did not constitute a nation. The idea that the Palestinian Arab population, especially its peasant farmers, had no real attachment to the land drew upon colonial racial views of non-European, indigenous populations as primitive and backward. Similarly, the idea of Palestine as a swampy wasteland inhabited by an unproductive people drew upon a broad European tradition of natural history and a modern paradigm of capitalist production in which subsistence habitats were empty landscapes awaiting discovery. The value of these habitats could only be realized by the production of surplus crops for the global market. Dismissing Palestinian Arab guardianship and attachment to Palestine, as scholar Ted Swedenburg has pointed out, defined Palestinian Arab "culture as being eternally destructive of the land."[142] As Ben-Gurion proclaimed in 1924, "We do not recognize the right of Arabs to rule the country, since Palestine is still undeveloped and awaits its builders."[143]

The Palestinian Arab farming population had its own way of imagining the land and its own attachment to it quite apart from Zionist imaginings. The idea of Palestine as a distinct country separate from other territories of the Ottoman Empire existed in the early twentieth century. This vision was accompanied by a sense of Palestinians as a national, political community that, by the Mandate period and in the face of the Zionist endeavor, developed rapidly.[144] The notions of *al-ʿard* (honor), *al-ard* (land), *sumud* (indivisibility of people from land), and a primordial connection to Palestine have been identified as central idioms in Palestinian nationalist discourse and memory. As Swedenburg observed,

141. Ben-Gurion, "Mi-tokh havikuah," *Anakhnu u-shkheneinu*, 184, as quoted in Shimoni, *Zionist Ideology*, 381.

142. Swedenburg, *Memories of Revolt*, 58.

143. Speech given on January 19, 1925, as quoted in Shabtai Teveth, *Ben-Gurion and the Palestinian Arabs* (Oxford: Oxford University Press, 1985), 38.

144. Rashid Khalidi, *Palestinian Identity: The Construction of Modern National Consciousness* (New York: Columbia University Press, 1997), 28–29, 108; Daniel Rabinowitz, "In and Out of Territory," in *Grasping Land: Space and Place in Contemporary Israeli Discourse and Experience*, ed. Eyal Ben-Ari and Yoram Bilu (Albany: SUNY Press, 1997), 178.

Old peasants saw their livelihood as clearly rooted in the land they had tilled. A man's position as a respected member of the community was viewed as closely tied to his possession of land, and hence the oft-repeated saying, "He who is without land is without honor." ... The primordialist claims regarding the Palestinians' primeval and prior roots in the land operated at the level of the collective.[145]

Furthermore, even though both the British and Zionists understood the Palestinian Arab economy as self-contained and perceived Palestine as lying in wait for European modernization and rejuvenation, recent scholarship has shown that the Palestinian economy was in fact already becoming integrated into the world market in the nineteenth century.

It is difficult to more deeply reconstruct a detailed description of the Palestinian imagination about different types of land or about malaria due to the scarcity of surviving Arabic health documents. Yet the situation of *fellah* holy sites near swamps and springs is evidence of a relationship between man and land. These sites of pilgrimage delineated a measure of ownership of that particular space.[146] This is not to say that every sector or individual of the Palestinian Arab community felt the same connection or degree of connection to the land (which can be said of the Zionist community as well) but only that an attachment existed.

Despite the difficulty in comprehensively discerning the nationalist meanings of disease and the environment for the Palestinian Arabs, values of nationhood, land and water use, and the meaning of malaria (and disease in general) intersected in the Zionist imagination and were evident in claims to Palestine as a Jewish national home and in plans to transform Palestine's physiography. Specifically, Zionist images of the land and the people and their dual transformation impacted the nature, implementation, and politics of antimalaria measures. But how severe was the malaria problem in Palestine?

145. Swedenburg, *Memories of Revolt,* 78–81. Both bedouin tribes and Arab villagers held possession or tenancy rights to the land; grazing rights of different bedouin groups were recognized and respected by the local sedentary population. Falah, "Pre-State Jewish Colonization," 291, 293.

146. Khalidi, *Palestinian Identity,* 115–116; Lockman, *Comrades and Enemies,* 16–17; Canaan, *Mohammedan Saints*; Eyal Ben-Ari and Yoram Bilu, "Saints' Sanctuaries in Israeli Development Towns," in *Grasping Land: Space and Place in Contemporary Israeli Discourse and Experience,* ed. Eyal Ben-Ari and Yoram Bilu (Albany: SUNY Press, 1997), 61, 74–75, 77.

PART I

Draining the Swamp to Heal the Land

Pathological Landscape

Epidemiology and the Medical Geography of Malaria in Palestine

Our people there are fighting disease while colonizing a wilderness, facing constant assault by the enemy while conquering the swamp and the desert. —Dr. Hershel Mayer (1940)[1]

Healing the pathological features of the land of Palestine was essential for healing the Jewish nation. Both were constitutive, interactive elements in the transformative project of Zionism. As the previous chapter showed, malaria and its control held a significant place in the ideology of labor Zionism. In practice, malaria was the most widespread infectious disease in Palestine before British occupation and the most extensively addressed public health initiative during British rule. Besides malaria, diseases with high prevalence rates in Palestine before and during the Mandate included trachoma, conjunctivitis, typhus, typhoid, tuberculosis, dysentery, pneumonia, cholera, sandfly fever, and smallpox. Different rates of infection, mortality, and fertility were evident between the Jewish and Palestinian communities. Factors that contributed to these unequal rates included, among others, the relative urban versus rural nature of each community, accessibility to medical services, rates of exposure, and funding for prevention and treatment.

1. Dr. Hershel Mayer, introduction, *Medical Leaves* 3, no. 1 (1940): 5, Ein Kerem Medical Library, Jerusalem.

TABLE 2.1   **Comparison of Birth, Death, and Infant Mortality Rates in Different Religious Communities of Palestine**

Year	Population Group			
	Jews	Christians	Moslems[1]	Total[2]
Population:				
1923	93,000	74,000	495,000	662,013
1924	94,669	73,533	504,960	681,245
1925	120,559	74,781	515,894	719,508
Birth rate:				
1923	35.2/1000	34.7/1000	50.8/1000	47.43/1000
1924	38.3/1000	40.4/1000	55.5/1000	51.53/1000
1925	33.2/1000	37.2/1000	54.7/1000	49.31/1000
Death rate:				
1923	14.1/1000	15.1/1000	29.1/1000	25.67/1000
1924	12.6/1000	16.8/1000	29.9/1000	25.04/1000
1925	15.1/1000	18.8/1000	31.2/1000	27.35/1000
Infant mortality:				
1923	125.8/1000	134.8/1000	199.3/1000	184.76/1000
1924	105.7/1000	151.9/1000	199/1000	184.80/1000
1925	131.2/1000	162.4/1000	200.5/1000	188.64/1000
1927	115.3/1000	187.2/1000	216.7/1000	200.5/1000

SOURCE: I. Kligler, *The Epidemiology and Control of Malaria in Palestine* (Chicago: University of Chicago Press, 1930), table 7, p. 17. Kligler's estimates for population numbers differ from those for similar years in Justin McCarthy, *Population of Palestine: Population History and Statistics of the Late Ottoman Period and the Mandate* (New York: Columbia University Press, 1990), 36–37.

[1] Exclusive of nomadic tribes.
[2] Including about 10,000 of other religions not delineated.

It is important to note that the campaign to combat malaria, a parasitic disease, was part of a much larger project to rid the area of infectious diseases, decrease infant mortality, and improve public health in general. Both Zionist and British officials during the Mandate period, especially at its beginning, recognized the eradication of infectious diseases and parasitic diseases as one of the most important, if not the most important, of all public health activities during the Mandate period.

In particular, Jewish health and settlement professionals concurred that Palestine was a land filled with malaria. The country possessed topographical and climatological features that contributed to the disease's transmission. Zionist colonization agencies increasingly understood that the land had to be "rendered fit for colonization."[2] Identifying the characteristics of malaria in Palestine was for them therefore essential to combating the disease and transforming both the land and the Jewish people. Ultimately, changes

2. Kligler, *Epidemiology and Control of Malaria in Palestine*, ix.

to these epidemiological and topographical features contributed to the vast transformation of Palestine's landscape from a perceived pathological one to one regarded as healthy.

The Historical Epidemiology of Malaria in Palestine

Understandings of Disease Etiology in Palestine

During the nineteenth century, most physicians and laymen in Palestine (and elsewhere) attributed malaria to disease-causing noxious vapors emitted from swamps that the body contracted while breathing. Indeed, the word "malaria" is taken from the Italian *mal aria*, or "bad air." Other doctors in Palestine believed that malaria was caused by filth and/or by eating raw fruit that was grown near the swamps.[3] Prescriptions for a cure of fever in Palestine during the nineteenth century included bathing in springs, eating an onion with either the word "Amen" or magical words inscribed in it, smearing cattle excrement on the body to ward off mosquitoes, and expelling mosquitoes from houses by use of smoke.[4] Blood-letting was also used as a treatment for malaria fever in the nineteenth and early twentieth centuries around the world. Additional indigenous remedies in Palestine for fevers included the use of plants known as *jadáa* (*teucrium polinum*), *shih* (*artemisia*), and by the bedouin, a plant known as *kolkh* or *gulh*. Spiritual healing and other herbal concoctions accompanied the use of these medicines.[5]

With the advent of bacteriology in the late nineteenth century, new explanations emerged in the European medical community about the etiology of malaria. Initial experiments and subsequent discoveries came as a direct result of and engagement with colonialism and colonial medicine,

3. Doctors in Palestine therefore prohibited eating such fruit without cooking it (especially melons). Treidel, "Technical Problems in Soil Drainage." Yosef Triedel (1876–1929) served as a scientific advisor in the German African colonies and did cartographical work for the Commission on the Exploration of Palestine (established 1903), directed by Otto Warburg. See Penslar, "Zionism, Colonialism and Technocracy," 150–151. Triedal immigrated to Palestine where the Jewish Colonization Association employed him. Penslar, *Zionism and Technocracy*, 67.

4. Zvi Saliternik, "Reminiscences of the History of Malaria Eradication in Palestine and Israel," *Israel Journal of Medical Sciences* 14, no. 56 (May 1978): 518; and Zvi Saliternik, *History of Malaria Control and Eradication in Palestine and in Israel* (Jerusalem: Israel Institute of the History of Medicine, 1980), 7.

5. Vartan Amadouny, "The Campaign against Malaria in Transjordan, 1926–1946: Epidemiology, Geography and Politics," *Journal of the History of Medicine and Allied Sciences* 52, no. 4 (October 1997): 461.

so malaria and its treatment cannot be separated from the colonial endeavor. It was considered the quintessential colonial disease.[6]

Scientific knowledge about the vector of malaria transmission filtered into Palestine through European missionary channels, through visiting European engineers and doctors hired by the Ottoman authorities, and through Palestinian Arab doctors trained in Western medicine. Palestinian Arab physicians also learned Western medicine through Jewish doctors living in Palestine. The latter, having mostly originated from Europe, maintained and kept up with Western medical developments.[7]

New medical knowledge of malaria's etiology, however, did not change the general way malaria eradication efforts were managed. Public health officials drained swamps both before and after the cause of malaria was discovered by Alphonse Laveran and Ronald Ross. Alphonse Laveran, a French physician who served as an army surgeon in colonial Algiers, found the parasite in human blood that causes malaria in 1880. Ronald Ross determined the mosquito as the vector for malaria in 1897 while serving in the Indian Medical Service in British India. Other scientists (Patrick Manson and Giovanni Battista Grassi) contributed to the discovery and explanation of malaria etiology and transmission.

With the discovery of the malaria vector, however, drainage planning became more scientific through a high reliance on statistical measurement and detailed knowledge about the mosquito, parasite, and swamps.[8]

6. Sheldon Watts, *Epidemics and History: Disease, Power and Imperialism* (New Haven: Yale University Press, 1997), 216. For other important figures in the science of malaria like Angelo Celli, Ettore Marchiafava, Camillo Golgi, Patrick Manson, Giovanni Battista Grassi, Emanuele Ficalbi, Amico Bignami, Giuseppe Bastianelli, and Sir Neil Fairley, see Harrison, *Mosquitoes, Malaria and Man*, 13–156; and Michael Worboys, "Germs, Malaria and the Invention of Mansonian Tropical Medicine," in *Warm Climates and Western Medicine: The Emergence of Tropical Medicine, 1500–1900*, ed. David Arnold (Amsterdam: Rodopi Press, 1996), 189–192. Also Dobson, *Contours of Death and Disease*, 349; Leonard Jan Bruce-Chwatt and Julian de Zulueta, *Rise and Fall of Malaria in Europe: A Historico-epidemiological Study* (Oxford: Oxford University Press, 1980), 131–145; Michael Worboys, "Manson, Ross and Colonial Medical Policy: Tropical Medicine in London and Liverpool, 1899–1914," *Disease, Medicine and Empire: Perspectives on Western Medicine and the Experience of European Expansion*, ed. Roy MacLeod and Milton Lewis (London: Routledge, 1988), 22; David Arnold, "Introduction: Tropical Medicine before Manson," in Arnold, *Warm Climates*, 3–6, 10–13; Haynes, "Social Status and Imperial Service: Tropical Medicine and the British Medical Profession in the Nineteenth Century," in Arnold, *Warm Climates*, 208, 220; and Ann Moulin, "Tropical without the Tropics: The Turning Point of Pastorian Medicine in North Africa," in Arnold, *Warm Climates*, 161.

7. See Karakara, "Ma'arechet habriut hamandatorit vehavoluntarit ve'aravi Eretz Israel: 1918–1948."

8. Arnold, introduction, in Arnold, *Warm Climates*, 1–19; and Watts, *Epidemics and History*, 258. For Algeria, Moulin, "Tropical without the Tropics," in Arnold, *Warm Climates*, 171; and

This scientific information made it possible to ascertain and describe the detailed interaction between topographical/climatological conditions, the mosquito, and the malarial parasite that facilitated the disease's transmission.

Malaria: Transmission and Symptoms

Colonial scientists' interests in the habits of mosquitoes and parasites, as well as the temperature, rainfall, and soil types of the land, reflected the need to understand the intricate process of malaria etiology and its epidemiologic outcomes as well as the nonhuman actors that came together to propagate the disease and determine its route of infection. Experts in Palestine continually amassed knowledge about these nonhuman elements by conducting experiments, producing maps, and by modifying their original plans as necessitated by setbacks in drainage projects in progress. Zionist antimalaria work followed what scholar Tim Mitchell has observed for engineering projects in colonial Egypt: despite political claims of mastering the environment, there occurred "a series of claims, affinities and interactions, all of which exceed[ed] the grasp or intention of the human agents involved." Human agency and intention, as he explains, are often incomplete outcomes of these environmental/human interactions.[9]

Jewish malariologists such as Israel Kligler recognized the human and environmental agencies involved in malaria transmission in Palestine: "The configuration of the country, the character of the geologic formations, the volume and distribution of rainfall, the temperature and its seasonal variations—these are the physical facts which determine, to a greater or lesser degree, the distribution, prevalence, and intensity of the disease. Of equal importance are the habits, customs, and economic conditions of the inhabitants."[10] Complex interactions among these factors shaped the changes in epidemiology of malaria in Palestine.

Anopheles *Mosquito and* Plasmodium *Parasite*

In addition to the characteristics Kligler noted above, the female of the *Anopheles* genus was (and remains) one of the primary nonhuman actors

---

Worboys, "Germs, Malaria and the Invention of Mansonian Tropical Medicine," in Arnold, *Warm Climates*, 198–199.

9. Tim Mitchell, *Rule of Experts: Egypt, Techno-Politics, Modernity* (Berkeley: University of California Press, 2002), 29, 34, 37, 53.

10. Kligler, *Epidemiology and Control of Malaria*, 1.

whose habits were of prime importance in determining malaria transmission, both in Palestine and throughout the world. In malaria transmission, the *Anopheles* serves as the vector, drawing the *Plasmodium* parasite from the blood of an infected person through her proboscis when biting, then transferring that parasite to a healthy person at the ingestion of the next blood meal.[11] Female *Anopheles* mosquitoes must have a blood meal in order to lay their eggs, so biting is an essential part of the propagation of the species. Another key nonhuman agent, the malarial parasite, is of the *Plasmodium* genus. The parasite first hibernates in the human liver, at which time the patient is infected but asymptomatic. From the liver, the parasite migrates to the red blood cells and feeds off cell hemoglobin. It multiplies asexually at regular intervals, causing increasing symptoms of chills and fever in the host. After multiplying several times in the red blood cells, the parasite begins to produce male and female gametocytes that are transferred to the mosquito when she bites the infected individual. Gametocytes must be in the host's cells for the infection to pass to the female *anopheles*, where they are then fertilized. After development, the gametocytes turn into a sporozoite that lodges in the salivary glands of the mosquito and is injected into the next human it bites.

Palestine possessed all three types of malaria fevers (benign tertian, quartan, and malignant tertian) and the corresponding parasites that cause these fevers. The most common forms of fever in Palestine were benign tertian fever, in which the victim's temperature peaks every third day and produces relapses but rarely causes death, and malignant tertian, the most lethal type of malaria, which is characterized by attacks that occur every two days. These two types of fever are caused by *Plasmodium vivax* and *Plasmodium falciparum*, respectively. *P. falciparum* frequently causes fatal malaria because it makes the red blood cells sticky, eventually blocking blood flow and depriving parts of the body of oxygen. If oxygen deprivation occurs in the brain or other major organ, the patient dies.[12] Blackwater fever, a very serious complication of falciparum malaria caused by the intravascular breakdown of red blood cells and blood in the urine, produced a mortality rate in Palestine of 30 percent (91 cases) between the years 1920–1940.[13]

---

11. Harrison, *Mosquitoes, Malaria and Man*, 115–120, 152–156; and Dobson, *Contours of Death and Disease*, 323.

12. Watts, *Epidemics and History*, 219.

13. Saliternik, *History of Malaria Control*, 9.

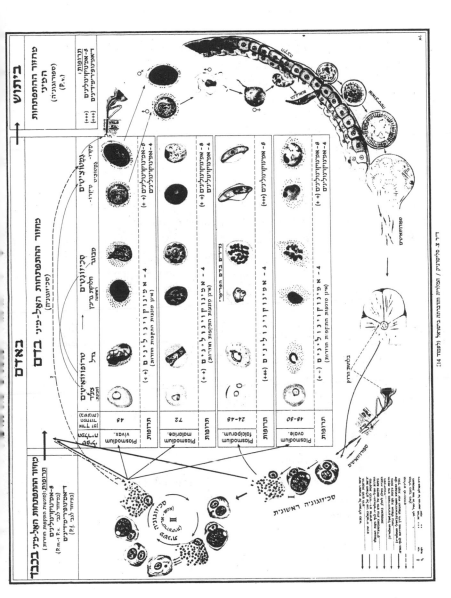

FIGURE 2.1. "Stages of the Malarial Parasite and the Influence of Specific Medicine." From Zvi Saliternik, *HaMalaria vehad-barata beIsrael* (Jerusalem: Rafael Chaim HaCoḥen, 1971), insert. The figure describes the development of the parasite in the *Anopheles* (sexual) and the development (asexual) of the parasite in human blood and liver. Corresponding medical treatment for malaria is given (aminoquinoline) for the period after the 1940s.

In Palestine and other colonies, *P. falciparum* was known to weaken those it did not kill, giving colonial officials the impression of "natives" as "listless and lazy."[14] The lethargic affect caused by malaria probably contributed to the Zionist malariologists' image of the *fellahin* (Arab peasant farmers) and bedouin in swampy areas as lazy, uncivilized, and primitive, a concept that we will revisit later in the book.

In contrast to benign tertian and malignant tertian fevers, quartan fever was less common in Palestine. Quartan fever is caused by *Plasmodium malariae*, occurs every four days, and is usually mild. *Plasmodium ovale* was not found at all in the country.

Most bouts of malaria last for one to two weeks, and if no drugs are taken, these episodes can reoccur for at least six months or more. In addition to fever and chills, malaria sufferers experienced symptoms such as anemia, enlargement of the spleen, and occasional jaundice. These symptoms, like malarial fever, could be acute or chronic and were often recurrent, depending on the type of malaria.

In the early twentieth century, physicians traditionally prescribed quinine to treat malaria. The drug's prophylactic use, efficacy, and potential danger, however, were topics of international medical debate, one in which Palestinian doctors actively took part.[15] In the 1940s newer antimalarials were developed, leading to the eclipse of quinine. In general, Jewish doctors in Palestine did not prescribe prophylactic quinine use, except in rare cases, where it was used until malaria control could be established. This practice changed in 1927, when the Keren Keyemet L'Israel (Jewish National Fund) and Zionist health professionals agreed that no settlement would be established until malaria control had already begun.[16]

Topography of Palestine

The topography of Palestine, another environmental factor, also contributed to malaria transmission. As Dr. Kligler noted, "The peculiar topography and the seasonal rainfall are largely responsible for the prevalence

14. Watts, *Epidemics and History*, 219.

15. Daniel Headrick cites quinine prophylaxis as a barrier to colonial exploration and control, *The Tools of Empire: Technology and European Imperialism in the Nineteenth Century* (Oxford: Oxford University Press, 1981), 73. For a critique, see Arnold, *Imperial Medicine*, 10; also see Sandra Sufian, "Colonial Malariology, Medical Borders and the Sharing of Scientific Knowledge in Mandatory Palestine," *Science in Context* (Fall 2006): 381–400.

16. League of Nations Health Organization Malaria Commission, *Reports on the Tour of Investigation in Palestine in 1925* (Geneva, 1925), 19.

of malaria throughout Palestine."[17] Paradoxically, the geological and climatological nature of Palestine that contributed to the high incidence of malaria was the same feature that made its reclamation easier: its porous soil, rainfall patterns, and low humidity.[18]

For purposes of malaria evaluation, malariologists described the country as having four main areas:

1. The coastal plain, extending from Haifa to the north to Gaza in the south. The coastal plain contained Pleistocene sand, gravel, and sedimentation from the hills. A chain of sand dunes and sandstone ridges blocked the main streams draining into the Mediterranean, resulting in stagnant lagoons, where *Anopheles* breeding was plentiful.

2. The valley of Esdraelon, the area where the coastal plain turns eastward to a plain that extends from the Mediterranean coast to the Jordan Valley. This area is a mixture of limestone and basalt where springs exist but cisterns were also prevalent as a water source.

3. The Jordan River Valley, which extends from a rift at the base of the Anti-Taurus range in Syria to the Dead Sea and Gulf of Aqaba in the south. This valley is set in a ravine (the *Ghor*) formed by a tectonic shift that resulted in a depression of a thousand or more feet below the surrounding hills. The Jordan River's course drops precipitously as it flows, features that make it difficult to access for irrigation.

4. The hill region, which rises from the western and northern plains through a series of foothills to the Judea, Samaria, and Galilee highlands, then declines on the east to the Jordan. These hills, mostly made of limestone, bisect the coastal and Jordan Valley plains, which have different climates and vegetation. There are few springs in the hill region, so residents depended on cement cisterns for their water source. These cisterns were dug in the limestone and made watertight, but when they were filled with rain water during the winter months, they made a perfect place for *Anopheles* breeding.

According to Israel Kligler, the country's drainage occurred on both the east and west: "The watershed [cut by the Valley of Esdraelon] lies along the ridge of the Judean, Samarian, and Galilean hills... On the west side of the ridge, there are a series of streams almost parallel and equidistant

17. Kligler, *Epidemiology and Control of Malaria*, 9.
18. League of Nations Health Organization Malaria Commission, *Reports on the Tour*, 38.

which drain into the Mediterranean. On the east side, the wadis drain into the Jordan."[19] During the rainy season, the streams carried a large volume of water, but once the rain stopped, the wadis became completely or partially dry. The western wadis carried water from the springs for a distance, but after leaving the foothills, these wadis turned dry, "except for holes caused partly by the erosion of the winter flood, and partly by the shepherds who dig watering holes in the wadi bed during the dry seasons."[20] Here Kligler acknowledged the mixture of human and nonhuman factors in the propagation of malaria.

On the east side of the ridge, the wadis plunged stepwise into the Sea of Galilee, the Jordan River, or the Dead Sea. The wadis south of Jerusalem were almost completely dry after the rainy season, while those north of the Holy City were mostly overgrown and perennial, with the majority carrying water from large springs year-round.

The Malaria Research Unit (MRU), the main Jewish antimalaria scientific body working within the Mandatory government, detailed the various topographical causes of malaria-infested areas across the country:

> The sources of mosquitoes and the consequent malaria in the various regions of the country—hills, foot hills and plain, are thus obvious. In the hills, e.g., Jerusalem, Safed, etc., the cistern is the principal cause of malaria. In the foot hills the numerous springs (only few of which are utilized) and their overgrown wadis constitute the chief cause of trouble (Rishon, Khulde, Karkur, Nahalal, Balfouria, Merchavia, Nuris, Beisan, Yavniel, etc.). In the flat plain along the coast the erosion holes in the wadi beds tapping ground water and the waters blocked by sand dunes (Kishon, Kabbara, Rubin) or sand stone outcroppings along the sea (Atlit, Tantura, Atta in Hadera) constitute the sources of malaria.[21]

Epidemic seasons varied with Palestine's geography. In areas such as Hadera and Zichron Yaakov, epidemics occurred in June and July. In

---

19. Kligler, *Epidemiology and Control of Malaria*, 3–4.

20. Kligler, *Epidemiology and Control of Malaria*, 5.

21. British Health Department of Palestine, Malaria Research Unit (hereafter MRU), Haifa, annual report for 1923, "Malaria Control Demonstrations: Breeding Places of Anopheles," part 1, p. 3, ISA M1670/130/33a/6420; League of Nations Health Organization Malaria Commission, *Reports on the Tour*, 36; and M. Osborne, "Resurrecting Hippocrates: Hygienic Sciences and the French Scientific Expeditions to Egypt, Morea and Algeria," in Arnold, *Warm Climates*, 80.

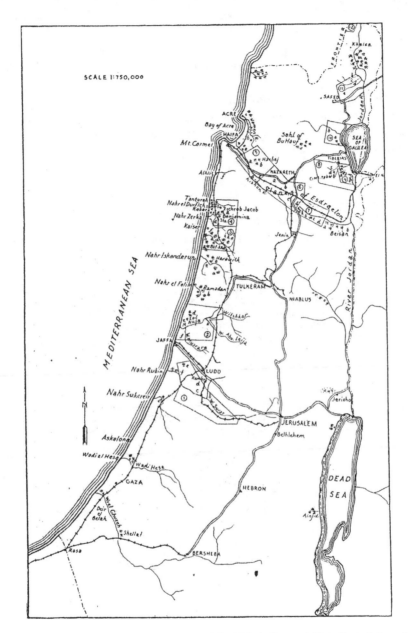

FIGURE 2.2. "Swamps, Streams and *Wadis* of the Malaria Research Unit Control Areas." From Israel Kligler, *Epidemiology of Malaria in Palestine* (Chicago: University of Chicago Press, 1930), 4.

other areas such as Yavniel, epidemics occurred in October or November. In areas such as Ein Harod and Kinneret, epidemics occurred in both periods.

Despite malaria's presence in most of Palestine, there was a relative absence of it in Acre and Nablus, which suggested to Palestine's malariologists that malaria conditions were not necessarily similar even when physical conditions were.

Rainfall

Like Palestine's topography, the rainfall patterns and the country's diverse climate contributed to the full but complicated picture of malaria

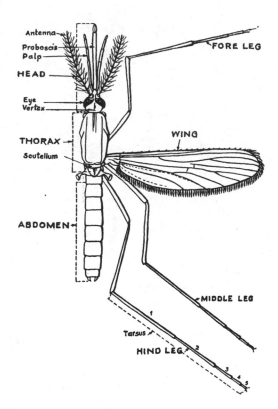

FIGURE 2.3A. "General morphology of a mosquito and adult *Anopheles*." From K. S. Krikorian and N. Bedrechi, *Atlas of the Anopheles Mosquitoes of Palestine* (Jerusalem: Palestine Department of Health, 1940), 4, Ein Kerem Medical Library.

חלקי האנופלס הבוגר (א· סופרפיקטוס) ושמותיהם

וראשי יתושים מבוגרים: כוליכס ואנופלס

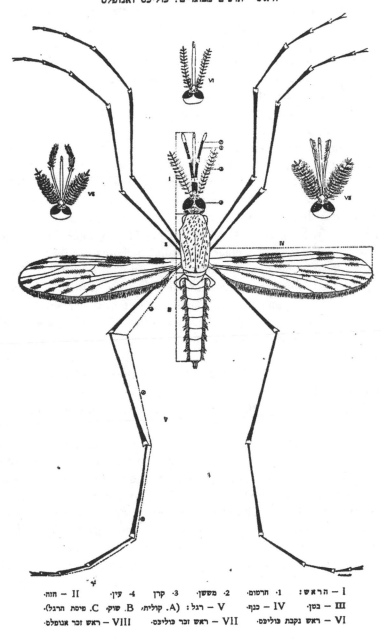

I – הראש : ·1 חרטום· ·2 מששן· ·3 קרן ·4 עין· II – חזה·

III – בטן· IV – כנף· V – רגל : (A)· קלית· B· סוק· C· מיסת תרנל)·

VI – ראש נקבת כוליכס· VII – ראש זכר כוליכס· VIII – ראש זכר אנופלס·

FIGURE 2.3B. General morphology of a mosquito and adult *Anopheles*. From Dr. Joseph Shapira and Zvi Saliternik, *HaMalaria beEretz-Israel: Sefer Shimushi* (Malaria in *Eretz Israel*: Practical book) (Jerusalem, 1930), 30–31, Ein Kerem Medical Library.

distribution and transmission. One reason for the variation is that the *Anopheles* sporozoite requires a particular temperature to develop. In addition, the specific temperature that facilitates development varies according to parasite type. While humidity also influences the transmission of malaria, it was of less significance in Palestine because humidity levels are relatively constant throughout the country.

Studies conducted by the Mandatory government Department of Agriculture showed that annual rainfall in Palestine averaged 25 inches, with larger amounts in the northern and western parts of the country. Despite irregular patterns in rain distribution in terms of geography and yearly fluctuations, the months of December, January, and February usually had the greatest rainfall. Malaria transmission depended upon rainfall volume and distribution in that both were factors in the creation of breeding places for mosquitoes. As Kligler wrote, "The primary physical factors which favor the development of *Anopheline* breeding places are the seasonal rainfall, the limestone formations, and the dunes."[22] *Anopheles elutus* (or *A. sacharovi*), the most important vector of malaria in Palestine, favored the stagnant or semistagnant waters covered with horizontal vegetation in the foothills and narrow valleys. *Myzomia sergenti* and *Myzomia superpictus*, two subspecies of anopheles, bred on the edges of moving streams and puddles. Winter floods that eroded the clay or gravel beds of the streams left behind large pools, which made breeding places in the plains as well.[23]

Whether or not the rain flooded an area or was distributed evenly affected the creation of breeding places. When flooding occurred, the water usually flowed down the limestone hills to the sea and pooled in landlocked areas, where little of it seeped into the rock to the source of the springs. When rain falls less tempestuously, the springs retain the water and the stagnant pools are reduced in volume. Again, Kligler pointed out the complexity of this nonhuman agent for malaria transmission: "When the total volume is above the average or average, the distribution factor is of utmost importance, because on it depends whether serious outbreaks of malaria are to be expected throughout the country or in the coastal sections alone. The character of the winter rains thus enables one to predict

---

22. Kligler, *Epidemiology and Control of Malaria*, 10.
23. Kligler, *Epidemiology and Control of Malaria*, 7; Saliternik, *History of Malaria Control*, 17.

with considerable accuracy the probable degree of mosquito prevalence in various parts of the country and to prepare the requisite measures of control."[24]

Malaria season in Palestine began with *Anopheles* breeding in late March or early April and ended in late December. From the middle of June to the end of July, most patients contracted benign tertian malaria. Kligler attributed this pattern primarily to the "residue of uncured malaria"; individuals with tertian malaria, however, often presented all year round. The next round of epidemics extended from the end of September to the middle of December. Patients in these months mostly exhibited tropical or malignant tertian malaria. Patients rarely presented with quartan malaria in Palestine, but scientists sometimes traced it in March and April.[25] Relapses of malaria, mostly tertian, occurred during the winter and spring months (February to May).[26]

In the three-month interlude between January and March, the British administration and Zionist agencies in Palestine regularly published educational propaganda (see chapter 5), conducted malaria surveys, designed and reorganized drainage plans, and performed medical examinations to prepare the population for the next season and prevent an increase in malaria, working within the vicissitudes of nature to design and disseminate their public health programs.

Climate

Variations in climate in Palestine also played a role in malaria transmission. Following general rainfall patterns, on average the coldest month in Palestine was January and the hottest month was August. Surprisingly, January had the highest humidity and May the lowest. Notwithstanding these average characteristics, each area of Palestine exhibited different temperature and humidity characteristics. According to the topographical regions delineated above, the coastal plain possessed a subtropical

---

24. Kligler, *Epidemiology and Control of Malaria*, 7.

25. MRU, "Malaria Control Demonstrations," 18; letter from Col. Heron to senior medical officers of all districts on antimalarial measures for 1935, March 29, 1935, 3, ISA MI503/1/86(59).

26. These seasonal trends certainly affected agricultural methods and planning. Israel Kligler, Joseph Shapiro, and I. Weitzman, "Malaria in Rural Settlements in Palestine," *Journal of Hygiene* 23 (1924): 288.

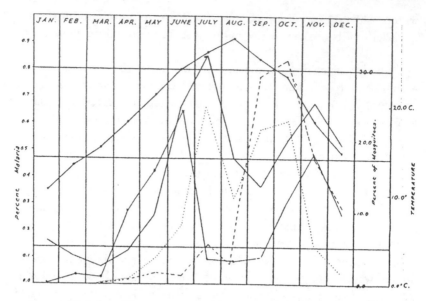

FIGURE 2.4. "Relation of Temperature, Mosquitoes, and Malaria." From Israel Kligler, *Epidemiology of Malaria* (Chicago: University of Chicago Press, 1930).

climate with high humidity throughout the year, while the hill country was "moderately temperate" with humidity that dropped after the heavy rains but then rose again after the summer. The Jordan Valley was mostly tropical with lower humidity than the other regions, although the Jordan region followed a similar drop and rise pattern in humidity as the hill region. Unlike the climate, winds in Palestine were relatively constant and generally blew either northwesterly or westerly. Occasional easterly winds (*hamsins*) were dry, hot, and sandy.

As we can see from the chart, the prevalence of *A. elutus* peaked in June and November, which accounted for the malaria outbreaks in Palestine in June-July-August (the hottest months)—especially along the seashore—and for those outbreaks in the middle of November and December, respectively. The outbreaks during November and December were found predominantly in the inland villages. In August, many adults infected with the *A. elutus* (or *A. sacharovi*) species died because of the extreme heat.[27] *M. superpictus* and *M. sergenti* both peaked in September–October (with *M. superpictus* causing one peak in July), accounting for

27. Saliternik, *History of Malaria Control*, 17.

the epidemics in Palestine during the fall season.[28] During the beginning of fall, the flight range of the mosquitoes in Palestine could reach fifteen kilometers. In the winter months, the female mosquitoes would stop developing eggs and gained weight.[29] The intimate relationship between mean yearly climate and species prevalence and behavior determined the timing and virulence of each malaria outbreak in each area of the country.

The Changing Profile of Malaria of Palestine

Certificate of Citizenship: Malaria before the Mandate Period

According to Dr. Puchovsky, chair of the Hebrew Medical Society of Jaffa from 1912 to 1913, contracting malaria was a "certificate of citizenship"; it was a rite of passage that indicated one's initiation into residence in Palestine. This initiation rite closely related to the notions of self-sacrifice and heroism so central to Zionist labor ideology, as delineated in chapter 1.

Dr. Puchovsky, who treated malaria in the late Ottoman era and Mandate Palestine, noted that prior to British rule, malaria incidence was so high that it was hard to find a person who had not suffered from it at least once. Arthur Ruppin, a key Zionist land and settlement expert during this time, agreed. Ruppin (director of the Zionist Executive's Settlement Office in Palestine) wrote that three-fourths of all illness in Palestine before and at the beginning of the Mandate was caused by malaria.[30] Dr. Kattan, delivering a paper entitled "Malaria in Palestine" at the Eighth Egyptian Medical Congress in June 1935, noted that from 1914 to 1916, almost half of all the patients attending the Polyclinic of the Sisters of Charity in Haifa suffered from malaria.[31]

---

28. Kligler, *Epidemiology and Control of Malaria*, 118; Saliternik, *History of Malaria Control*, 11.

29. Saliternik, *History of Malaria Control*, 17.

30. Arthur Ruppin, "Sanitation of Palestine," reprinted in *Palestine* 4, no. 20 (n.d.): 2, CZA J113/1430.

31. Malaria is believed to have existed in Palestine for over 2,000 years and even frustrated Crusader attempts to take Jerusalem in the eleventh century. Alfred Crosby, *Ecological Imperialism: The Biological Expansion of Europe, 900–1900* (Cambridge: Cambridge University Press, 1986), 66–67. For the prevalence of malaria in Palestine in ancient times, see these Palestine governments pamphlets: HMG, *The Campaign against Malaria in Palestine* (Jerusalem: Government Printing Press, March 1936), Public Records Office (hereafter PRO), London, Colonial Office Papers (hereafter CO), 733/345/10; HMG, *Campaign*

Other prominent Zionist activists also worried about the disease's effects. Dr. Hillel Yofe (1864–1936), a well-known Zionist malariologist in Palestine, concurred with Puchovksy, Ruppin, and Kattan. Yofe noted that he did not know of one place free from malaria before and during the early part of the Mandate period. Moshe Smilansky (1874–1953), a Zionist active in land purchase and a leader of Jewish agricultural settlement, worried about the high incidence of malaria for future generations: "[A]lmost every child has a swollen abdomen like a pregnant woman. What a generation of laborers and colonizers will grow up from them!!!"[32] Malariologists of the time did not deem case fatality rates as excessive among adults as they were among children. They believed, however, that high rates of malaria in Palestine warranted thorough control measures.[33]

Consistent with a widely accepted belief of the late nineteenth and early twentieth centuries that fevers constituted one form of miasmatic disease, both Jewish and Arab doctors in Palestine commonly diagnosed malaria for any ailment with symptoms of fever, including influenza and typhus. Although in the early 1890s scientists had not yet discovered the etiology of malaria, they differentiated it from other fevers as an intermittent fever and febrile disease.[34] Physicians' continued identification of malaria as part of the general rubric of "fevers" was likely a vestige from earlier diagnostic practices, evidence of the time lapse between scientific discovery and the actual integration of new knowledge into medical practice and consciousness. Diagnosing malaria as "fever" in Palestine did not differ from courses of action in other countries of the period. Most physicians in Europe were also unaware of the new etiological definition of malaria presented during the late nineteenth century.[35]

---

against Malaria in Palestine (Jerusalem: Government Printing Press, March 1936), PRO CO 733/345/10; Nissim Levy, "Zichron Yaakov-mirkaz harefuah harishon betzfon Eretz Israel" [Zichron Yaakov: The first medical center in the north of Palestine], HaRefuah 108, no. 2 (January 1985): 90–99.

32. An enlarged spleen causes the abdomen to swell, which indicates a history of malarial infection. As quoted in Saliternik, "Reminiscences," 518. Such statements were influenced by eugenic concerns. See Alison Bashford, Imperial Hygiene: A Critical History of Colonialism, Nationalism and Public Health (Houndsmills: Palgrave Macmillan, 2004).

33. League of Nations Health Organization Malaria Commission, Reports on the Tour, 43.

34. Worboys, "Germs, Malaria and the Invention of Mansonian Tropical Medicine," 186–187, for difference between zymotic and febrile diseases; Osborne, "Resurrecting Hippocrates," in Arnold, Warm Climates, 82.

35. In fact, most medical textbooks did not change their accounts of malaria until shortly after the 1880s, when the etiology of malaria was discovered. Worboys, "Germs, Malaria and

The medical narratives of Puchovsky further clarify the quinine fetish of the time. Puchovsky wrote in his memoirs that doctors were "taken by 'malariomania'" in Palestine; they overdiagnosed malaria for ailments with fever. As a result, they wrongly prescribed quinine, considered a tonic in the nineteenth century, for most feverish illnesses.[36] Puchovsky's first story describes a young girl, daughter of one of the important merchants in Jerusalem, who fell ill during World War I and was diagnosed "as usual" by her doctor as having malaria. Her doctor prescribed quinine despite the fact that "the microscopic examination of her blood was negative for malaria." Even after the quinine didn't help, the doctor increased the quinine dose, only to be countered by a consulting doctor from Kushta, who determined that all of the symptoms pointed to typhus. Puchovsky met with the Jerusalem doctor and asked, "If the blood test was negative and the quinine didn't help, what basis did you have to think that this was malaria?" The doctor, still insistent on his original diagnosis and thus revealing societal "malariomania," resentfully answered, "But in any case, it was malaria!"[37]

Palestine's residents followed the "malariomania" trend of medical professionals. Patients took quinine so frequently that quinine sulphate was considered a household remedy. Everyone knew it by the name "sulphate" or *sulfata* in Arabic. It was used more than any other drug in the country and was administered either by mouth or intramuscularly.[38]

Quinine was distributed without standardization of dose. According to Puchovsky, "The Arab population also took quinine, but their beloved medicine was quinine wine—not so much from quinine as from wine. Every sickness was treated with quinine among the Arabs." S. Friedman of Haifa produced quinine wine (*sharb al-qana* in Arabic) under the name

---

the Invention of Mansonian Tropical Medicine," in Arnold, *Warm Climates*, 188; Worboys, "Manson, Ross and Colonial Medical Policy," 23; for British India, see Arnold, *Colonizing the Body*, 36.

36. Doctors prescribed it for several conditions besides malaria, usually in conjunction with a cathartic. Puchovsky, "Mezichronotav shel rofe vatik," 3, Histadrut Archives, IV-104-853-24-2; and Rosenberg, *Explaining Epidemics*, 15.

37. Puchovsky, "Mezichronotav shel rofe vatik," 3. The girl's Vidal reaction (a test for typhus) was done by the Pasteur Institute in Jerusalem. Pasteur institutes were established across the Ottoman Empire, including Istanbul. Moulin, "Tropical without the Tropics," in Arnold, *Warm Climates*, 168.

38. "Reconstruction in Turkey" (1919), 3, CZA J113/1430. For a discussion of kina in North Africa, the ancestor to colonial quinine, see Moulin, "Tropical without the Tropics," in Arnold, *Warm Climates*, 165.

Quina-Laroche S.C.,[39] and Dr. Hillel Yofe granted official license for its production and use on May 20, 1895.[40]

Puchovsky's second memoir related the extent of public insistence upon using quinine even when it was not indicated. He described an Arab patient who came to his practice with advanced pulmonary tuberculosis. Although Puchovsky prescribed him quartacol capsules, the patient refused to take them, explaining that he had already taken a lot of quinine without any benefit. "As much as I tried to explain to him that this was a different medicine entirely, nothing helped and he stubbornly reiterated that he would not take any more *sulfata* and that he could not understand that the capsules could include another medicine except quinine."[41] Although quinine may have been overprescribed or misused in early twentieth-century Palestine, Puchovsky's narrative suggested a high enough incidence of malaria to justify its misuse.

Combined with the abandonment of villages and colonies, the high number of new infections threatened the survival of Jewish settlements, especially before the Mandate period. The settlement of Hulda, for instance, consistently had a population infected with malaria, which rendered it unfit for work. To combat this situation, one-third of the Hulda annual budget was used for medical aid but often at least half of the settlers remained "invalided."[42] Newer Zionist settlements such as Nahalat Yehuda, Ein Ganim, Shuni, Degania, Merchavia, Kinneret, and Gan Shmuel had the same difficulties. Older settlements such as Yessod, Mishmar and Athlit, Ekron, Yavniel, and Hadera also suffered greatly from high rates of malaria.

Many malariologists expressed frustration at the lack of action taken by the Ottoman government to alleviate the malaria and general sanitation problems. Arthur Ruppin's plan for the sanitation of Palestine after World War I, however, revealed that Turkish laws regarding the health of the population already existed. British and Zionist reports of the Mandate period commonly overlooked or disregarded these laws and efforts,[43] probably because of a belief that the Ottoman government did not

39. Personal communication with Muhammed Karakara, Jerusalem, June 1996.

40. If Puchovsky is right in his assessment, then this may also tell us about the extent of some of the Arab population's religious practices. Levy, "Zichron Yaakov," 93.

41. Puchovsky, "Mezichronotav shel rofe vatik," 4.

42. Kligler, *Epidemiology and Control of Malaria*, viii.

43. Secondary sources, like Naomi Sheperd, *Ploughing Sand: British Rule in Palestine, 1917–1948* (Rutgers University Press: New Brunswick, 1999), 131–132, accept this claim

do *enough* for sanitation rather than that they did nothing. Such beliefs most likely stemmed from a desire to see themselves as the harbingers of progress.

Yet late Ottoman reports like Ruppin's actually provided the background and landscape upon which the British and Zionists then worked in the field of reclamation and sanitation during the Mandate period. In his report, Ruppin recommended that all landed property in Palestine, if swampy or damp soil, should be considered as a potential breeding ground for mosquitoes and therefore subject to sanitation and drainage. The legal basis for this sweeping action, Ruppin remarked, was

> afforded by the Turkish expropriation law, which makes it possible for the authorities to take in the public interest all necessary measures for the health of the population. The authorities would then have to request the owners of the farms to carry out the sanitation within a fixed space of time in accordance with plans drawn up by the health authorities. Should this not be done within the required time, steps should be taken to carry out the work of sanitation at the expense of the owner, the land serving as a pledge for the costs of the work and being liable to be partly expropriated and sold, if necessary, in order to cover the expenses of sanitation. In the case of land, lakes and rivers belonging to the State, their sanitation must be carried out at the expense of the country.[44]

The spirit of Ruppin's plan was more or less implemented thereafter in the British Mandate government's Antimalarial Ordinance of 1922. Ruppin's plan proposed that large sanitation schemes should ultimately be the responsibility of the state.

Before British rule, physicians raised the question of the state's duty to carry out malaria eradication. Following a lecture on malaria eradication given by Hillel Yofe to the Hebrew Medical Society in June 1912, Dr. Stein—like Ruppin soon after—promoted the central role of government in ridding the country of endemic disease. He noted that in Jaffa in 1894, one could not find even one family whose members had not contracted malaria. The city, Stein claimed, was one of the main sources of malaria in the area. Because of the severe situation, the Ottoman government

---

uncritically. Peter Muhlens, "Bericht ueber eine malaria expedition nach Jerusalem" *Zentralblatt für Bakteriologie* 69 (1913): 41–85, as cited in Fritz Yekutiel, "Masterman, Muhlens and Malaria: Jerusalem 1913" (unpublished paper, 1997), 10.

44. Ruppin, "Sanitation in Palestine," 3.

decided to drain the swamps surrounding the city, thereby significantly improving the Jaffa health situation and making it "more healthy" than Jerusalem or even other settlements. Stein's comments provided additional evidence of Ottoman efforts toward malaria eradication and helped describe the epidemiology of malaria in late Ottoman Palestine.[45]

Throughout the Mandate period, the Mandatory government and the *Yishuv* intensely debated the designation of responsibility for malaria control. This repeated exchange was tied to a larger issue of British colonial developmental policy in Palestine that sought to advise on infrastructural projects but not to solely undertake or finance them. Although British officials like High Commissioner Herbert Samuel acknowledged the problem of malaria in his plan for developing Palestine at the beginning of the Mandate, it was unclear who held the main responsibility for the projects' implementation and funding.[46]

Malaria in World War I

Malaria continued to pose a very serious problem in Palestine during World War I. In cities like Jerusalem, 60 percent of the Jewish population was said to have contracted malaria during the war.[47] Between 1918 and 1919, 50 percent of children in Jewish settlements suffered from chronic malaria.[48]

As in most colonies, conquering malaria became part of the military conquest of Palestine. The disease significantly affected British military strategy and operations. British military officials considered compiling data for malaria control as an essential step before being able to engage in hostilities against the Turkish Army. In fact, General Edmund Henry Hynman Allenby prepared for the campaign in Palestine by reading Crusader accounts of the Levant that included discussions of malaria.

45. Hillel Yofe, "Tafkid harofe bemilchama 'al haendemiut shebearetz" [The role of the doctor in the war against endemicity in Palestine], twelfth meeting of the Hebrew Medical Society, June 6, 1912, *Zichronot HaDavarim* 203 (May 1913): 52, CZA Library. See also Puchovsky's lecture "Havra'at al davar haripui shel kadachat" [The question concerning treatment of malaria], twelfth scholarly meeting of the Hebrew Medical Society, November 31, 1912, *Zichronot HaDavarim* 203 (May 1913): 91, CZA Library.

46. See Biger, *Empire in the Holy Land*, 81–87; and Smith, *Roots of Separatism in Palestine*, 12, 38, 57.

47. Jerusalem Jewry and the War (Special Correspondent) 3.1, February 9, 1918, CZA J113/1430.

48. HMG, Campaign against Malaria in Palestine, 8.

Despite such preparation, there were 8,500 primary cases (new cases, not relapses) among the frontline British troops in Palestine between April and October 1918. From October to December 1918, officials recorded over 20,000 cases.[49] The high incidence of malaria among the British Expeditionary Forces in 1918 and among German and Turkish prisoners of war (which reached 60 percent) induced British forces to set up "antimalaria squads" and "sanitary squads" in every military unit. These squads identified mosquito-breeding places near the unit's camp, canalized streams, drained pools, and covered cisterns and wells. A brigade malaria officer supervised the work. Still, in 1919, British army camps close to Haifa, Jenin, and Beisan had to be evacuated due to the high incidence of malaria.[50]

Before they could undertake squad tasks, British medical officers gathered data about malaria-infested areas of Palestine and conditions within their military camps.[51] Taken together, British and Zionist studies provided an important basis on which to proceed with drainage necessitated by British military needs. Malariologists during the Mandate period then used these studies in antimalaria planning.

The reports categorized each area as intensely malarious, malarious, or slightly malarious. A concomitant study by Hillel Yofe of thirty-four Jewish colonies, including the upper and middle Jordan Valley, the Esdraelon and Yavniel Valleys, and the coastal plain, categorized areas similarly.[52] In addition to the study of British military camps during World War I, entomologists of the Royal Army (Barraud, Austin, and others) carried out a partial survey in 1917 and 1918 that determined which species of *Anopheles* mosquitoes were prevalent in the area, concluding that *A. bifurcatus* was the prime species in the cities, while *A. elutus* and *A. superpictus* were more common in rural areas. The entomologist of the newly established British Health Department of Palestine completed the survey between 1921 and 1923. In many cases, such as the Auja area in 1918, the Egyptian Labor Corps provided the manual labor for malaria eradication.

49. HMG, *Campaign against Malaria in Palestine*, 2; Crosby, *Ecological Imperialism*, 66; Jewish Agency for Palestine (hereafter JA), memorandum submitted to the Palestine Royal Commission, "Antimalaria and Drainage Work by Jewish Bodies" (December 1936), 4, CZA Library.

50. League of Nations Health Organization Malaria Commission, *Reports on the Tour*, 12–13.

51. Saliternik, "Reminiscences" 518.

52. Yekutiel, "Masterman, Muhlens and Malaria," 14; JA, "Antimalaria and Drainage Work by Jewish Bodies," 4–5.

The British Mandate government also established a sanitary engineering section within the British Health Department of Palestine that assisted in some large drainage projects.[53]

In July 1920, the British Civil Administration found that the fatality rate of malaria in Palestine was sixty-eight per thousand people, affecting both infants and children in a community still feeling the ravages of war. In the village of Beit Jibrin, for instance, about one-sixth of the population died in three months due to the scourge.[54]

Malarial Trends during the Mandate Period

Although antimalaria measures were undertaken prior to the Mandate, British and Jewish health officials considered these efforts neither comprehensive nor conclusive. Malaria remained a serious health threat. One difference between the pre-Mandate and Mandate periods regarding malaria control was a shift from focusing on minor control work (pouring oil in the swamps, sparing Paris Green to kill larvae, constructing some irrigation) to addressing problems in large areas by organizing and executing systematic swamp drainage plans with highly trained drainage personnel.[55]

British and Zionist antimalaria control activities were performed and funded separately. This arrangement mirrored the larger organization of public health and clinical systems in Mandatory Palestine. The Mandatory government took care of malaria control and education in Arab areas of residence, while the Zionist agencies largely concerned themselves with Jewish settlement areas. Zionist agencies, however, sometimes implemented swamp drainage measures in nearby Arab villages that, in turn, affected the health of their own settlements. Despite divisions in implementation, exchange of scientific information between the government and Zionist antimalaria agencies was frequent.

The most systemic, serious malaria problems during the Mandate period occurred in the 1920s, especially in the rural areas. Sporadic epidemic trends in the 1930s and 1940s followed. At the beginning of the Mandate period, malaria control work showed quick and drastic results. In 1922, for instance, a minimal annual malaria incidence (new cases) ranged from

53. JA, "Antimalaria and Drainage Work by Jewish Bodies," 8–9; Arnold, *Imperial Medicine*, 14.

54. League of Nations Health Organization, Malaria Commission, *Reports on the Tour of Investigation in Palestine in 1925* (Geneva, 1925), 9.

55. "Twenty Five Years of KKL: 5662–5687," 40, CZA A246/441.

50 to 100 percent in many studied areas.[56] Monthly prevalence data (new and existing cases) showed that an average of 5.7 percent (out of approximately 620,000 residents, for a crude estimate of 35,340 people) of the population had malaria in 1922. With the initiation of the MRU control demonstrations in 1923, there was a decrease in malaria incidence everywhere in Palestine except Hadera and Rosh Pina. This decline was greatly attributed to large-scale malaria control projects undertaken by Jewish, government, and other engineers at the beginning of the Mandate and in 1922.[57] The MRU, for instance, carried out their work by dividing the country into ten, later fourteen, districts, focusing on the Jewish settlements. It extended its work to neighboring Arab villages where possible. A decline to 2.9 percent average monthly prevalence resulted in 1923, and by 1925, the average monthly percentage rate was 0.8.[58]

Even after some malaria control work in the early years of the Mandatory period, the disease persisted as a serious concern. Spleen rates (a measure of parassitemia, or parasites in the blood) and blood exams still showed the presence of malarial infection in many areas of Palestine. Aref al-Aref, administrator of the Beersheba district, noted that even in the 1940s, 90 percent of the bedouin in that region examined in his investigation had swollen spleens.[59]

Occasional difficulty securing quinine exacerbated the situation. In an effort to alleviate the malaria situation, the Mandatory government sold quinine packets at cost at post offices in the cities. For those in villages, systematic free distribution of quinine was offered, even in schools. Each packet contained five grains of quinine sulphate. A packet of twelve doses cost 3.5 piasters. For the villages, the subinspector would collect the malarious patients, record their names, and distribute a diluted quinine solution for the patient to take for four days. Upon initial treatment, the patient would be given a calomel tablet with quinine in it to swallow in front of the subinspector to insure compliance.[60] In 1923, free distribution of quinine was discontinued for financial reasons but was available in

56. MRU, "Malaria Control Demonstrations," 17.

57. MRU, "Malaria Control Demonstrations," 19; and "Proceedings of the Antimalarial Advisory Commission," November 22, 1923, 11, JDC Archives, file 280.

58. Kligler, *Epidemiology and Control of Malaria*, ix.

59. "Proceedings of the Antimalarial Advisory Commission," May 22, 1924, 4, JDC Archives File 280; Aref al-Aref, *Bedouin Love, Law and Legend: Dealing Exclusively with the Bedu of Beersheba* (Jerusalem: Cosmos Publishing Co., 1944), 155.

60. League of Nations Health Organization Malaria Commission, *Reports on the Tour*, 17–19.

emergency situations. Inspection and treatment were made easier in the Jewish colonies by the presence of a resident doctor or nurse who would notify the MRU of malaria cases. The MRU would then resolve any problems. Even when quinine was prescribed in the Jewish settlements, compliance was evidently quite low. Despite the urging of nurses in the colonies, settlers threw pills away because of their bitter taste and bad smell as well as the loss of appetite and nausea they caused. A. D. Gordon, the father of *Dat ha'avoda*, when asked why he did not take the prescribed quinine, answered, "Why should I take quinine if I shiver with quinine as much as without it!"[61] In 1925, overall malaria incidence reached the lowest level since British occupation. There were no malaria epidemics that year, in part because of dry weather but also because of broad antimalaria measures like swamp draining (with some impact from quinine treatment or prophylaxis) of that and previous years.

Assessing the Malaria Situation: Problems with Data

Although the Mandatory government's annual health reports describe the yearly malaria situation in Palestine and the efforts to improve it, these documents very likely underestimated the situation on the ground. Underestimation probably resulted from inaccurate census numbers for the whole population of Palestine. Census data for the general population would form the denominator for general prevalence and incidence rates and were found in separate tables from malaria statistics. Compounding the problem, census data were given variously as end-of-year or midyear numbers.

Moreover, the number of existing malaria cases for all of Palestine for 1922–1924 in the British Health Department of Palestine reports were very similar, if not identical, to data given for dispensary visits for malaria from fifty clinics. This suggests that the population data on malaria were based on dispensary data alone and thus that the data presented are epidemiologically biased. These dispensary statistics likely missed those people who had malaria but did not attend dispensaries and thus probably underestimated malaria morbidity in the country at the time. In addition, dispensary data included only a fraction of the total number of dispensaries in Palestine at the time.

61. D. Kohn, Z. Weiss, and E. Flatau, "The History of Malaria in the Jezreel Valley in the Years 1922–1928 in Ein Harod," *Korot* 8, nos. 3–4 (1982): 177–186; quotation on 184–185.

TABLE 2.2 **Overview of Malaria Morbidity and Mortality from 1922–1933 in Mandate Palestine**

Year	Total Number of Patients in Dispensaries (about 50 dispensaries)	Number of Patients Suffering from Malaria	Number of Malaria Deaths	Prevalence Rate (per 1,000)	Mortality Rate (per 1,000)
1922	283, 156	20, 297	67	71	3.3
1923	270,945	13, 280	78	49	5.8
1924	295,215 (in 1925 report) 285,215 (in 1927 report)	11,732	33	41	2.8
1925	339, 795 (1925) 344,150 (1927)	7,249 (1925) 7835 (1927)	13	21 (1925) 23 (1927)	1.8
1926	388,729	7,956	n/a	20	n/a
1927	394,932	9,069	n/a	23	n/a
1928	410,854	7,207	n/a	17.5	n/a
1929	404,188	11,404	n/a	28.2	n/a
1930	406,623	11,503	n/a	28.2	n/a
1931	458,941	8,147	n/a	17.7	n/a
1932	495,583	2,984	n/a	6	n/a
1933	528,186	3,330	n/a	6.3	n/a

NOTE: Based on available dispensary data.
SOURCE: Taken from Department of Health annual report for 1925, 16; annual report for 1927, 25; and annual report for 1933, 32, all from ISA M4475/06/1.

Malaria rates among different segments of the population are even more difficult to discern with British Health Department of Palestine report statistics.[62] Instead, one must go through each village, discern which community predominantly lived in each area, and then calculate comparative rates between Jews and Arabs, adjusting for village population. Malaria statistics for districts where Jews and Arabs lived side by side (more than a few) are difficult to disaggregate for the Jewish and Arab populations. Accurate and reliable malaria statistics are therefore almost impossible to discern: numbers given here are best estimates of incidence, prevalence, and mortality rates derived from the available data. Israel Kligler did provide some information about Arab and Jewish blood and spleen rates in his study of the epidemiology of malaria in Palestine. His information seems to be the closest, most complete data available for comparing malaria rates in communities in Palestine.

Across the board, the percent that indicated enlarged spleen and parassetemia, indicators of eventual malaria, among adults and children in the Arab population in these regions were higher, sometimes much higher, than that of the Jewish population in the early years of the Mandate,

62. These would be the documents that would supply such aggregate information.

TABLE 2.3 **Blood Rates in Adjacent Jewish and Arab Villages with Equally High Malaria Prevalence (1922–1923)**

Place	Number Examined Adults/Children/Total	Number Positive Adults/Children/Total	Percent Positive Adults/Children/Total
Ekron:			
Jews	215/110/325	17/11/28	7.9/10.0/8.6
Arabs	52/139/191	6/42/48	11.5/30.3/25.1
Gedera:			
Jews	77/43/120	3/4/7	3.9/9.3/5.9
Arabs	2/46/48	0/5/5	0.0/10.9/10.4
Ein Ganim:			
Jews	255/118/373	10/6/16	3.9/5.1/4.3
Arabs	14/28/42	1/7/8	7.1/25.0/19.0
Merhavia:			
Jews	n/a/n/a/162	n/a/n/a/16	n/a/NA/9.9
Arabs	n/a/n/a/185	n/a/n/a/54	n/a/n/a/29.2
Yessod:			
Jews	155/83/238	2/9/11	1.3/10.8/4.7
Arabs	36/38/74	3/9/12	8.4/23.7/16.2
Ayeleth:			
Jews	50/0/50	1/0/1	2.0/n/a/2.0
Arabs	13/13/26	1/3/4	7.7/23.0/15.4

SOURCE: I. Kligler, *The Epidemiology and Control of Malaria in Palestine* (Chicago: University of Chicago Press, 1930), 91.

TABLE 2.4 **Spleen Rates in Adjacent Jewish and Arab Villages with Equally High Malaria Prevalence (1922–23)**

Place	Number Examined Adults/Children/Total	Number Positive Adults/Children/Total	Percent Positive Adults/Children/Total
Ekron:			
Jews	160/85/245	14/8/22	8.7/9.4/9.0
Arabs	43/139/182	39/131/170	90.7/94.2/93.4
Gedera:			
Jews	73/38/111	27/14/41	37.0/36.8/36.9
Arabs	5/47/52	4/43/47	80.0/91.5/90.4
Ein Ganim:			
Jews	247/116/363	95/50/145	38.5/43.1/40.0
Arabs	14/28/42	14/27/41	100.0/96.4/97.6
Merhavia:			
Jews	95/32/127	24/4/28	25.3/12.5/22.0
Arabs	58/58/116	10/30/40	17.2/51.7/34.5
Yessod:			
Jews	155/83/238	55/28/83	35.5/33.7/34.9
Arabs	15/39/54	11/29/40	73.3/74.4/74.0
Ayeleth:			
Jews	50/n/a/50	3/n/a/3	6.0/n/a/6.0
Arabs	8/12/20	4/6/10	50.0/50.0/50.0

SOURCE: I. Kligler, *The Epidemiology and Control of Malaria in Palestine* (Chicago: University of Chicago Press, 1930), 91.

TABLE 2.5 **Blood and Spleen Indices among Different Population Groups**

	Blood Samples Examined	Positive for Malaria	Percentage	Spleens Examined	Positive for Malaria	Percentage
Group:						
Jews	6,738	310	4.6	5,486	1,104	20.1
Arabs	1,746	219	12.5	1,220	763	62.5
Bedouins	616	83	13.5	385	226	58.8

NOTE: These figures represent the total findings during the general examinations of population groups prior to starting control in the various areas throughout the country. These examinations were generally made in March and April, before the epidemic season. I. Kligler, *The Epidemiology and Control of Malaria in Palestine* (Chicago: University of Chicago Press, 1930), 60.

when malaria infection was not yet under control. The bedouin population rates were even higher than the sedentary Arab populations.

Another chart provided by Kligler shows similar differences in indices of infection. Kligler's prose, however, paints a different picture than the numbers infer: "As seen from this table, both the blood and spleen rates are only one-third as high in the Jewish group as in the others. Nevertheless, the bulk of the Jewish rural population, being relatively new and settled in the most malarious areas, suffers as much as, [or] usually much more than, their neighbors."[63]

Kligler, Weitzman, and Shapiro, in a scientific article about the use of spleen and blood rates to determine malaria prevalence, explained the varying numbers. Reviewing data from demonstration areas in Palestine, these authors concluded that different living conditions, access to treatment, age composition, and use of quinine prophylaxis were also important factors for consideration in the interpretation of results of malaria examinations.[64] They considered that the "native population does not readily submit to examination either of blood or spleen.[65] Furthermore, the examination of Muslim women was prohibited, so Arab data were based on men and children. As a result, the prevalence rates (carrier rates) for the Arab population were often overestimated.[66] Compounding the problem of data collection was the fact that *fellahin* men were frequently away in the fields working for days or weeks and were therefore unavailable for examination. Finally, successful medical examination of Arabs

63. Kligler, *Epidemiology and Control of Malaria,* 92.
64. Kligler, Shapiro, and Weitzman, "Malaria in Rural Settlements in Palestine," 297.
65. Kligler, Shapiro, and Weitzman, "Malaria in Rural Settlements in Palestine," 291.
66. Kligler, Shapiro, and Weitzman, "Malaria in Rural Settlements in Palestine," 295.

involved social ritual and the acquaintance and trust of the village leader, according to Kligler and his colleagues, who described that, "[b]lood and spleen examinations in a Moslem village in Palestine is a special event which begins with sitting about in the Sheikh's house, talking, drinking coffee and talking some more. Only after these preliminaries are over and after the first man, brave enough to have himself stabbed for blood, has been found, does the work begin in earnest. The first 'victim' is always sure to bring others."[67] In contrast, the investigators were able to examine what they considered representative groups of men, women, and children among the Jewish population. The specificity of the society's customs and norms required certain methods of examination and scientific approaches, like restraining from spleen examinations of Arab women for purposes of modesty.

Epidemic Episodes and Efforts to Combat Malaria in Palestine

Despite incomplete statistics, the Mandatory government saw the necessity of addressing the malaria problem. It enacted the Antimalarial Ordinance of 1922, which required individual settlers and landowners to share the work and expenditures of malaria control. The law limited some cultivation and swamp usage by prohibiting growing rice in the swamps, for instance, "within three kilometers of any boundary of any municipal area." Three kilometers was the range of mosquito flight, according to the scientists whose knowledge the government and the KKL used to specify agricultural methods and Jewish settlement restrictions. In addition to prohibiting rice paddies, participation of colonists and villagers in malaria control was compulsory. Failure to comply with the ordinance made a landowner, peasant farmer, or settler subject to a fine and a short-term prison sentence.[68] As governmental and Zionist engineers undertook further drainage schemes, the scope of the ordinance was expanded to include other areas of Palestine. In 1923, for instance, the attorney general amended regulation 10 of the ordinance after an objection to the original decree was made. The amendment stated,

67. Kligler, Shapiro, and Weitzman, "Malaria in Rural Settlements in Palestine," 291.
68. League of Nations Health Organization Malaria Commission, *Reports on the Tour*, 29–34.

In the case of irrigated lands or gardens, every occupier or owner shall keep any canals, irrigation channels, water courses or drains, which are on his land or in respect of which he enjoys a servitude or easement, clear and free from obstruction and repair the banks thereof, etc.

With the enforcement of such regulations, the British intended to expedite work in public water areas; however, the extent to which the government enforced this ordinance is questionable. We do know that owing to the Anti-Malarial Ordinance, there were 713 prosecutions in 1922, as compared to 200 in 1921. Despite an upsurge in prosecutions, Colonel Heron, the director of the British Health Department of Palestine, cited potential difficulties with regard to the implementation of this ordinance. In a letter to Bernard Flexner of the Joint Distribution Committee, Heron anticipated resentment by Arab *fellahin* for exerting legal pressure on "his land or his water supply which he has held for ages" but noted that no resentment had been recorded in the major projects completed.[69] In the last decade of the Mandate, under the Public Health Ordinance of 1940, revisions of antimalarial rules addressed "certain aspects of irrigation" and "artificial fish ponds."[70]

Malariologists and sanitary engineers conducted hundreds of large and small antimalaria projects in urban and rural areas during the Mandate period. In 1921, for instance, antimalaria projects included those in Retaniah, Menachamia, Deganiah, and Betaniah, in the Kinnereth area, and the entire Migdal area, including Yemma, Bet Gan, and Um-El-Alek.[71] Urban and rural antimalaria efforts were part of a vast infrastructure project that included the building of railroads, roads, and electrical, water, and telegraph systems. British, Zionist, and Palestinian Arab parties

69. Government comments prefacing Heron's quote state that the passage was omitted from the memorandum submitted by the Jewish Agency in 1937 to the Royal Commission. This intentional omission is recognized by the government on page 5l since the quote gave clear acknowledgment of Arab cooperation. Comments on Memorandum submitted by the Jewish Agency to the Royal Commission, Antimalarial and Drainage Works by Jewish Bodies, 4, PRO CO733/345/10; and "Proceedings of the Seventh Meeting of the Antimalarial Advisory Commission," May 24, 1923, 3, JDC Archives File 280; and "Proceedings of the Eighth Meeting of the Antimalarial Advisory Commission," November 22, 1923, 13, both in JDC Archives, file 280.

70. British Health Department of Palestine, annual report for the year 1941, 1, JDC Archives, AR 21/32/291, microfilm 40.

71. I. Kligler, "Report of the Anti-malarial Work for May and June 1921" and "Progress Report of the Anti-malarial Demonstration for the Months July and August 1921," RF, RF5/2/61/398.

all participated in and contributed to this wider program. Their respective involvements, however, were tinged with different meanings and purposes. Like the malaria educational propaganda that will be discussed in chapter 6, British and Zionist health officials employed the message of responsibility for one's own health and sanitary surroundings in recruitment efforts for antimalaria projects. Arab purposes and meanings are harder to discern due to a lack of available documentation and the wide disparity between rural and urban resident living. From the materials that exist and/or are accessible, it does not seem that malaria projects were connected in any signficant way to Palestinian nationalism in discourse or practice. Among other things, this is likely because the British were in charge of their malaria control and as such there were no Arab antimalarial agencies, or trained malariologists during this time. However, because the Arab community's health affairs came, in general, under the jurisdiction of the British Mandatory government, we can presume that they were exposed to, and perhaps influenced by, British messages of self-responsibility for one's health.

In addition to lessening their own expenditure on health-related and public works endeavors, the government urged cooperation of all inhabitants in drainage projects, both Arab and Jewish. This was allegedly "to awaken the interest of the inhabitants in their malaria problem" and to develop a consciousness and responsibility for disease eradication, thought to be a sign of civilization and progress.[72] The documents, however, suggest that the government promoted contribution to antimalaria work for reasons of cheap labor and the defrayal of government costs rather than for the social welfare and independence of the population.

The distinct nationalist flavor and ramifications of Zionist self-help and public health activities for the nation-building project distinguished Zionist efforts from those of the British. The theme of self-help figured prominently in all Zionist disease eradication campaigns, tying into the larger notion of *Kibush hakarka* (Conquest of Land), wherein the Jewish settler was seen as an active agent in shaping himself in Palestine and in constructing an intimate tie to the *Eretz Israel* rather than what was considered the fatalistic, detached Diaspora Jew or Palestinian Arab.

The self-help ethos of the Zionists translated into concrete programs to address the malaria problem. By 1934, an outbreak of malaria following a heavy rain and a lack of strict antimalaria control caused alarm in the

---

72. MRU, "Malaria Control Demonstrations," 2; Moulin, "Tropical without the Tropics," in Arnold, *Warm Climates*, 170, for issue of medical transformism as activism, a sentiment considered by colonial officers as opposite of the fatalism of Muslims.

Yishuv. Communities sent letters seeking help to the Vaad Leumi, the entity chosen to administer the affairs of Palestine Jewry and represent the Jewish community during the Mandate period, as well as to the Mandatory government. At the same time, Kupat Holim (the Sick Fund of the *Histadrut,* or "Federation of Laborers," the Zionist labor union) gathered statistics showing a significant increase in new malaria cases among their members.[73] The outbreak compelled the Vaad HaBriut (the health council of the Vaad Leumi) to take quick action in several areas to prevent further malaria problems in 1935. Such measures included drainage work, hiring additional inspectors, and putting pressure on the government to do same. The Vaad HaBriut called for a Central Bureau for Malaria Control, a course in malariology for physicians, increased visitation to various districts, and intensified education work.[74] A special committee consisting of Professor Kligler, chairman of the Health Council, Dr. Noack, medical officer of the Vaad Leumi, and Dr. Mayer, medical director of Kupat Holim, toured the malaria-infected areas and discussed implementation of intensified malaria supervision with local committees.[75]

In the wake of the Arab Revolt (1936–1939), a sustained Arab strike and violent uprising against British support for a Jewish national home in Palestine, malaria increased because of inconsistent distribution of quinine and the curbing of drainage measures. World War II conditions also reduced quinine availability, but Palestine saw a general leveling off of new malaria cases with an occasional epidemic resurgence during the 1940s.[76] The British Health Department of Palestine described the improved malaria

73. Letter from the Hadera Local Council to the high commissioner, May 14, 1935, CZA J1/1303; a copy of the same letter was sent to Vaad Leumi; letter from Kibbutz Ein Shemer in Karkur to the Jewish Agency Secretariat, February 5, 1935, CZA S25/377; Kupat Holim report on "Milchama bemalaria," CZA S25/377.

74. "Prevention of Malaria in 1935: Resolution Adopted by the Vaad HaBriut at Its Meeting of March 5, 1935," ISA M1503/1/96 (59); and "Impending Malaria Season, communicated by Health Section of Vaad Leumi," April 29, 1935, CZA S25/377; letter from Heron to secretary of Executive Committee, General Council of the Jewish Community of Palestine, on antimalarial measures, April 12, 1935, ISA M1503/1/86(59); letter from Katznelson to government director of medical services, on antimalaria measures, May 28, 1935, ISA M1503/1/86(60); Fifth Meeting of Vaad HaBriut, April 9, 1935, CZA J1/3571.

75. Health Section of Vaad Leumi, "Impending Malaria Season," April 29, 1935, CZA S25/377. This document also discusses problems of the Vaad Leumi with the government on antimalaria matters. Vaad HaBriut, official communique no. 22/35, CZA S25/377.

76. Atabrine was considered as a replacement for quinine. Manufacturing it in Palestine was deliberated. Letter from Dr. Jonas S. Friedenwald to Mrs. A. P. Schoolman of Hadassah, June 3, 1942, Hadassah Archives, RG 1/box 100, folder 2; letter from Schoolman to Yassky, September 29, 1942, Hadassah Archives, RG 1/box 100, folder 2; memorandum to Mrs. Schoolman from Denise Tourover, August 6, 1942, Hadassah Archives, RG 1/100/2.

situation as "almost negligible in most of the urban areas...and low in most of the rural."[77] Malaria incidence had decreased tremendously by the end of the Mandate era. In the 1940s, a gradual decline in adult and infant mortality rates accompanied a decrease in malarial incidence.[78]

In 1945, heavy winter rains brought the threat of another malaria outbreak to the southern part of the Huleh area (between Lake Tiberias and Huleh Lake). Dr. Abraham Katznelson of the Vaad Leumi, at a press conference that reviewed the *Yishuv*'s health services, remarked:

> We are on the brink of a malaria outbreak which may exceed even the difficult times two decades ago, when Jewish settlers first started fighting the disease here. Last year there were some two thousand Jewish cases, half of which were new patients.[79]

Jewish settlers lost 15,000 workdays because of the 1944 outbreak in the Huleh Valley.

Palestine joined the worldwide shift to dichloro-diphenyl-trichloroethane (DDT) use for mosquito control during and after World War II.[80] Zionist health officials were optimistic about the effects of DDT for the ultimate eradication of malaria. But in 1945, civilian use of DDT had still not been officially approved. In addition, spraying DDT in large areas was expensive and required airplanes, a disincentive for civilian use while war expenditures were a concern. The director of the Health Department commented that

> I must strike a warning note with regard to DDT. With all its properties it is not likely to be a cheap panacea for the control of malaria and the people must not expect, as a result of its use, to be exempt from the necessity of doing antimalaria work and to be free to allow irrigation water to flow about anywhere without control.[81]

77. British Health Department of Palestine, annual report for 1941, 11.

78. British Health Department of Palestine, annual report for 1941, 4.

79. Abraham Katznelson, "Danger of Malaria Epidemic," *Palestine Post*, June 25, 1945, 3, CZA J/2284; document suggesting use of DDT in the Huleh area, CZA J/2284; letter from Katznelson to editor of *Palestine Post*, July 8, 1945, CZA J/2287.

80. See Frank Fenner, "Malaria Control in Papua New Guinea in the Second World War: From Disaster to Successful Prophylaxis and the Dawn of DDT," *Parassitologia* 40, nos. 1–2 (June 1998): 55–64.

81. Letter from MacQueen to Katznelson, July 6, 1945, CZA J/2287.

Dr. Katznelson disagreed, citing the need for additional measures (besides the governmental ones of canalization, cleaning, and chemical treatment) such as DDT to wipe out malaria in highly infested areas like the Huleh. Katznelson also argued for DDT by pointing out its effectiveness on war fronts and in Palestine. He wrote, "I fully agree, however, that DDT in no way renders routine antimalaria work unnecessary."[82]

Notwithstanding all the work done during the twenty-five years of the Mandate, an article entitled "Fewer Cases of Malaria This Year," published in the *Palestine Post* on June 28, 1945, noted alarmist reports of the potential for a countrywide malaria epidemic. The article mentioned the canalization of 1,000 kilometers of streams, funded by the government, to prevent stagnant waters and the existence of special mosquito-catching stations across the country. Despite its financial concerns, the Mandatory government agreed to release a small quantity of DDT for experimental purposes in the Huleh in order to address the malaria threat. The quantity was not sufficient, so the Vaad Leumi looked to Hadassah in the United States to secure the rest. By November 1945, the Shell Oil Company was ready to supply the Vaad Leumi with a new product of gas oil and green concentrate that they had tested in the Suez and which they said killed mosquitoes.[83] In order to continually press the government for more DDT, the Vaad Leumi (via Dr. Katznelson) stressed its need for both the Jewish and Arab populations, particularly noting the existence of malaria in mixed Arab-Jewish districts.[84] Katznelson addressed the sympathies of the government by emphasizing coexistence, odd in a period that had recently seen the three-year Arab Revolt, the separation of Arab and Jewish economies, and was headed toward the dissolution of positive Arab-Jewish relations in the 1948 war and thereafter.[85] In 1946, the government began to spray DDT in 122 villages situated north of Haifa (total population 36,000 persons, Upper Galilee area).[86] New malaria cases

82. Letter from Katznelson to editor of *Palestine Post*, 2.

83. Letter from Katznelson to editor of *Palestine Post*, 2; "Fewer Cases of Malaria This Year," *Palestine Post*, June 28, 1945, CZA J1/2284; letter from Katznelson to Editor of *Palestine Post*; letter from Birnbaum to Katznelson, January 7, 1945, CZA J/2284; letter from Shell Oil Co. to Katznelson, November 6, 1945, CZA J1/2287.

84. Letter from Katznelson to MacQueen, June 27, 1945, CZA J1/2284.

85. Such an emphasis on coexistence can also be seen in proposed cooperative antimalaria initiatives (see chapter 7).

86. Letter from J. MacQueen to Katznelson, January 29, 1946, CZA J1/2287; Saliternik, *History of Malaria Control*, 13.

dropped from 703 in 1945 to 180 in 1946.[87] By 1947, total malaria cases decreased significantly, from 1,652 out of approximately 1.69 million people in 1944 to 528 cases out of approximately 1.84 million people in 1946, a decrease largely attributed to the use of DDT.[88] In addition, DDT was used to kill flies, bedbugs, cockroaches, and fleas.[89] In 1948, DDT was distributed to most Jewish settlements in Palestine.[90]

In spite of all these epidemiological shifts, malaria was still an endemic disease at the end of the Mandate period. According to Zvi Saliternik, a prominent Zionist malariologist during the Mandate and afterward, the disease was highly endemic in areas like Beisan, Huleh, the Jordan Valley, and Haifa Bay. The general rate of first infections in these areas in 1949 with *P. vivax* was 80.3 percent, with *P. falciparum*, 19 percent, and with *P. malariae*, 0.1 percent.[91] The interaction of humans and the environment in constructing and reconstructing the landscape of Palestine certainly effected change but did not rid the country of the disease entirely. As we will see in the concluding chapter of this book, negotiation between humans and the environment continue to this day.

87. Saliternik, *History of Malaria Control*, 13.

88. "Large Decrease in Cases of Malaria," *Mishmar*, May 29, 1947, CZA S71/402; Justin McCarthy, *The Population of Palestine: Population History and Statistics of the Late Ottoman Period and the Mandate* (New York: Columbia University Press, 1990), 68–69.

89. Letter from P. Noack to all [of Vaad Leumi], May 13, 1947, CZA J1/8400.

90. Document on DDT and the war against malaria, 1948, CZA J1/8400.

91. Saliternik, "Reminiscences," 519, and idem, *History of Malaria Control*, 13.

Potential Landscape

Swamp Drainage Projects and the Politics of Settlement

Colonization will be impossible unless the disease [malaria] be brought under control.
—Justice Louis Brandeis, 1919[1]

In the nineteenth and early twentieth centuries, the presence of endemic malaria in Palestine and the Arab-Jewish settlement patterns that malaria morbidity produced were among the primary factors making Jewish land purchase and colonization possible. Getting rid of malaria made the next step of Jewish settlement possible.

Due to a variety of reasons, including Ottoman land reform, Arab patterns settlement gradually expanded to the valleys and coastal regions in the mid-to-late nineteenth century. Such disproportionate speculation and purchase in these areas is evidenced by the fact that by 1945, 66 percent of Jewish land was concentrated in the valleys and coastal region. By the end of the Mandate period, Jews owned 23 percent of the coastal plain, 30 percent of the northern valleys, and only 4 percent of the hill country.[2] Despite a growing Arab inhabitancy, these lands remained more vulnerable to Jewish land speculation during early Zionist settlement and

---

1. Paraphrase of Justice Brandeis's comment in 1919, presumably written by I. Kligler, "A Battle for Health" (December 8, 1943), 5, Hadassah Archives, RG 1/100/2.

2. Shafir, *Land, Labor and the Origins of the Israeli-Palestinian Conflict*, 18, 40–41, 183; Kligler, *Epidemiology and Control of Malaria*, viii, 103; Metzer, *Divided Economy of Mandatory Palestine*, 86.

thereafter. Nearly all of the Jewish colonies before and during the Mandate period were located in highly malarial areas of the valleys and the coastal region of Palestine since large landowners held these lands and they were more readily purchased because of their relatively sparse populations.[3] During the nineteenth century, Palestinian Arab residency was concentrated in the hilly regions. Scholars largely explain this pattern as a reflection of the peasants' attempt to evade heavy taxation imposed by the Ottoman regime and as a method to avoid bedouin raids. Unlike Palestinian Arabs who lived in the cities and villages, the Arab bedouin wandered throughout Palestine with their herds.

But this settlement pattern was also a way to avoid malaria. Although the hills of Palestine (other than Nablus) were not completely free of malaria, the valleys and coastal region had a much higher prevalence of the disease.[4] Given this profile of morbidity, Arab cultivators living in the nearby hills would often descend to the plain during the day, when mosquitoes were not as plentiful, to tend to their lands.[5]

Discussions regarding the politics of Zionist land speculation and settlement—from attention to land perception and recruitment of prospective immigrants to debates about land reform and policies—all featured the notion of "potential." Zionist leaders and medical professionals believed that the land possessed inherent, vast possibilities for Jewish settlement. They expected to "heal" the land by making it productive. The potential utility of different types of land in Palestine, including swamps, shaped British land policy, and it also influenced the extent of Zionist purchase and impacted demographic changes during this period. Indeed, the complex dynamics between malaria and the government's land tenure policies together advanced and often constrained the Zionist pattern of land acquisition.

Swamp drainage projects—as a way to rid Jewish areas of malaria—were used as practical instruments to realize what Zionist malaria and settlement officials saw as the land's productive capacity, but equally important, as rhetorical tools in policy debates with the clear objective of expanding Jewish immigration and close settlement in Palestine. These projects not

3. JA, "Antimalaria and Drainage Work by Jewish Bodies," 6–7; Metzer, *Divided Economy of Mandatory Palestine*, 7.

4. Kligler, *Epidemiology and Control of Malaria*, vii; Firestone, "Land Equalization," 818, 818n15.

5. The Zionist immigrants did not do this—when they settled, they lived near watercourses and marshes. Kligler, *Epidemiology and Control of Malaria*, 103.

only forged tangible connections between disease, land transformation, and Zionist redemption but also were semantically bound with nationalist agendas. In this way, Zionist malaria control efforts were not simply a war against disease but perhaps even more a campaign to create more room for immigration and settlement: to further the building of a Jewish national home. Particularly during difficult economic and political periods for the nationalist movement, Zionist medical and land reclamation agencies viewed swamp drainage as one way to maximize immigration and generate greater Jewish manpower. Leaders of these agencies also saw swamp drainage as a step towards exploiting the land for commercial products and food. In their broadest sense, the connections made between disease eradication, immigration, and settlement here are not unique; such agendas were used in other colonial contexts to facilitate European colonial settlement.[6] What makes the Mandate Palestine case unique is how these particular connections play out in this contested, colonial territory.

Settling on a Sanitary Land: Malaria, Settlement, and Potential Immigration

The notion of the potential of the land and of the Jewish people in Palestine is implicit in *havra'at hakarka vehayishuv* (healing the land and the nation). The idea of healing the land and the Jewish settlers in Palestine suggests that both had the capacity to be cured if only reformed and managed appropriately. Beyond being seen as beneficial from a medical perspective—to rid the people of the malaria scourge—antimalaria measures were framed by Jewish scientists in Palestine within a discourse of the land's potential. Antimalaria measures were thought to let loose the land's productive capacity, reactivate it by external means, and make the land active and healthy. Zionist settlers undertook antimalaria measures to build and prove the productivity of both the Jewish people and the land. These conditions then were thought to facilitate the redemption of both.

Seeing and describing nature's untapped potential was a common outlook in colonial science. Kavita Philip has shown how colonial botanical and agricultural research administrators replicated this viewpoint in

6. For Morocco, see Will Swearingen, *Moroccan Mirages: Agrarian Dreams and Deceptions, 1912–1986* (Princeton: Princeton University Press, 1987), 48; and Paolo Palladino and Michael Worboys, "Science and Imperialism," *Isis* (1993): 84–97.

settings like India. These experts, like their Zionist colleagues in Palestine, believed that to free nature's wealth one had to know how to manage and regulate a colony's chaotic landscape.[7] Draining swamps was one way to achieve that objective. Like its corollary in the persona of the new Jew, the land and its potential were mentally constructed by Zionist medical professionals as dichotomies: productive/unproductive, real/potential, diseased/healthy. These qualities—like the Diaspora Jew—were seen as mutable, if manipulated "rationally." As one early pamphlet states:

> But are not marshes the negation of the desert? The existence of marshes in Palestine is proof of the abundance of unused springs and misdirected torrents. Their rational utilization will rid the country of the scourge of malaria and increase production by irrigation.[8]

Unleashing the land's potential could only be done through the intervention of humans, and the Zionists at that: "The treasures of Palestine have remained hermetically sealed for centuries; it is now only that the open sesame of these treasures has been discovered: Jewish devotion and Jewish idealism."[9] This statement produces a certain history of Palestine whereby the land lay unattended until the arrival of the Zionists. They are seen as the only actors who could cause the land's productive forces to gush forth.

According to Zionist malariologists, malaria and marshes impeded the realization of the land's potential and the prospective success of the Zionist endeavor. As we will see, although malaria morbidity paradoxically *enabled* Jewish land purchase in the late nineteenth century due to the demographic patterns it produced, it also operated as "an effective bar to *settlement* of large tracts of fertile land";[10] that is to say, it was a key obstacle to permanent close settlement of Jewish immigrants in Palestine. Such con-

---

7. Kavita Philip, *Civilizing Natures: Race, Resources and Modernity in Colonial South India* (New Brunswick: Rutgers University Press, 2004), 89–90.

8. "Palestine: The Real Country," *Palestine: Organ of British Palestine Committee* 2, no. 1 (July 28, 1917): 3, CZA J113/1430.

9. "Palestine: The Real Country," 6.

10. Yaakov Goldstein, "Tochnit zioni rishona leyibush bitzot haHuleh" [The first Zionist plan for draining the Huleh swamps], *Cathedra* 45 (1985): 164, emphasis added. Lord Passfield also considered drainage and irrigation as solutions to the problem of *fellahin* demographic congestion in the hill district. This, along with intensive cultivation, was thought to provide more land on which the *fellahin* could reside. Letter no. 487, from Passfield to chancellor, high commissioner of Palestine, June 26, 1931, PRO CO 733/221/9.

tiguous and sustainable settlement was one of the major goals of the Zionist endeavor.[11] Justice Louis Brandeis, a prominent Zionist and U.S. Supreme Court judge, was one of the many Zionist leaders who understood the connection between health and Zionist regeneration through settlement. Making the connection between land potential and human potential through immigration, he argued that without improvements in the sanitary and hygienic conditions of Palestine, especially with regard to malaria eradication, no significant Zionist immigration and inhabitation would be likely to take place.[12]

Henrietta Szold, founder of the Hadassah Organization, echoed the concerns of Justice Brandeis in 1920 when she identified the malaria problem as a potential cause for the collapse of Zionism. She declared: "I fear the events of next spring [expected malaria epidemic]. There is bound to be a collapse, either the physical collapse of the bands of the young men coming here, or the moral collapse of the Zionist idea...if we wish to forestall the mosquito we must begin at once."[13] Four years later Dr. Hillel Yofe, an early Zionist malariologist, made the treatment of malaria the top priority of the Jewish health establishment, over and above all other diseases, because it was "one of the greatest obstacles to creating settlement in many places in our land, especially those with the most fertile soil."[14]

Jewish settlements like Hadera (1890) and Atlit (1903), for example, had exceptionally high case fatality rates. Other colonies that were established, like Petach Tikva (1878), also suffered greatly from the malaria scourge.[15] Tantura and Yesod Hama'ala (1883) in the Huleh Valley were also deemed significantly malarious. Like the Arab villages on those sites that had been abandoned, some of these Zionist settlements had to be abandoned due to the severity of the malaria situation.[16] In other places, like Khulde in 1921, presence of the disease impeded permanent

11. See Kitron, "Malaria, Agriculture and Development," 297.

12. Goldstein, "Tochnit zioni," 164, CZA Yellow Subject Folder on Swamp Drainage.

13. Letter from Henrietta Szold to Judge Mack, December 26, 1920, 4, CZA J113/554.

14. Hillel Yofe, "Leshe'elat haholim bemalaria chronit," *Briut ha'Oved* (Tel Aviv: Kupat Holim, 1924), 6.

15. JA, "Antimalaria and Drainage Work by Jewish Bodies," 3; Levy, "*Zichron Yaakov,*" 97.

16. Shafir confirms malaria as being one of the causes for the loss of villages between the mid-sixteenth century and early nineteenth centuries. See his description of a shift in settlement patterns to the valleys and coastal zone by the mid-nineteenth century. Shafir, *Land, Labor and the Origins of the Israeli-Palestinian Conflict,* 38–40.

settlement.[17] Clearly, Zionist drainage projects not only tried to alleviate the malaria problem but perhaps just as important, they sought to facilitate immigration and permanent settlement.

The triangular relationship between swamp drainage, immigration, and labor for the Zionist project did not go unacknowledged by Palestinian Arab intellectuals. Dr. Aftim Acra, a professor of environmental science practicing as a pharmacist in Palestine during the Mandate period, recalled: "This project [of swamp drainage] was believed by the Arab populace to be tinged with the Zionist objective of clearing the area for the establishment of some Jewish settlements. Their suspicion proved to be true as time went on." Acra's statement was confirmed in interviews with two other Palestinian Arab health professionals who also grew up and worked during the Mandate period.[18]

During of the Thirteenth (1923, Carlsbad) and Fourteenth (1927, Vienna) Zionist congresses, delegates acknowledged the association between successful settlement and malaria morbidity. Resolutions of the Thirteenth Zionist Congress, confirmed at the Fourteenth Congress, ordered the consideration of the sanitary and hygienic needs of the settlements (including drainage), particularly when new areas were to become occupied. Interests of not only health but also economy motivated such resolutions; the goal was to maximize the labor force in each settlement and to minimize the amount of sick days used by Jewish residents.[19]

If Jews were considered the only actors who could unleash the land's potential, then the Zionist project required considerable Jewish recruitment and migration to Palestine. In order to facilitate maximum immigration, the Zionist leadership had to make immigration attractive. Indeed, between October 1921 and September 1939 a total of 311,868 Jews immigrated to Palestine.[20] Arthur Ruppin, head of the Settlement Office of

17. Letter from Henrietta Szold to Vaad Hatzirim, September 4, 1921, CZA KKL3/99b; ESCO Foundation for Palestine, *Palestine: A Study of Jewish, Arab, and British Policies* (New Haven: Yale University Press, 1947), 1: 391.

18. Personal written correspondence with Dr. Aftim Acra, September 26, 1998; confirmed in oral history interview with Dr. Majaj, June 18, 1996; and oral history interview with Said Rabi, vice-head of Department of Health Education, Israeli Ministry of Health, July 1, 1996.

19. Sanitary Sub-commission of the Fourteenth Zionist Congress, January 19, 1926, attached letter from Kisch to members of Palestine Zionist Executive (PZE), CZA J1/2678.

20. A. Ulitzer, *Two Decades of Keren HaYesod* (Jerusalem: Eretz Israel Foundation Fund, Keren HaYesod, 1940), reprinted in *Rise of Israel*, 14 (New York: Garland Publishing, 1987), 371.

the Zionist Organization, addressed this issue when he spoke to the Fifteenth Zionist Congress in 1927. He recognized that immigrants would not come to Palestine if they could not protect themselves from hunger and disease or if education facilities were not available to their children. If they came to Palestine and did not receive these essentials, they would leave the country. The immigrants, he claimed, must be provided these necessities so that they could devote their time to cultural affairs, so as to make Palestine the center of Jewish culture.[21] Antimalaria projects contributed toward that end by facilitating a disease-free environment and by enabling intensive agriculture.

Despite efforts to weaken the correlation between malaria prevalence and immigration patterns through swamp drainage, this relationship remained strong. Dr. Israel Kligler, the most esteemed Jewish malariologist of the time and director of the MRU, noted early in the Mandate period that "[t]he seriousness of the malaria problem is indicated by the fact that over 40 percent of the immigrants are infected during the first six months and over 60 percent in the course of the first year. Aside from the material loss, the loss in energy, vitality and morale are incalculable."[22]

Zionist settlement policy and preparation was strongly influenced by Kligler's opinions about malaria. As we shall see, Kligler strongly objected to the settlement of immigrants and existing Jewish residents on malaria-infected lands bought by Zionist land purchase and settlement agencies before drainage had taken place.[23]

The converse association between malaria rates and immigration also held true: positive immigration trends correlated with decreasing malarial incidence and the completion of antimalaria work. In 1935, malaria outbreaks resulted from heavy rains. The British Mandatory government noted that large immigration during this time made possible active settler cooperation in fulfilling antimalaria work. However, immigrants performing swamp drainage work did not expect to be responsible for this endeavor. Newcomers living in temporary or semipermanent conditions in coastal swamp areas expected the British Health Department of Palestine

21. Ruppin is espousing *Ahad Ha'Am*'s idea of transforming Palestine into a Jewish cultural center. "A Period of Crisis," reprinted in Ruppin, *Three Decades of Palestine* (Jerusalem: Schocken, 1936), 165.

22. I. Kligler, "Report on Preliminary Survey of Malarial Situation among Selected Groups of Immigrants and *Chaluzim*," February 24, 1920, 2, CZA J15/7212. The report included the areas surrounding the roads of Haifa/Gedda and Afule/Nazareth.

23. See chapter 5.

to undertake the malaria eradication in the area, given the Department's efforts elsewhere. They emphasized their point: "[I]n villages where the inhabitants have suffered malaria for generations, Government undertakes to wipe it out, while in areas of new settlement the pioneers must themselves clear the vicinity of malaria [seems] paradoxical."[24] These settlers noted that the Mandatory government carried out land reclamation in Arab villages but in the case of Jewish settlements, the government only helped plan drainage activity. Implementation rested upon the Jewish settlers.

The Mandatory government's differential treatment of the Arab and Jewish communities regarding the execution of swamp drainage reflected the broader arrangement of separate medical and public health services during this time. Linked to their critique of the government's policies on the disposal and improvement of state lands, the health activists of the *Yishuv* disagreed with the government's approach towards antimalaria work. They claimed that while Zionist agencies accepted the responsibility for draining swamps in areas they bought, such consent did not preclude governmental assistance in reclaiming areas in the vicinity of Jewish land. The Mandatory government responded to the activists' demands by invoking the Anti-Malarial Ordinance (1922). This ordinance placed responsibility for drainage upon the Jewish settlers and Arab residents.[25]

Despite the persistent tension between the *Yishuv* and the British Mandatory government on this issue, ridding areas of malaria enabled significant Jewish settlement. Moreover, by transforming the landscape of Palestine, swamp drainage projects changed the image of certain areas as diseased or death-ridden to ones that were considered healthy, thus contributing to the larger goal of *havrảat hakarka ve haYishuv*. As one Keren Keyemet L'Israel publication exalted: "Settlements established on the spots where previously disease and death prevailed, can now be counted among the healthiest in Palestine."[26]

Drainage measures facilitated transformations not only in landscape perception but also in intensive agriculture and in the irrigation of dry land. Drainage, irrigation, and settlement issues merged together in the theory that more settlement would provide greater utilization of the springs and therefore greater regulation of the waters, at least in areas where springs

24. "Government and Malaria Danger: Conditions in Jewish Settlements Call for Action," *Palestine Post*, June 14, 1935, ISA M1503/1/86(60).
25. Letter from Heron to chief secretary, on the topic of antimalarial measures, April 14, 1935, 2, ISA M1503/1/86 (59).
26. "Twenty-Five Years of KKL: 5662–5687," 40, CZA A246/441.

FIGURE 3.1. Emek Hefer: girl throwing larvacide. Courtesy of the Keren Keyemet L'Israel Photo Archive.

were the main cause of malaria. Zionist medical professionals believed that greater demands for drinking water and greater irrigation would significantly decrease the amount of unused water in any one given area and in turn, give rise to demands for subsoil drains and canals.[27] The more people to serve, the more reason for investment in drainage and irrigation operations and the greater the need for increased immigration.

Invest they did. Having the luxury of financial investment from abroad, in 1924 the MRU, an organization sponsored by the American Jewish Joint Distribution Committee and various other Jewish institutions, spent 14,492 Egyptian pounds on malaria control, or sixty-six piasters per capita (equal to $3.30 per capita).[28] This sum was slightly more than forty-six times the government's (and the affiliated Malaria Survey Section, a Rockefeller Foundation initiative) expenditure per capita on malaria control for the same year. The government spent an average of 1.42 piaster

27. MRU, "Malaria Control Demonstrations," 8.

28. For a discussion on the use of the Egyptian pound at the beginning of the Mandate period in Palestine, see Smith, *Roots of Separatism in Palestine*, 26–28.

per capita of malaria control for the Arab sector.[29] Such a huge discrepancy in control expenditure between the government and the MRU, working primarily on behalf of the Jewish population, significantly affected the respective communities' malaria rates.[30]

In addition to concerns about the potential for immigration and settlement, malaria morbidity frustrated efforts to achieve maximum labor productivity. Apart from the great suffering malaria caused to settler health, Zionist medical professionals seconded labor's concerns about malaria-related labor and agricultural losses. Letters to the Jewish Agency from workers in several kibbutzim (communal settlements), including Kibbutz Shacharia, Kibbutz haNoar Ha'Ivri-Petach Tikva, Gordonia, HaSadeh in Rishon leTzion, and Tiberias, complained that they did not have sufficient supplies, beds, clothes, money and space. All of these conditions, they claimed, led to diseases such as malaria because they increased the susceptibility to and transmission of disease among residents. A letter from laborers in the HaShomer HaTzair Kibbutz Gimel used the act of draining swamps to justify their worthiness of better general accommodations: draining swamps showed their adoption of the Zionist notion of self-help and conquest of labor. "After seven years we established [ourselves] in Palestine, years of hard work and different kinds of very hard labor of *kibush ha'avoda* (conquest of labor) in the city and in the village, draining the swamps and paving the roads and all that accompanies this."[31] Conditions in some of the workers' camps, especially in areas with malaria, were so bad that one doctor described them as "bedouin"-like, referring not only to the temporary tents in which the new immigrants lived but to the overall unsanitary conditions.[32]

Despite harsh living conditions and malaria morbidity, drainage work supplied needed employment for Jewish laborers, especially in periods of economic difficulty. Though antimalaria work proved arduous for most new immigrants, the Keren Keyemet L'Israel recognized it as a good skill

29. Kitron states that depending on the year, up to 20–40 U.S. dollars per capita were spent on malaria control in certain Jewish areas. Kitron, "Malaria, Agriculture and Development," 304; League of Nations Health Organization Malaria Commission, *Reports on the Tour*, 44.

30. See chapter 2 and the diagnosis of the League of Nations in chapter 5.

31. Letters from Shacharia, etc., to the Jewish Agency, CZA J1/2233; letter from HaShomer HaTzair Kibbutz Gimel to Jewish Agency, May 23, 1932.

32. Memoranda: The Situation in the workers' camps in the settlements. Cooperative visit from the Jewish Agency, Vaad Leumi and Kupat Holim in the settlements of Judah and Sharon, November 22, 1931, CZA J1/2233.

FIGURE 3.2. Yagur (Yajur) swamp draining, 1924. Courtesy of the Keren Keyemet L'Israel Photo Archive.

to have for future employment opportunities since it was thought to nurture perseverance and technical acumen.[33]

Laborers engaged in antimalaria activity were responsible for maintaining and repairing the drains and pipes installed during the original swamp drainage project. This work was necessary to prevent seepage or blockage of watercourses that could create stagnant pools, which could eventually lead to new malaria cases.[34] But people did not always comply. By 1931, for instance, drainage works at Nuris installed in 1923 needed repair since damage had caused the spread of malaria. A letter from Mr. Granovsky of the KKL to the Jewish Agency Health Department noted that the drainage had been fixed several times but that the populations of Kfar Yezekiel and Kibbutz HaGiv'ah refused to accept help for their drainage problems, eventually causing a malaria problem in Nuris.[35]

33. Minutes of the Central Executive of the KKL, Jerusalem 1921–1923, 2, CZA A246/441.

34. Letter to KKL from unknown, May 29, 1922, 4 CZA KKL3/57b; letter from KKL to Jacob Zwanger, May 26, 1922, CZA KKL3/57b.

35. Other acts of noncompliance were recorded. For instance, after working six hundred days on antimalaria schemes, residents then refused to help a government doctor with other work. A legal case was brought against them for their refusal. Letter from Granovsky of KKL

Poor supervision by doctors and medical officers posed another maintenance problem for the settlements and schools. For example, in the late
1930s Zvi Saliternik, a central figure in Jewish antimalaria measures during the Mandate, noted that the person in charge of drainage maintenance
in Nuris had too much area to cover and did not fulfill his obligations successfully. Residents' resistance did not help the situation. Those in Nuris
ridiculed, "You can take us to court—we will not pay more than half a
lira which is much cheaper than repairing the situation." In response to
such defiance, Saliternik felt it was time to impose fines in order to gain
compliance; the time had passed, he argued, for using moral arguments to
influence the settlers. A similar situation occurred in Nahalal.[36] To complicate the situation, as new cases of malaria decreased, supervision of
antimalaria treatment and work was not seriously adhered to.[37]

And despite a downturn in malaria incidence, by the late 1930s about
one-third of Jewish land purchases still needed swamp drainage.[38]

Potential Settlement and the Politicization of Reclamation and Redemption

The Politics of Settlement: Land Categories and Land Transfer

As discussed in chapter 1, Zionist labor discourse and planning placed a
premium on agricultural settlement and development. Renewing rural areas was seen as a vital component for the revitalization of the Jewish people. Consistent with this vision, the World Zionist Organization directed
most of its investments towards rural development while private capital
brought in by immigrants provided the bulk of urban-industrial investment. As such, general Zionist policy did not necessarily guide private
investments; these investments depended principally upon the market.[39]

The choice of the rural Jezreel and Huleh drainage projects presented
in chapter 4 as key case studies reflects not only the rural emphasis of

---

to JA Health Department, April 28, 1931, CZA J1/2233; letter from Dr. Hurvitch from Kupat
Holim of Emek to Vaad Leumi Health Department, July 21, 1931, CZA J1/2233.

36. Letter from Dr. R. Katznelson, head of Kupat Holim Amamit, to doctors, November
1937, CZA J113/511.

37. Letter from Saliternik to Kupat Holim, with a survey of the Nuris area attached, April
21, 1937, 10, CZA J1/1726.

38. Yehuda Reinharz and Ben Halpern, *Zionism and the Creation of a New Society* (Hanover: Brandeis University Press, 2000), 276.

39. Reinharz and Halpern, *Zionism and the Creation of a New Society*, 274; for the rural,
private industry of citriculture, see Karlinsky, *California Dreaming*.

Zionist malariologists, land settlement experts and the labor Zionist leadership in general, but also the organizational division of malaria work in Palestine in the early part of the Mandate. The Mandatory government carried chief responsibility for cistern supervision in municipalities, while nongovernmental agencies undertook most of the large, usually quite costly, rural schemes.[40] Furthermore, British land registration, surveys, and title discernment focused on the rural plains, where most Jewish purchase and settlement occurred and where the Arab and Jewish citrus industry was focused, rather than on the hills.[41]

Urban antimalaria measures were implemented first because of high population density.[42] Urban antimalaria projects primarily involved antilarval measures; cistern covering, cleaning and repair; and distribution of quinine, as well as installing wire netting on house doors and windows. These activities did not radically transform the visible outer landscape of the city. The building of modern water supply systems during this time certainly altered the underground landscape of Palestine's urban centers.

Rural projects included the same measures as urban programs but necessitated other, more complex technological operations and required a larger financial investment. The methods for swamp drainage in rural areas varied according to place and *Anopheles* species. They included and/or mixed methods of subsoil drainage, concrete drains, river regulation, drainage channels, and irrigation ditches.[43] Rural projects most clearly expose the physical and ecological transformations that large anti-malaria works effectuated on the landscape of Palestine.[44]

Zionist leaders and physicians emphasized the land's potential for productivity when addressing settlement issues because land for sale and intensive cultivation during the Mandate was generally limited. This territorial reality contrasted with a belief in the Diaspora that land was plentiful in Palestine.[45] Real limitations on land acquisition primarily stemmed

40. Kitron, "Malaria, Agriculture and Development," 298.

41. Warwick Tyler, *State Lands and Rural Development in Mandatory Palestine, 1920–1948* (Brighton: Sussex Academic Press, 2001), 173; Karlinsky, *California Dreaming*.

42. For urban malaria in Palestine, see HMG, *Campaign against Malaria in Palestine*, 4–5.

43. League of Nations Health Organization Malaria Commission, *Reports on the Tour*, 61–63.

44. "Proceedings of the Eighth Meeting of the Antimalarial Advisory Commission," 2; and Goubert, *Conquest of Water*, 68.

45. See, for instance, Elan Zaharoni, "Bitzot she baEmek," 88, CZA Yellow Folder on Swamp Drainage.

from six factors: high prices, limited funds for Jewish purchase, a developed sense of ownership by the local population, a larger indigenous population than expected, extant settlement on arable land, and a lack of great commercial potential on the land. In addition to a narrow margin of transferable land, the total proportion of unused land was quite low.[46] Such conditions caused uneven Jewish land speculation practices.

Once they acquired marshland, the Zionists utilized it to meet their national goals. Zionist engineers and antimalaria workers, while correcting the general conditions of the land and the people, also sought to change or correct the extent of land for settlement through swamp drainage. Their act of healing or correcting the land was not only intended for the ideological purpose of Jewish redemption but also for the practical corollary of settlement expansion and more immigration.

Land tenure practices and agricultural reform also affected the limited availability of land. Jewish settlement patterns and land purchase practices must therefore be first situated within this broader issue. Knowledge about Palestine's land categories as well as the disposal and acquisition of different types of land helps us understand the larger topographical and political terrain in which swamps and their reclamation took place.

Like shifting Arab settlement patterns at the turn of the century, land tenure issues had their roots in the Ottoman Land Code of 1858. The British drew and adapted from the Code when they occupied Palestine. This Code was part of a larger set of reforms called the Tanzimat reforms that sought to modernize the Ottoman Empire in order to compete in the global economy. The main purpose of the Land Code of 1858 was to gain full control of certain lands and to stimulate agricultural production in order to maximize tax returns.[47] Although the process and nature of the Code's implementation was location-specific, a major effect of the Code was to shift the idea of property rights and relations from one describing the distribution of access to surplus to one denoting access to land;

---

46. Marshland in Palestine can be included among what scholar Baruch Kimmerling has called potential frontierity—the total amount of land in a territory having no relation to its accessibility or ability to be immediately used. The existence of swamps in rural areas exacerbated an already substantially low "frontierity" of land in Palestine that frustrated Zionist land agencies in their quest for land purchase. Baruch Kimmerling, *Zionism and Territory: The Socio-Territorial Dimensions of Zionist Politics*, Institute of International Studies (Berkeley: University of California, 1983), 10. Ussishkin perhaps recognized potential frontierity when he noted that everything could increase: population, productivity, and wealth, but not land. M. Ussishkin, "Call of the Land," *Rise of Israel*, 14: 322.

47. Kenneth Stein, *The Land Question in Palestine, 1917–1939* (Chapel Hill: University of North Carolina Press, 1984), 11, 13, 282.

tax revenue claims made by multiple parties accompanied by cultivators' usufruct rights changed to actual ownership of land by a single title holder where revenue claims were limited to the centralized state alone.[48]

Under Ottoman law, land was divided into six categories: *mulk* (freehold land), *waqf* (religious endowment), *miri* (commonly but incorrectly translated into "state" land), *mewat* (uncultivated land), *matruka* (communal land), and *mahlul* (vacant *miri* land).[49] Two other types of land ownership, called *musha'a* and *mudawwara* (*jiftlik*) were also common.

Miri lands were taxable under Ottoman law and composed a large portion of agricultural land in Palestine. Before the Code, these lands were "state" lands insofar as the Ottoman government held a monopoly over deciding access to their surplus. The "owner" of the *miri* plot did not own the title deed but had heritable rights (*tasarruf*) and rights of usufruct (boundaries determined by the state) so long as he continued to pay the tithe and engage in perpetual cultivation. The ruler held the title to the land (*raqaba*). One stipulation of the 1858 Land Code was that it brought land uncultivated for three years (*mahlul*) into the government's hands. In addition to resolving disputed land claims, the code awarded land ownership if the claimant had paid the required taxes and if he had cultivated *miri* land for more than five years. It also prohibited transfers of land due to peasant indebtedness and tried to limit trespassing. Despite these provisions, peasant dispossession as a result of the code did occur. Finally, the code established the government's singular claim to revenue where previously numerous tax farmers had access to surplus.[50]

Originally, *mulk* lands were not absolute private property as such. The designation of *mulk* entitled the grantee to the tax revenues and the fruits of the land by an official document (*berat*) issued by the Ottoman state. The grantee could transfer his right to revenues to his heirs or convert it to a religious endowment. Like *miri*, the *raqaba* (final ownership) rested with the centralized state.[51] Eventually Land Code specifications of lease contracts, registration, and issuance of title deeds were extended to *mulk* lands in the late nineteenth century.

48. Huri Islamoglu, "Property as a Contested Domain: A Reevaluation of the Ottoman Land Code of 1858," in *New Perspectives on Property and Land in the Middle East*, ed. Roger Owen (Cambridge: Harvard Middle Eastern Monographs, 2000), 36; and Roger Owen, introduction, in Owen, *New Perspectives*, xi–xii.

49. Tyler, *State Lands*, 7–8.

50. Owen, introduction, in Owen, *New Perspectives*, vvii; and Islamoglu, "Property as Contested Domain," in Owen, *New Perspectives*, 31–32, 35.

51. Islamoglu, "Property as Contested Domain," in Owen, *New Perspectives*, 18, 27.

Mahlul lands were vacant *miri* lands that were left uncultivated because of neglect or an absence of heirs.[52] This designation affected some swamp areas. Similarly, *mudawwara* (also called *jiftlik*) lands were originally freehold but were gradually registered under the Sultan's name. *Jiftlik* also denoted a tract of land cultivated yearly that required a pair of oxen to work it. Despite registry designation, the owners and their families continually thought of themselves as the legal owners. Upon their occupation of Palestine, the Mandatory government acquired *mudawwara* lands like the Beisan district in the Jezreel Valley.[53]

Matruka lands were those left by the state for public use, like communal pasture. Besides some swamp areas, forests, wadis, irrigation channels, rivers, and roads fell under this category.[54] *Mewat*, or "dead lands," belonged to the Ottoman state. They were unoccupied, uncultivated, were not left for the public good and were usually a specific distance from a village. Desert, stony land and some swampy land, sand dunes and scrub were commonly considered *mewat*. Unlike the usual Western and colonial categorization of swampland as exclusively "waste" land, it seems that swamp areas came under various land categories according to their use. Trees were treated in a similar fashion. This variable designation shows that swamps were not in and of themselves always wasteland in Ottoman understanding.[55] They fell under different land categories according to utilization, surrounding lands, and who had access to their revenue.

Among these categories, *musha'a* land was held collectively by the village and divided amongst them every year or two for crop rotation. Several tracts of land with varying soil quality and water resources were allocated according to each village family's relative ownership of animals and

52. Zureik, *Palestinians in Israel*, 39–40. Although most secondary sources state three years, Zureik says that the land had to remain uncultivated for five years in order to be declared *mahlul*.

53. For case studies of *mudawwara* lands and colonial treatment of disputes, see Martin Bunton, "Demarcating the British Colonial State: Land Settlement in the Palestine *Jiftlik* Villages of Sajad and Qazaza," in *New Perspectives on Property and Land in the Middle East*, ed. Roger Owen (Cambridge: Harvard Middle Eastern Monographs, 2000), 121–158; Geremy Forman and Alexandre Kedar, "Colonialism, Colonization and Land Law in Mandate Palestine: The Zor al-Zarqa and Barrat Qisarya Land Disputes in Historical Perspective," *Theoretical Inquiries in Law* 4, no. 2 (July 2003): 492–540.

54. Tyler, *State Lands*, 7.

55. Islamoglu, "Property as Contested Domain," in Owen, *New Perspectives*, 31; and Bunton, "Demarcating the British Colonial State," in Owen, *New Perspectives*, 153n25.

number of males.[56] During the Mandate period, the Zionists and most British officials criticized the *musha'a* system, claiming that because tracts repeatedly changed hands, there was no incentive to make investments that would improve the land and promote intensive cultivation. This allegation was consistent with the broader British and Zionist land tenure approach that favored conversion to private property and the transfer of such property to "enterprising" parties who would produce crops for the global market.[57]

Under the land code, the Ottoman government forced title registration of freehold owners to attain land consolidation and use. The state itself did not have to register its holdings. When the registration effort failed, the law was changed to allow foreigners to own land. This provision ultimately led to the alienation of *miri* land from state control with a resultant increase in private property held mainly in the hands of large landowners. Some reasons for the failure of the initial Land Law include the peasants who, for fear of military draft, registered their land in the name of the head of the tribe or others' names.[58] These changes jeopardized peasants' rights to cultivate and pass on the land to their heirs.

Although the Ottoman Land Code of 1858 was slowly put into effect in Palestine, it affected the rural valleys more significantly than the hill regions, since the land in the rural valleys largely held *miri* status. These areas often had large swamp areas and consequently, high prevalence rates of malaria. Much of this land was owned in large tracts by absentee landlords who, for various reasons, sought to dispose of these lands at the end of Ottoman rule and into British occupation. The creation of a land market resulting from Ottoman land reform made the eventual purchase of these sizeable tracts much easier for the Zionists; in effect the Tanzimat reforms unwittingly facilitated Zionist colonization.[59]

Still Zionist land purchase was not trouble-free. Besides financial and other concerns, the issue of land transferability should be emphasized.

56. Under the Ottomans, each village was collectively responsible for the payment of taxes and tithes. Tyler, *State Lands*, 172.

57. Bunton, "Demarcating the British Colonial State," in Owen, *New Perspectives*, 121–122.

58. Stein, *Land Question in Palestine*, 11; and Shafir, *Land, Labor and the Origins of the Israeli-Palestinian Conflict*, 34. Khalidi argues that there were more absentee land sales than Stein admits. *Palestinian Identity*, 94–95, 113.

59. Smith, *Roots of Separatism in Palestine*, 96–97; Shafir, *Land, Labor and the Origins of the Israeli-Palestinian Conflict*, 24. Also Zureik's quote of Warriner in Zureik, *Palestinians in Israel*, 43.

First, *mulk* (private) land, which seemingly could be more easily transferred, was mostly located in urban areas and was not a key target of Zionist agencies. They were intent on building an agriculturally based homeland. *Waqf* land was almost entirely nontransferable, and *matruka* land was highly unattainable because the government used it for public purposes.[60]

At the beginning of the twentieth century, *musha̓a* land formed the main type of cultivated land in Palestine. The Zionists found acquisition of *musha̓a* areas equally difficult but as the Mandatory government gradually surveyed the territory, this type of communal tenancy was transformed into individual titles, allowing a larger percentage of it to be eventually sold. Several Arab villages also unofficially switched from periodic redistribution to the permanent partitioning of land. Such land was called *mafruz*. Due to this conversion, 25 percent of Arab land in Palestine was *musha̓a* land by 1940, in contrast to an original 70 percent at the beginning of the British civil administration in 1917.[61]

Warwick Tyler, whose recent book analyzes the dynamics surrounding state lands, claims that despite the transition of communal land to partitioned land, which was supplemented by a greater trend toward land consolidation, peasant agriculture was not drastically transformed. Partition of land as an attempt at privatization suffered from extensive fragmentation where shareholders owned several plots scattered throughout the village, reportedly hindering intensive agriculture and utilization of water resources.[62] Although this transition eventually freed some land for Jewish purchase, Zionist land agencies still preferred to buy consolidated tracts (and did for those that became consolidated). Individual and secure title made those lands more attractive to Jewish speculation, which in turn frequently increased land prices and heightened the incentive for *fellahin* (Arab peasant farmers) to sell. Not surprisingly, Jewish purchase of land from middle and small Arab landowners increased in the 1930s when land from large absentee landowners dwindled and when the shift towards increased privatization of land was well underway.

Until the middle of the Mandate period, however, to a great extent, the land available for transfer was *miri* land owned by the state or by large landowning elites both within and outside of Palestine.[63] The limited

---

60. For more on *waqf* land, see Yitzhak Reiter, *Islamic Endowments in Jerusalem under British Mandate* (London: Frank Cass, 1996).

61. Tyler, *State Lands*, 173.

62. Tyler, *State Lands*, 172, 174.

63. For alienation of land from peasants to large landowning elites, see Tyler, *State Lands*, 13–14.

availability of other lands meant that Zionist land purchase as the Mandate period proceeded was contingent not only upon developing phases in land speculation but, at least in theory, upon increasingly stringent Mandatory policies on land transfers.[64] By the middle of the Mandate period, many large estates owned by elite Arab families were either partly or completely sold to Jewish land agencies. The elite families were motivated by the high prices paid by the Zionists, diversion of capital from the sale to more profitable purposes, and in the case of absentees, the desire to relinquish administration of properties after the breakup of the Ottoman Empire. Large purchases by Zionist land agencies, according to Kenneth Stein, showed not only a commitment to establishing a Jewish homeland but "inevitably meant having to give considerable thought to tenants and to decisions about whether the vendor or purchaser would see to their future location. The acquisition of larger land areas also required protracted planning because of the large amounts of money needed to effect a transfer."[65]

Tyler believes that one of the fundamental reasons that gave rise to the tenant relocation issue was the disinclination of the British Mandatory government to challenge the authority of these large landowning Palestinian Arab elites.[66] He argues that its reluctance to produce tense relations with Palestinian Arab elites, alongside a desire to perhaps constrain investment in the development of the Arab sector, led the British to avoid implementation of significant land and agricultural reforms. Instead, in their attempt to uphold obligations towards the Arab population set forth in the Mandate, they enacted band-aid measures to protect peasant cultivators.

One of these stopgap measures was the Land Transfer Ordinance of 1920 that permitted the sale of Arab lands of small size to Jews as long as the area did not exceed 300 Turkish dunams of agricultural land. Other provisions included the transfer of property as along as the new owners reserved a maintenance area for the peasant families, the owners were residents in Palestine and as long as they intended to cultivate the area immediately.[67] According to Martin Bunton, in addition to protecting the status of tenants, this ordinance was ultimately meant to encourage a market

64. Tyler reminds us, however, that there were always Arab landowners willing to circumvent government restrictions in order to sell their land. Tyler, *State Lands,* 9.

65. Stein, *Land Question in Palestine,* 47.

66. See Tyler, *State Lands.*

67. Stein, *Land Question in Palestine,* 46. The dunam was an aerial unit and, in the nineteenth century, took the place of the *cift,* the area that could be plowed with two oxen. See Islamoglu, "Property as Contested Domain," in Owen, *New Perspectives,* 53n46.

in land and facilitate economic activity. The Ordinance required that the administration approve all transactions in writing. This stipulation tried to prevent an excessive rise in land prices and inhibit the consolidation of large estates.[68] An amendment in 1921, which resulted in large part from opposition to the 1920 Ordinance by active Arab politicians and landowners who rejected the proviso of written consent, eased restrictions on owner-occupiers to sell land but still sought to protect tenant farmers. The Mandatory government did not fully enforce the ordinance, and circumvention was a common occurrence. For example, landowners could first evict tenants—sometimes offering them compensation—and then sell their land before notifying the government.[69]

Leading up to and after the 1929 riots, a period of violent demonstrations and rioting that involved Arabs, Jews, and British police and that began as a dispute over access to the Western Wall in Jerusalem, the government became keenly aware of the conflict between their desire to facilitate land transfers and the social dislocation that resulted from the gradual shift to individual property rights. This phenomenon occurred not only in Palestine but also throughout the empire. Increasing dislocation questioned the assumption that individual title enabled development and facilitated social progress. It led to legal measures, which lacked coherence yet tried, often unsuccessfully, to reconcile this programmatic tension.[70] Ordinances like the 1929 Protection of Cultivators Ordinance, a 1931 amendment, and another 1933 ordinance tried repeatedly to prevent eviction of the peasantry after land transfer. Instead of providing a subsistence area, the 1929 regulations required landowners to offer monetary compensation to tenants when they had to relocate. The ordinance, however, protected peasant farmers who cut reeds or had grazed or watered animals in the area for at least five years from eviction. Length of tenancy was hard to prove, however, since there was no record of tenancies in Palestine.[71] In the wake of substantial tenant displacement, the 1933 Ordinance sought to prevent the eviction of "statutory tenants" (those

68. Jews were not the only ones involved in land speculation and transactions. In the 1930s, Arab capitalists also bought land, adding to the land transfer dilemma. Bunton, "Demarcating the British Colonial State," in Owen, *New Perspectives*, 123, 155n46.

69. Stein, *Land Question in Palestine*, 48; and Bunton, "Demarcating the British Colonial State," in Owen, *New Perspectives*, 125.

70. Bunton, "Demarcating the British Colonial State," in Owen, *New Perspectives*, 122, 124, 148–149.

71. Tyler, *State Lands*, 176; M. Bunton, "Demarcating the British Colonial State," 133–134.

actively cultivating land) by requiring landlords to provide a subsistence area when a transfer was transacted. Since swampland commonly formed part of the topography of land that was transferred, these issues are commonly raised in drainage concession transactions and in swamp drainage plans. We will see below how these protective terms were managed in the Jezreel and Huleh Valley drainage projects.

By 1936 recommendations were set forth in a Lot Viable Ordinance to the High Commissioner Wauchope that required small landowners in areas like the Jezreel and Huleh Valleys to retain a *lot viable* (minimum subsistence area) for themselves and for their family. The definition of lot viable found in the draft of the ordinance read: "The size of a lot viable will vary from place to place and from time to time, according to the locality, the type of land and the mode of cultivation. The lot viable is now taken to be 73 to 150 dunams and in some cases 200 dunams in unirrigated land, 30 dunams in irrigated but extensively cultivated land and 10 dunams in irrigated under intensive cultivation."[72] Land exempt included the whole of the land in the Beersheba subdistrict, land under citrus cultivation, and urban land. Due to the Arab Revolt of 1936–1939 this ordinance was not passed. Only the 1940 Land Transfer Restrictions tried to safeguard small landowners by restricting Jewish land purchases to a small free zone.[73]

For the Zionists, a careful balance had to be struck between these ordinances and conditions to facilitate Zionist land purchase so that Jewish immigration could occur. Not only did the British have Mandatory obligations set forth in the League of Nations Mandate towards the Zionists in this regard (most notably to encourage close settlement in article 6 of the League of Nations Mandate charter, which granted control over Palestine to the British), but perhaps more importantly, they relied upon Jewish capital for the "surrogate colonization" of Palestine.[74] The Jewish Agency presented such arguments when contesting these protective measures. For instance, the agency challenged the extent of the lot viable, determined as 130 dunams for tenants in the Johnson-Crosbie Report as opposed to an existing subsistence area of 88 dunams calculated in the 1931

72. "Brief on Government's Proposals for the Protection of Small Owners by the Reservation of a Subsistence Area" (no author but assumed to be Sir Arthur Wauchope's staff), May 2, 1936, CO 733/290/75972 (2). Thanks to Ken Stein for finding this document for me.

73. Stein, *Land Question in Palestine*, 192; Tyler, *State Lands*, 175–176.

74. Tyler, *State Lands*, 12. The term is borrowed from Scott Atran, "The Surrogate Colonization of Palestine: 1917–1939," *American Ethnologist*, special section entitled "Tensions of Empire," 16, no. 4 (November 1989): 719–744.

census estimate report for the central range of Palestine.[75] It claimed that the average subsistence area needed for a rural Arab family should be 75 dunams, a significant decrease from the Johnson-Crosbie figure and the lower range of the Lot Viable Ordinance draft figure but an increase from what the agency alleged was an existing average of 44 dunams (for unirrigated land). The Lot Viable Ordinance and other land transfer restrictions that drew upon the lot viable estimates would, it feared, keep significant portions of agricultural land from being put on the market.[76] It would require, on average, more land to be set aside in reserve when land transactions and concession negotiations were worked out. The lower figure they proposed would presumably free up more land for Jews to ameliorate and ultimately enable more permanent Jewish settlement.

These debates became especially acute during the middle of the Mandate period when, as discussed above, most Zionist land purchases shifted from large tracts of *miri* land owned by absentee landowners to Palestinian *effendis* or bankrupt smallholders and middle-sized landowners who needed the money to repay their debts. Indeed, as Stein states, cultivable land per capita—at least as the government defined it—was severely reduced throughout Palestine in the 1930s. This development was accompanied by Palestinian Arab migration to the hills, to the outskirts of Jewish settlements, to neighboring Arab villages or to the cities. Such migration resulted, in part, from Jewish land purchases in the 1920s and created population congestion and labor and agricultural issues in the thirties.[77]

In addition to Arab *miri* land, Zionist land agencies could acquire *miri* land from the government. The British inherited approximately a million dunams of potentially cultivable *miri* land, most of it also located in the valleys and plains west of the Jordan River. Another several million dunams of wasteland were also in government possession. Some of this land, however, was actually fallow land owned by Arabs. Another portion of the reclaimed wastelands went unregistered so as to avoid paying a registration fee and annual tithes. Because there was no reliable register of state lands at the time of the British occupation of Palestine, the extent and borders of *miri* land were unknown. These measurements would have to be determined through tedious land surveys and drawn-out land proceedings (both of which were never fully completed), a process that

75. Stein, *Land Question in Palestine*, 186–187.
76. Stein, *Land Question in Palestine*, 191–192.
77. Stein, *Land Question in Palestine*, 185–187.

unintentionally allowed for gradual Arab habitation on and subsequent claims to state land while delaying a possible lease to the Zionists until title had been established.[78] The Mandatory government used the term "land settlement" to denote the process of determining land boundaries, rights, and ownership to land as well as registering land. This term should not be confounded with settling people on the land, the meaning of "land settlement" more commonly used by the Zionists.[79]

Misunderstandings about the meaning of *miri* land and the extent of cultivable lands at the beginning of the Mandate period commonly led to exaggerations about the availability for disposal and purchase.[80] Some British officials often confused the definition of *miri* land, equating it with the more familiar notion of crown or public domain land. Derived from a misunderstanding of *miri* land, Bunton argues that a new category called state land was created during this time.[81] As we will see, such misperceptions led to frustrations clearly expressed in Zionist exchanges with the Mandatory government.

Once a *miri* purchase (from the government or from large landowners) was made, Zionist land agencies often had to invest further to improve "nonarable" lands, such as swampland. These additional reclamation expenses were extremely costly. They also tied up funds that could have been used to acquire more land. From 1921 to 1939, land purchase made up 11 percent (LP 9 million) of all Zionist investment in Palestine with an added five percent spent on reclamation activities.[82] Perhaps recognizing the leasing of differential quality of state lands to Arabs and Jews—mostly fertile to the former and nonarable to the latter—the British Mandatory government acknowledged the connection between technological improvements and Zionist objectives: "The extent of this [Jewish] immigration [for the National Home] depends almost entirely on schemes of irrigation, water-power development and public works, opening up for settlement land in Palestine which is now uninhabited and uncultivated." The Mandatory government explicitly recognized plans for irrigation of

---

78. In 1937, the Peel Commission noted the incomplete survey of state land. By the end of the Mandate, only 20 percent of state land had been transferred. Tyler, *State Lands*, 41.

79. *Survey of Palestine*, 1: 234.

80. Tyler, *State Lands*, 15, 40–41. For transfer of the Atlit, Kabbara, and Caesarea Concession as well as the sale, rather than lease, of Beisan lands to Arab cultivators, see ibid., 23–24.

81. Bunton, "Demarcating the British Colonial State," in Owen, *New Perspectives*, 127.

82. Tyler, *State Lands*, 9–10, citing Kimmerling, *Zionism and Territory*, 12.

the Negev and drainage of the Huleh, for example, as facilitating more land for Jewish settlement.[83]

Semantic Transformation: The Debate on Absorptive Capacity and Cultivability

Releasing the land's potential for "redemption" through malaria eradication went hand in hand with its role in helping to rectify Zionist acquisition of largely nonarable land. Such objectives were ostensibly achieved on a practical level, through actual drainage measures, but also accomplished on a semantic level in policy debates. Jewish swamp drainage projects were used as strategic devices to challenge political terms in British land-transfer control policies. These stipulations frustrated the quick extension of Zionist purchases. In effect, the use of drainage measures in policy debates would serve to reconfigure British land concepts for the benefit of Zionist settlement agendas.

Before land-transfer control policies were enacted, the first high commissioner, Herbert Samuel, sought to provide long-term leases on state lands. He established a Land Commission in 1920 to ascertain the extent, quality, and availability of state lands, to establish mechanisms to protect the rights of Arab tenants on state lands, and to recommend ways to increase the land's productivity, according to British notions of productivity as deriving from intensive cultivation.[84] While taking steps to protect Arab tenants, his administration also sought to prevent what was considered illegal encroachment upon state lands through a 1920 Ordinance and the 1921 *Mewat* Land Ordinance.[85] The 1920 Ordinance, eventually proven ineffective, required that all those who had encroached on *mahlul* (vacant) land inform the administration immediately, while the *Mewat* Land Ordinance made illegal cultivation of *mewat* (dead) land liable to charges of trespass. The terms of these ordinances contrasted with the

83. See Din veheshbon me'et halishka hareshit shel KKL [Minutes by the head council of the KKL], 1922–1927, CZA A246/441; extract from semiofficial letter from Sir Arthur Wauchope to Sir Philip Cunliffe Lister, March 13, 1935, PRO CO 733/271/7.

84. Samuel introduced the concept of "moral rights" as so as protect Arab cultivators who had traditionally grazed on certain areas but lost their rights to the land under Ottoman rule and were therefore not covered under extant law.

85. The Mewat Land Ordinance made it possible for the government to opt to lease land to those considered illegally cultivating *mewat* land. The Woods and Forests Ordinance (1920) sought, among other things, to prevent ecological damage of the land by grazing animals. Tyler, *State Lands*, 21–22, 28.

previous Ottoman treatment of *mahlul* and *mewat* land, where cultivators who paid a tithe received a grant to use the land. The Ottoman state could not expropriate land that was under cultivation. The British ordinances, on the other hand, shifted ultimate decisions over land use and control to the government. This transformation allowed more land to be registered and directly vested with the state. These legal changes ultimately narrowed the land rights of cultivators.[86]

Other, more restrictive British land regulations directly affecting Zionist agendas and malaria eradication plans shortly followed. According to these policies, Zionist immigration and settlement could continue as long as it did not violate the "economic absorptive capacity" of the land. This phrase clearly drew upon colonial notions of nature's untapped potential. The idea of increasing the productive capacity of the land grew out of a trend in the nineteenth century toward intensive forms of taxation (extracted from land) that coincided with global commercial expansion.[87] The concept of absorptive capacity was first proposed in the Churchill White Paper of 1922. Its logic became an essential condition for the fulfillment of the Jewish National Home.

The Zionists quickly realized that in order to increase the number of immigrants allowed to enter Palestine, and therefore the amount of land settled by the Jewish population, they had to increase the economic absorptive capacity of the land. Drainage measures attempted to do just that. In their tactical function to advance Zionist political and land purchase agendas, swamp drainage projects inevitably became an emblematic tool embroiled in policy debates with the Mandatory government. Swamp drainage was cited in political struggles over the policy definition of "economic absorptive capacity" and over the term "cultivable land."

References to drainage projects as instruments to increase the economic absorptive capacity abound:

> This improvement work was an essential preliminary step to the settling of the colonists, and at the same time enormously improved the land prospects of the settlers. The cultivable area was at once extended by the drying up of the swampy ground.[88]

86. Bunton, "Demarcating the British Colonial State," in Owen, *New Perspectives*, 127–128. For colonial changes to Ottoman land law, see this article.

87. Islamoglu, "Property as a Contested Domain," in Owen, *New Perspectives*, 21.

88. E. Epstein, *Subduing the Emek* (Jerusalem: Head Office of the Jewish National Fund, 1923), 8.

Arthur Ruppin, the leading Jewish settlement and agricultural expert, addressed this issue at the Eighteenth Zionist Congress in 1933: "But the Palestine of today is far superior in this absorptive capacity to the Palestine of 1922. The fact that we have in Palestine 220,000 Jews and not 85,000 is itself an important factor in the increased absorptive capacity of the country."[89]

Zionist spokesmen used drainage projects to claim the flexibility of the concept "economic absorptive capacity." They also employed these efforts to challenge what they saw as the false rigidity of the "cultivable" land category. They contested the British definitions of these terms and redefined them to fit Zionist land and settlement agendas.[90] The latter term, "cultivable," figured principally in the state lands and land-transfer controls debate. In its land policy set forth in the Hope-Simpson Report of 1930, the Mandatory government defined cultivable land as " land which is actually under cultivation or which can be brought under cultivation by the application of the labor and financial resources of the average Palestinian cultivator." This definition excluded marsh areas.[91] Hope-Simpson's estimate of the total cultivable land available in Palestine was based on this definition. The administration's definition was objectionable to the Jewish Agency, which considered cultivable land as that area capable of cultivation regardless of reclamation costs. To this end, agricultural methods and reclamation schemes were recognized as one way to increase the economic absorptive capacity of the land. They were also invoked to illustrate the Zionists' share in the country's economic development, a claim made to further legitimate the movement's commitment and connection to the land. This assertion reasoned that as opposed to an extensive form of cultivation used by the Arab peasantry—deemed by the Zionists as "degenerate agriculture"—the Jews undertook productive, intensive cultivation on their land.[92] The agency further argued:

> It is hardly necessary to point out that in the view of the Jewish Agency the area of cultivable land in Palestine can in any event be increased considerably

89. Arthur Ruppin, "Settling German Jews in Palestine," Eighteenth Zionist Congress, Prague, August 1933, reprinted in Ruppin, *Three Decades of Palestine*, 275–276.

90. Barbara Smith notes that the idea of economic absorptive capacity was distorted since it was only applied to the economy of the Jewish community. For her discussion of economic absorptive capacity, see Smith, *Roots of Separatism in Palestine*, 68, 72–85.

91. Stein, *Land Question in Palestine*, 105.

92. JA, "Antimalaria and Drainage Work by Jewish Bodies," 11.

by the use of advanced agricultural methods and improvement of the soil [e.g., drainage] entailing investment on a scale not within the means of the average *fellah*.[93]

Invocation of swamp drainage projects was an excellent way to achieve both a semantic refocusing of the land debate and the actual transfiguration of certain areas into cultivable land. Their use exploited the notion of potentiality by emphasizing that characteristic in the term "cultivable," rather than taking the latter term as fixed and dependent upon the Arab *fellah*. It highlighted the transformability of the land, rather than a relatively inherent quality based on majority agricultural practices. It was intended to prove once again—yet with concrete political ramifications—the creative, human contributions of the Zionist enterprise and the path toward *havra'at hakarka vehayishuv* (healing the land and the nation).

Reworking the definitions of "cultivable land" and "economic absorptive capacity" through the example of swamp drainage enabled the Zionist leadership to argue that there was enough land in Palestine for intensive cultivation that would satisfy the needs of the Jewish national home without prejudicing the rights of the Arab population. This claim would ostensibly prevent a breach of provisions made in the Balfour Declaration and the British White Papers; it also addressed the concerns of article 6 of the League of Nations Mandate charter that promised "close settlement of Jews on the land" by the British but that the Zionists disputed as unfulfilled. Like most Zionists, Abraham Granovsky of the Jewish National Fund understood this provision to mean that the British would make available to them all unoccupied or unused state and wastelands not needed for public purposes. In addition, he believed that encouragement of intensive cultivation would reduce Arab unit holdings, thereby releasing superfluous land for Jewish purchase. Frustrated with the lack of British initiative to fulfill article 6, Chaim Weizmann, president of the World Zionist Organization, criticized the Colonial Office in 1923 for failing to honor its pledges. The leader pointed out that the only land available at fair prices was "government Waste Land—sands and marshes." Why should Jewish agencies buy from absentee landowners, Weizmann contended, when there was a large amount of land that was lying fallow and in many cases were "sources of disease and infection as in the case

---

93. Jewish Agency for Palestine, *The Area of Cultivable Land in Palestine* (Jerusalem, 1936), 13.

of marshes...."? The *Yishuv* was ready to invest "labour, money and energy into this waste land and transform" it into "real soil."[94] Arthur Ruppin disagreed with Weizmann's approach, recognizing that the only state lands left after Ottoman land reform were those of the poorest quality—like swamps—which would necessitate large reclamation costs. It was worth it, he thought, to expend more money on fertile land from private owners.[95] The Jewish Agency would later reproach the government, claiming that it leased inferior land to the *Yishuv* and cultivable land to the Arabs. It would complain that the government showed an increasing reluctance to lease land to the Jews at all. Finally, the Jewish Agency would criticize the government for failing to develop its retained lands for public benefit. But at this point, Zionist spokesmen repeatedly argued that the Mandatory government needed to quickly and seriously address their inaction. Palestinian Arab leaders, on the other hand, opposed transfer of state land to Jews under article 6. They invoked the same phrase from the Balfour Declaration in order to claim that such actions would "prejudice" the rights of the Arab people.[96]

The notion of cultivability became especially pertinent after the Hope-Simpson Report of 1930 that reemphasized absorptive capacity and the associated limits to immigration. Although it refuted allegations that the British were failing to dispose of state lands possible for Jewish settlement, the report reiterated the commitment of His Majesty's Government (HMG) to pursue article 11 of the Mandate charter. This article promised to promote an active policy of agricultural development with close settlement and intensive cultivation by both Arabs and Jews.[97] Hope-Simpson stated the government's duties under the Mandate and the connections between immigration, land development, and close settlement in his 1930 report:

> It is the duty of the Administration, under the Mandate, to ensure that the position of the Arabs is not prejudiced by Jewish immigration. It is also its duty under the Mandate to encourage the close settlement of the Jews on the land,

94. Tyler, *State Lands*, 22–23; Chaim Weizmann, report on a visit to Palestine, November–December 1922, enclosed in a letter from Weizmann to Shuckburgh, marked "secret," February 15, 1923, CO 733/62, as quoted in Tyler, *State Lands*, 24.

95. Tyler, *State Lands*, 26–27.

96. Tyler, *State Lands*, 17.

97. The Hope-Simpson Report was published at the same time (October 21, 1930) as the 1930 Passfield White Paper. Tyler, "The Huleh Lands Issue," 356; Stein, *Land Question in Palestine*, 105–118.

subject always to the former condition. It is only possible to reconcile these apparently conflicting duties by an active policy of agricultural development, having as its object close settlement on the land and intensive cultivation by both Arabs and Jews. To this end drastic action is necessary.[98]

Other parts of the report dealt with the emergent Arab landless problem in Palestine, with the creation of a development department to manage the economic development scheme (which was subsequently postponed), and most important for our purposes here, the extent of cultivable land. The Passfield White Paper of 1930 restated the report's stance but threatened Jewish immigration further by stipulating that Jewish immigration would not be dependent upon economic absorptive capacity but upon the unemployment rate in Palestine.[99]

After fierce opposition by the Jewish Agency, the Mandatory government, with the help of Zionist leaders (who helped draft the document), published the MacDonald Letter, which effectively annulled both the Hope-Simpson Report and the Passfield White Paper. The debate that ensued until that time, however, was telling.

In this exchange, the Zionists countered Hope-Simpson's claim that there was no margin of land left for agricultural settlement by noting the positive ramifications of their settlement and drainage projects:

Thus even if we ignore for a moment the fact that by drainage and other improvements, a substantial portion of the land which, according to the Government, is not cultivable may be made available for cultivation and settlements (for many of the Jewish colonies have been built on desert and waste lands and have subsequently been turned into flourishing gardens), that area of 2,500,000 or 3,000,000 dunams of cultivable but uncultivated land could support 20,000 families and—given modern irrigation methods and intensive cultivation— even 50,000 families and more.[100]

The Jewish Agency refuted Hope-Simpson's recommendations on other grounds as well. They claimed that Hope-Simpson's estimate differed

---

98. The Hope-Simpson Report, as quoted in Bunton, "Demarcating the British Colonial State," in Owen, *New Perspectives*, 135.

99. Stein, *Land Question in Palestine*, 117–1181; JA, *The Area of Cultivable Land in Palestine*, 1; JA, "Memorandum Submitted to the Permanent Mandates Commission of the League of Nations, June 1930," reprinted in *Rise of Israel*, 19: 290; and Metzer, *Divided Economy of Mandatory Palestine*, 181.

100. JA, "Memorandum," 19: 248.

from figures that were previously supplied by HMG to the Jewish Agency. Significantly, the Jewish Agency claimed that surveyors for the Hope-Simpson Report classified land as cultivated or uncultivated and not cultivable or uncultivable. Cultivability, they pointed out, could vary with rainfall (and rainfall affected malaria incidence) and capital. In a separate publication, the agency claimed that Jewish ownership of fertile land only amounted to one-fifth of the total of Palestine and that much of the fertile land in its possession had to be rendered as such through drainage.[101]

Furthermore, the Zionists protested Hope-Simpson's estimates of cultivable land (six and a half million dunams)—which were admittedly pure conjecture—on the basis that they were based on aerial photographic surveys and on the rural property tax rolls.[102] The rural property tax policy did not focus on taxing land ownership per se but rather on land cultivation and revenue from it. Therefore wastelands (which included rocky areas, desert land, and sometimes swamps), if not cultivated, were not taxed. Once reclamation projects on wastelands began, taxation measures set in. The Jewish Agency complained: "He who has improved his land, has sunk effort and capital in it and increased its yield, is obliged to pay."[103]

The Jewish Agency argued that understanding the financial disincentives of the provision, the local Arab leaders who assessed the land areas for the Palestine administration allowed their villagers to categorize land as uncultivable so that they didn't have to pay taxes. Such a provision likely deterred a push for *fellahin* to undertake drainage schemes despite the Anti-Malaria Ordinance of 1922, which legislated that landowners undertake needed drainage in designated areas.[104] To advance their case, the Jewish Agency claimed that by changing the rural property tax categories, the Mandatory government could reap tax benefits and the amount of cultivable land would increase. A new system that would tax wastelands and poorly cultivated areas would provide an incentive to adopt intensive agri-

101. Jewish Agency for Palestine, *The Area of Cultivable Land in Palestine*, 2, 3, 8; Jewish Agency for Palestine, *Statistical Bases of Sir John Hope-Simpson's Report on Immigration, Land Settlement and the Development in Palestine* (London, 1931), 4, 5, 7, 15, 22; Jewish Agency for Palestine, *Jews and Arabs in Palestine: Some Current Questions* (Swann and Ibbott, 1936), 17.

102. Stein, *Land Question in Palestine*, 105–107.

103. The payments began with a tithe on the first crop even though most farmers did not receive a return on the crop during the first few years of cultivation. JA, "Memorandum," 290.

104. Forests were considered cultivable land but were classified as uncultivable for taxation purposes, perhaps because they often came under the *matruka* land category in the Ottoman Land Code. This confusion illustrates the flexibility and weakness of some colonial categories. Jewish Agency for Palestine, *Area of Cultivable Land in Palestine*, 1.

culture, which would then increase the absorptive capacity of the land and allow for more Jewish immigrants to enter Palestine.

As Stein notes, "The Jewish Agency had to discredit Hope-Simpson' estimate if the greater question of continued Jewish settlement was not to be jeopardized . . . [T]o admit to the accuracy or even the semi-accuracy of Hope-Simpson's estimate would have politically endangered the entire Zionist enterprise."[105] Despite the agency's claims, the Cabinet Committee decided that Jewish immigration would be restricted to a quantity that could only be settled on Jewish land reserves. This provision further elevated the importance of increasing the absorptive capacity of the land and using land reclamation schemes on purchased land to do so. Given the Government's limitations, exploiting the land's potential would become even more indispensable for the success of Zionist settlement. Despite the annulment of these policies with the publication of the MacDonald Letter, the strategic value and practical utility of swamp drainage projects could not be ignored.

Debates between *Yishuv* leaders and the Mandatory government about these policy frameworks and their implications continued throughout the Mandate period. In 1937, the issue of "economic absorptive capacity" arose again. The Peel Commission, appointed in May 1937 to investigate the underlying reasons for the 1936–1939 Arab Revolt, recommended a partition of Palestine. It also recommended that economic absorptive capacity be considered when calculating the (future) rate of Jewish immigration.[106] The Jewish Agency refuted this condition, arguing that "economic absorptive capacity" was not a fixed category that could determine the extent of immigration. On the contrary, they claimed, economic absorptive capacity varied and was determined by the process of immigration (with its subsequent labor force). In the end, the British administration withdrew the partition proposal and the commission recommended that immigration should be limited to 12,000 people per year for five years.[107] It also proposed the creation of public utility companies that would provide development schemes in rural areas in the interest of both Arab and Jewish communities. These

105. Stein, *Land Question in Palestine*, 106.

106. Metzer suggests that the political issue of economic absorptive capacity grew more acute when there was a decrease in immigration due, in part, to the decrease in capital inflow that then decreased labor demand. Metzer, *Divided Economy of Mandatory Palestine*, 137.

107. HMG, *The Political History of Palestine under British Administration: Memorandum by His Brittannic Majesty's Government presented in 1947 to the United Nations Special Committee on Palestine* (Jerusalem, 1947), 23.

companies were to promote drainage, contribute money, and investigate irrigation possibilities. The Peel Commission concluded that partition of Palestine was the only way to end Arab-Jewish unrest and resolve the conflicting promises made by the British in the Mandate charter.

Ultimately, the subsequent White Paper of 1939 limited immigration to 75,000 for the next five years with further immigration pending Arab consent; it also curtailed land sales. This policy marked a significant shift in British policy toward the Zionist project and led to illegal Zionist immigration known as *Aliyah Bet*. Due to these measures, the Zionist labor movement thereafter changed its tactics from territorial expansion to increasing the demographic density in already bought lands. As a result, the importance of reclamation works that would facilitate dense settlement was reinforced. The new approach had to take advantage of the settlement conditions produced by completed swamp drainage projects.

Malaria and drainage projects, as components in political debates about state lands and land transfer controls, were central in the discursive manipulation and transformation of the land of Palestine. Malaria's presence conditioned Jewish land purchase. Its eradication served as a political justification for increasing immigration and settlement and for the fulfillment of British obligations toward building the Jewish national home. Drainage projects provided a strategic tool for managing any limitations imposed upon the Zionists by the British during times of increased political tension. The political use of drainage measures for manipulating the terms "absorptive capacity" and "cultivable land" in Mandatory debates supplemented the physical, technological reclamation of Palestine with a semantic re-cla(i)mation by making malleable political definitions that described Palestine's topography. In this way, the Zionists reconfigured the notion of the land's potential for practical settlement agendas.

Land, Potential, and the Ethos of Development

In addition to employing drainage schemes to maneuver the terms "economic absorptive capacity" and "cultivable land," the Zionists also used them, and their resultant improvement of health conditions, to make a general political claim that they were taking a leading role in the development of the country. They argued that their contribution to development demonstrated a presence in Palestine that was worthy and desirable.[108]

108. For Palestinian Arab response to this claim, see chapter 7.

FIGURE 3.3. Haifa-Afik HaFuwarra, September 1929. Courtesy of the Keren Keyemet L'Israel Photo Archive.

An "ethos of development" formed an essential element of the Zionist endeavor and was used as one of its moral and practical justifications for colonization in Palestine.[109]

Especially during times of political tension within the country, which were accompanied by the commissions and reports just reviewed, the Zionists tried to persuade British leaders of their positive sociopolitical and economic role. They identified swamp draining and malaria eradication as proof that they were making Palestine a healthier place; they were realizing the land's potential. A memorandum submitted in December 1936 to the Palestine Royal Commission by the Jewish Agency of Palestine, "Antimalaria and Drainage Work by Jewish Bodies," clearly makes this claim. The memorandum begins:

It is the object of this Memorandum to show:

1) that Jewish colonization was carried out chiefly in areas infested with malaria where cultivable areas had become waste marsh land due to centuries of neglect;

109. Levine, "Discourse of Development in Mandate Palestine," 101.

2) that Jewish agencies spent large sums of money—hundreds of thousands of pounds—to reclaim these areas and render them cultivable and free of malaria;

3) that this process of reclamation has favored Arab as well as Jewish lands and that the malaria control financed by Jewish bodies benefited the Arab as well as the Jewish cultivators; and

4) that whereas Jewish bodies contributed to Government for control and also towards drainage of Arab lands, neither Government nor Arabs shared in any way in the heavy expense incurred by the Jews.[110]

These points are raised again later in the text to emphasize the constructive results of the Zionist endeavor in Palestine. Points three and four assert the benefit of Zionist drainage for the Arab population, a topic taken up in chapter 7 below.

In addition to the economic expense and value of Zionist antimalaria works, the Jewish Agency underlined the theme of heroism, claiming that pioneers who settled in malarious areas suffered in their fight against the scourge. This, they asserted, showed the extent to which Jews would sacrifice for the good of the agricultural and economic development of the country and for the benefit of incoming immigrants and settled residents.[111]

The memorandum concluded by restating its objectives: that the Jewish settlers acquired malarious wastelands and, by persistent and "rational means," rendered them productive and healthy; that this work highly benefited the "infected and debilitated" Arab residents living near the marshes; and that the Mandatory government contributed almost nothing to these schemes but gave full attention to Arab lands on which Jews also expended money.[112] By diminishing the role of the indigenous Palestinian Arab population and the Mandatory government in promoting and undertaking reclamation projects, the Zionist leadership implicitly cast itself as the main actor in the agricultural and economic development of the country.

The Mandatory government did not uncritically accept the agency's assertions. Understanding that this memorandum was written for political purposes with likely propagandistic exaggerations, the government sought further information to confirm Jewish contentions, especially with

---

110. Note in the second point the indirect manipulation of the term *cultivable*—that land can be rendered cultivable. JA, "Antimalaria and Drainage Work by Jewish Bodies," 1.

111. JA, "Antimalaria and Drainage Work by Jewish Bodies," 6

112. JA, "Antimalaria and Drainage Work by Jewish Bodies," 24.

FIGURE 3.4. Emek Hefer: men on boat spraying larvacide. Courtesy of the Keren Keyemet L'Israel Photo Archive.

regard to material benefit bestowed upon Arabs residents.[113] To refute the claim, the government asserted mutual influence: antimalarial work, it remarked, carried out by a Jew favored the Arab and that done by an

113. Questionnaire by Sir Morris Carter re: memorandum entitled "Antimalaria and Drainage Works by Jewish Bodies," PRO CO 733/345/10; and letter from L. Andrews, development officer for commissioner on special duty to secretary, Palestine Royal Commission, February 3, 1937, PRO CO 733/345/10.

Arab favored a Jew—"fortunately the results of such work are like the sunshine and the rain shared in equal measure by all."[114] The Mandatory government's notes about the Jewish Agency memorandum specified blatant manipulations made by the agency to promote a political agenda. The government challenged the Jewish Agency's designation of Jewish organizations' expenditures on reclamation works. They compared claims published in the memorandum against figures given in MRU files, citing Jewish Agency magnification of figures for some works ran anywhere from five to ten times the cited MRU costs.[115] In contrast to the Jewish Agency's contentions, the British Health Department of Palestine downplayed the alleged heroism and martyrdom of Jews who settled in malarious areas, charging the Anti-Malaria Ordinance rather than Zionist initiative as the main motivator in malaria prevention. It also refuted Zionist complaints that Jewish organizations contributed to work in instances that were essentially government functions.

[D]isaster was inevitable had an attempt been made to settle a non-indigenous population on such lands without complying with the law in regard to antimalarial measures . . . While, therefore, it is an admirable attitude of mind to make a virtue of necessity, compliance with the law cannot be construed as a grievance against the government more especially since the law was drafted in accordance with the recommendations of the Antimalarial Advisory Commission composed of members of both Arab and Jewish communities and received the approbation of the medical advisers of the Jewish Agency at that time. In no cases has a Jewish body been required to expend funds by Government otherwise than in compliance with the law. In the isolated instances in which contributions have been made towards antimalarial work in other than Jewish lands these have been entirely voluntary and with a specific object in view, such as the safeguarding or promotion of a commercial undertaking.[116]

114. Palestine Health Department, comments on memorandum submitted by the Jewish Agency to the Royal Commission, "Antimalaria and Drainage Work by Jewish Bodies," 1937, p. 9, PRO CO 733/345/10.

115. Compare figures on pp. 30–31 in JA, "Antimalaria and Drainage Work by Jewish Bodies" against comments on memorandum, 7 PRO CO 733/345/10. See also Ein Mayeti in the Jezreel Valley as case in point about inflation of figures.

116. Palestine Health Department, comments on memorandum, 2–3. See also Scott Atran, "The Surrogate Colonization of Palestine, 1917–1939," *American Ethnologist*, special section, "Tensions of Empire," 16, no. 4 (November 1989): 719–744.

Moreover, the Mandatory government denounced the Jewish Agency's omission of quotes from letters that gave clear recognition to Arab co-operation and participation in MRU drainage works. By censoring such statements found within a letter from Colonel Heron, head of the Palestine Health Department to Bernard Flexner, head of the health committee of the Joint Distribution Committee that funded the MRU (while including other parts of the same letter), the Jewish Agency tried to heighten the Jewish contribution in the MRU's work while obscuring, even erasing, the significant participation of other parties. For instance, the Jewish Agency for Palestine memorandum stated:

> It is evident from this map [table 6] that, with few exceptions, which will be referred to below, Jewish effort has been chiefly responsible for rendering large sections of uncultivable, malaria-infested regions healthy and habitable.

The Mandatory government responded:

> In referring to the useful work accomplished by the Malaria Research Unit under Government direction it is regrettable that the Jewish Agency thought fit to conceal the fact that successful cooperation had been obtained between Arab villagers or land owners and the Jewish settlers in work which was to the mutual benefit of both communities.[117]

Challenging number one of the objectives of the Jewish Agency's memorandum, the government argued that

> few major schemes of drainage involved reclamation of considerable areas of land previously entirely uncultivated. There is always a considerable variation in the extent of a marshy area at the end of winter and the beginning of the next wet season and it was customary for the Arab to bring into cultivation for a late or summer crop the land surrounding a marsh as the winter floods receded and the land became cultivable during the seven or eight months of the dry season. There was little land of this nature surrounding perennial swamps which was not used for crops of one variety or other while in Arab ownership.[118]

117. Palestine Health Department, comments on memorandum, 5. The omission is made on pp. 12–13 of "Antimalaria and Drainage Works."

118. Palestine Health Department, comments on memorandum, 5–6.

The government was referring to the *fassil* system used by many *fellahin* and bedouin. In this rotational scheme, water distribution alternated from area to area every twenty to thirty days. The *fassil* method was accompanied by a flooding technique that used gravity to irrigate lower spots, often resulting in the creation of pools or manmade swamps. These methods suited agricultural practices based on seasonal crops where irrigation was only required periodically every year. At the beginning of the Mandate period, Arab agriculture generally focused on the production of dry cereals, which were dependent upon rainfall. Crops sowed in the plains were generally divided into winter and summer crops. Depending on the type of soil, winter crops included wheat, barley, or lentils. After harvest, the land was let fallow. Summer crops included millet, melons, and other fruits and vegetables as well as the sesame. The *fellahin* used the swamps to produce rice, reeds for baskets, papyrus, and as locations for animal grazing. In the hills, mixed farming was practiced, including growing olives and nuts. Intensive cultivation methods used by Jewish settlements and promoted by the government for the development of Palestine demanded continuous water availability and distribution.[119] On the whole, Jewish agriculture focused on indirect or processed products for the market like dairy and poultry farming, honey, eggs and fish. Direct soil crops included grapes (wine), citrus, and a variety of other vegetables and fruits. In contrast to Arab agriculture, Jewish intensive farming consistently utilized fertilizers and grew fodder crops. By the end of the Mandate, Arab agriculture had shifted to include more vegetable production, including tomatoes, potatoes, garlic, cabbages, eggplants, onions, and cucumbers. Poultry farming in the Arab sector also grew at this time.[120]

The government also argued that the Zionist memorandum was misleading in taking full responsibility for drainage projects that were not entirely financed by them alone but rather involved Arab and government contributions. The Jewish Agency memorandum, however, claimed that the "great majority [of large-scale drainage and reclamation schemes] were carried out by Jewish bodies entirely at their own expense."[121] These claims refer specifically to parts of the Hadera works, Rizikat marsh, and Wadi Hawarith lands.

119. Tyler takes a modernization approach to land tenure issues in Palestine and therefore adopts British and Zionist notions of progressive or backward agriculture in his discussion. *State Lands*, 69–71. For conflict over water rights, see chapter 7.

120. Tyler, *State Lands*, 163–171.

121. See JA, "Antimalaria and Drainage Works by Jewish Bodies," 14, 20–21, and tables 6–7 on pp. 30–32; and p. 6 for the counter claim, Comments on memorandum.

By revealing the cultivation methods of the *fellahin* vis-à-vis the swamps and by showing that the land was put to use, the government implicitly countered Zionist claims that the land was total waste only to be made cultivable with Jewish hands. It stated that "the Jewish Agency ignores or is entirely ignorant of the great mass of antimalarial canalization and control work done yearly in streams and wet areas throughout the country by Arabs at the request of the government. It is therefore not in a position to assess the work done, judge its comparative value to one community or another or to advise what further works are required."[122]

Conflict between the government and the *Yishuv* concerning financial contributions to various drainage measures continued throughout the Mandate period alongside the state lands matter. As mentioned above, the *Yishuv* claimed that the government refused to contribute to important schemes for the development of the country and in doing so released itself from obligations to close Jewish settlement prescribed in article 6 of the Mandate charter. The Mandatory government argued that it was not obliged to actually carry out or finance schemes but only to advise and supervise under the Anti-Malaria Ordinance. *Yishuv* health activists also complained of few government-employed Jewish doctors and sanitary officers as well as the nonprovision of free malaria pharmaceutical supplies to Jews as they were to Arab communities.[123] These tensions never subsided, persisting until the end of the Mandate period.

122. Comments on memorandum, 8.

123. For the organizational correspondence on the matter, see Kupat Holim, "Milchama bemalaria," CZA S25/377.

CHAPTER FOUR

Technological Landscape

The Jezreel Valley and the Huleh Valley Projects

K avita Philip, in her writing on the historical ecology of India, asserts the interdependence of ecology and political economy: ecological projects can be driven by culture-specific factors and can result in permanent culture-specific adaptations. With so much at stake, players possess their own passionate interests in ecological projects. So, too, in Palestine, Zionist malariologists, engineers, and land settlement agencies recognized their unique social, political/national, and economic conditions and goals when planning malaria eradication schemes.[1]

Forgotten Landscape: Antimalaria Agencies and a Land before Drainage

Topographical and meteorological knowledge and measurements significantly aided sanitary planners in the exact physical manipulation of the land for malaria control. Sanitary engineers recognized that the swamp's character determined what type of mosquito would breed there. Such knowledge was important for understanding certain habits of mosquito species (where they bred, how far they flew, their life cycle, etc.) and for designing schemes sensitive to the mosquito species.

---

1. Philip, "Global Botanical Networks," 125–130.

In 1907, Dr. Malcolm Watson designed an approach, called "species sanitation" in colonial malariology circles while doing work in Malaya.[2] The implementation of species sanitation clearly shows that scientists and physicians recognized their "conquest of the land" as highly dependent upon the very specific habits and fluctuations of nonhuman factors within various areas. Using the species sanitation method, scientists in Palestine found that *Anopheline* larvae lie parallel to and just beneath the surface of the swampy water. They deemed techniques, like oiling and spraying Paris green, as effective forms of antimalaria measures. In 1922, a Mandatory government entomologist found that the most common species of *Anopheles* carrying malaria was the rural group that did not enter houses: *A. hyrcanus*, *A. algeriensis*, and less so, *A. mauritianus*. However, rural species like *A. superpictus*, *A. elutus*, *A. sergenti*, and *A. multicolor* entered houses and tents. They, along with the urban malaria vectors of *A. bifurcatus* and a very rare *A. pharoensis*, were considered the most important to eradicate.[3]

With this knowledge in mind, the organization primarily responsible for Jewish drainage, the Keren Keyemet L'Israel (KKL/JNF, the Jewish National Fund),[4] set out in the late 1920s to eliminate any swamp that stood three kilometers to the west and two kilometers to the east of any Jewish settlement. These measurements were calculated according to summertime wind direction and strength, which extended the flight distance of the *Anopheles*.

In addition to considering the flight distance of the *Anopheles* mosquito, the KKL made sure to divert rainwaters within ten to twelve days after the rainy season so that no formed swamp area could remain. This length of time was based upon scientific data that asserted that it took that specific amount of time for *Anopheles* to develop.[5] The KKL plainly conceded the importance of considering the intermingling of nonhuman factors in eliminating the disease.

Zionist agencies like the Palestine Land Development Company, American Zion Commonwealth, the Jewish Colonization Agency (ICA)/ PICA,

2. Harrison, *Mosquitoes, Malaria and Man*, 138.

3. British Health Department of Palestine, annual report, 1922, part C: "Antimalarial Service," 2, ISA M1669/130/18.

4. Walter Lehn and Uri Davis, *Jewish National Fund* (London: Kegan Paul International, 1988); Shilony, *Ideology and Settlement*. For a Zionist historical perspective on the KKL, see Abraham Granovsky, *Land Policy in Palestine* (Westport: Hyperion Press, 1940).

5. "Metoch din veheshbon shel hanhala mirkazit leKKL" (1924–1925), 1.

and the Palestine Salt Company also took part in the "healing of the land" (*Havra'at haaretz*). The American Zion Commonwealth undertook drainage schemes near Balfouria on Wadi Muveily and Rub el-Nasara swamp. The Palestine Salt Company dealt with drainage at Atlit, including drainage of the coastal swamps, drainage of Wadi Doustros and the filling in of railway borrow pits. Reclamation meant to convert these areas into sedimentary and evaporating basins for salt manufacture.[6] Similar commercial concerns dominated the motivation for draining the coastal swamps with malaria eradication as a secondary goal.

The MRU, the main Jewish antimalaria agency funded by the American Joint Distribution Committee, supervised antimalaria work on Jewish land and undertook initial malaria surveys to prevent malaria outbreaks or delays in the malaria control work. The MRU often conducted surveys with the Malaria Survey Section, a private agency funded by the Rockefeller Foundation, which collaborated with the Palestine Mandate government. Their cooperative reports were regularly submitted to the British Health Department of Palestine. The information presented here about the Jezreel Valley (Nuris and Nahalal) and Huleh Valley drainage projects are mainly based upon MRU malaria surveys.

Ethnography and Malaria Surveys

In contrast to other colonies, sanitary engineers in Palestine, as part of the MRU or other agencies, regularly conducted preliminary malaria surveys for creating drainage schemes. Professor Swellengrebel, head of the League of Nations Malaria Commission to Palestine, praised these surveys for their scientific value. In this way, he disagreed with Ronald Ross, the discoverer of the vector of malaria, who argued that postponing practical measures in order to carry out detailed preliminary research was a mistake. To Ross, it implied the "sacrifice of life and health on a large scale" without promising valuable results.[7]

Besides their contemporary scientific value to some malaria researchers, malaria surveys are rich sources for present-day scholars. They supply information not only about "indigenous beliefs and practices from the

6. "Drainage Operations of Various Jewish Agencies," part of British Health Department of Palestine, MRU annual report, Haifa, 1923, 11, 13, ISA 1670/130/33a/mem/6420.

7. League of Nations Health Organization Malaria Commission, *Reports on the Tour*, 37–38. See Pratt, *Imperial Eyes*, 28, for discussion of nature (and therefore disease) as narratable.

earliest stages of a European presence" but also offer useful demographic, economic, epidemiological, climatological, and topographical mappings of the country before drainage.[8] As such these malaria surveys are as much technological documents as they are ethnographies. A retrospective analysis of these reports can help us envision the topographical and ecological changes that transpired with drainage technology. Equally important, a critical assessment of these surveys reveals the biases and ideologies that inform the writing of the hydraulic and sanitary engineers with regard to the land and its inhabitants.[9] Because engineers' observations span both the physical and cultural landscapes of Palestine, they can and should be used critically for writing social, cultural, medical, and political histories of Palestine.[10] As mini-ethnographies in a historical archive with scarce resources on the sociomedical conditions of the indigenous population, and with little knowledge of Palestine's environment before drainage projects erased certain physical characteristics of the land, malaria surveys and other publications on the subject give the historian insights—albeit specially constructed—into forgotten landscapes, practices, and people.

In his work about malaria and its role in depopulation in 1928, Dr. Israel Kligler, for instance, traced what he saw as the pathological features of the land. Like other epidemiological descriptions of the time, Kligler's thick description reads very much like an ethnographic account with details about the lifestyles and habits of Palestine's populations. Readers are able to visualize Kligler's impressions of the substantial consequences of malaria upon the landscape of Palestine:

> [T]here is little doubt that the static condition of Palestine during the last several centuries is due almost entirely to malaria. One has but to pass over the country and see the numerous *khirbet*, or remains of deserted villages, in the vicinity of the marshes to realize the role this disease has played in its depopulation. The once famous city of Beth Shan, standing in the midst of an intensively

8. For historical utility of medical documents, Arnold, "Introduction: Tropical Medicine before Manson," in Arnold, *Warm Climates*, 14.

9. See Bernard Cohn, *Colonialism and Its Forms of Knowledge: The British in India* (Princeton: Princeton University Press, 1996), 5–8, 11–12, for historiographic, survey, and enumerative investigative modalities of colonial knowledge; also, Arnold, *Colonizing the Body*, 23. A detailed analysis of how the British used these surveys in ruling Palestine is beyond the scope of this book but seems worthy of inquiry.

10. Many of these reports are written in English in their original or are translated into English because they were submitted to the Mandatory government. Their accessibility to a wide scholarly community is therefore possible.

irrigated plain, was in the course of time completely surrounded by water-logged marshes, became one of the most malarious points in a highly malarious area, and dwindled to a small village. The same is true of Jenin—the Ein Ganim of Jewish history. Even the religious zeal and devotion of the German Templars could not withstand the ravages of the disease, and their colonies at Migdal and Semonia had to be abandoned.

If we pass from Galilee and the Valley of Esdraelon to the coastal plain, the effect is even more striking. One sees large stretches of richly watered, potentially cultivable land inhabited only by a few Beduoin tribes, all infected with malaria, and eking out a precarious existence from the proceeds of baskets made of marsh reeds and from the milk of the buffaloes which wallow in the marshes. The entire hinterland of the famous city of Caesarea is a waste of dunes and marshes, inhabited by about seven hundred souls. One passes a Circassian village half encircled by marshes—one of the outposts colonized by the Turks about forty years ago with hardy stock from Circassia, nine hundred strong. Now there are scarcely fifty souls left in the village, mostly native stock and all stamped with the effects of malaria. This condition is repeated at Ramadan, Jelile, at Nebi Rubin, in short, all along the coast.

Five or six years ago many of the Jewish villages were mournful spectacles. Almost without exception the Jews had chosen to settle in the most malarial areas. The swampy lands were the least populated and most readily bought; besides they were, on the face of it, the best watered, and water is scarce in Palestine. Probably these were the prime considerations in the minds of the purchasers of the land of the early settlements; of the dangers that lurked in the swamps they were no doubt totally ignorant.[11]

Malaria here is described as contributing to the "static" condition of Palestine, a condition that could be changed, in Kligler's mind, by the application of science. Kligler continued by pointing out the burden malaria caused for the workers in the settlement of Hulda before the Mandate period where "half the working force was invalided."[12] Other settlements like Degania, Kinneret, Merhavia, Nahalat Yehuda, and Ein Genim suffered as well. Kligler also points here to the logic of Jewish settlement in malarial areas as based not only on low cost of land but also upon the perceived abundant availability of usable water in these areas. As opposed to other kinds of settlements, his description of Jewish settlements implies a substantial transformation of the land during the Mandate period from

11. Kligler, *Epidemiology and Control of Malaria*, vii–viii.
12. Kligler, *Epidemiology and Control of Malaria*, viii.

one abandoned by others due to the land's "devastation"—a land in a state of "potential" cultivability—to one, according to Kligler, made productive and stable particularly by and for the Zionist settlers. We will continue to see this emphasis upon the land's transformation throughout the book.

In addition to a discussion of climate and topography, Kligler's *The Epidemiology and Control of Malaria in Palestine* (1930) included factors like diet and general health conditions as central to understanding the contours of the endemicity and epidemicity of malaria in Palestine and to assessing the state of sanitation in the country during the Mandate period. For these reasons, and for its more serious presence among poorer countries, Kligler thought that malaria could be classified as a social disease in Palestine.[13] He dedicated a full chapter of his account to the details of the habits and differences among and between the numerous populations of Palestine.

Just like his delineation of variations among nonhuman agents, Kligler denoted the various "gradations of civilizations" among the human actors in Palestine for which he commented that no other place could boast of a more heterogeneous population in the world. Palestine was made up of Oriental and Occidental Jews, city Arabs, Arab peasants, and roaming bedouins, Christians, Muslims and Jews, industrialized and pastoral populations. Kligler characterized the Jews in both the cities and in the rural settlements as highly literate, with lifestyles of the "more civilized peoples of the West." In contrast, the urban Arab population was described as largely illiterate, the rural Arab population as almost entirely illiterate. "Their standard of life in the city is low, and in the rural districts lower still."

Nutritional levels also varied according to the urban/rural divide and according to the communal divide. Poor nutrition put individuals at greater risk for malaria infection and helped facilitate the sustained, high endemic level of malaria.[14] Kligler noted that the urban Arab diet was described as mixed and balanced while the poorer, rural residents depended more upon availability and affordability of foods. Their diet was therefore usually uneven. When food was available in sufficient quantities, though, the "native population," according to Kligler, adopted a diet well suited to the "climatic conditions of the country."[15]

The Jewish population, he stated, also varied in nutritional levels. Although the Jewish settlements generally had better food conditions than

13. Kligler, *Epidemiology and Control of Malaria*, 12, 14–15.
14. Kligler, *Epidemiology and Control of Malaria*, 18.
15. *Kligler, Epidemiology and Control of Malaria*, 15.

those of the rural Arab population, availability also factored into food preparation. Jewish settlers in the newer settlements and larger cooperative settlements, according to Kligler, had very bad, unbalanced diets, partly due to persistent efforts to retain European food habits.

Kligler described farming methods as archaic; manure was unheard of and irrigation was "primitive." Urban Arabs and many Oriental Jews lived in crowded, poorly ventilated areas. In the rural areas, older Jewish settlers lived in ventilated, well-built houses. The new settlers, however, were particularly at risk for infection since they had never been exposed to malaria before. They often spent one to three years in tents or temporary wooden huts. Their housing situation contributed to malaria morbidity.

In contrast, according to the malariologist, the rural Arab population lived in mud houses covered with cow-dung roofs and no windows. According to Kligler, the rural "natives" lived with their family, "animals, fleas, lice and other bugs," and commonly slept on reed mats on the floor. For their part, the bedouins roamed westward and northward in the spring and summer and eastward/southward in the fall before the rains. Kligler wrote that their migration helped spread the disease.[16]

Patterns of rural living influenced the state of malaria in the country. According to the scientist, both Jewish and Arab agricultural populations moved in groups. They largely lived in villages in order to protect the population from bedouin raids. Villages were usually established near a stream or spring when possible. Their location made residents close to water, one of the main environmental sources of malaria.

Available water supplies in the cities influenced the use of cisterns and consequently the sources of malaria. By 1928, central water supplies only existed for Jerusalem, Tel Aviv, and an area of Haifa. The Mandatory government even found these water supplies inadequate and supplemented them with cisterns. Where central supplies were unavailable, residents widely used individual wells and cisterns or fetched water in vessels from the major springs (i.e., Nablus and Jenin).[17] Sewage disposal systems were virtually unknown at this time, leading to the use of cesspits, which regularly bred mosquitoes.

In addition to natural causes of malaria, malariologists cited manmade contributions to swamp formation in practically every malaria survey report conducted in Palestine. According to scholar Mark Harrison, scien-

16. Kligler, *Epidemiology and Control of Malaria*, 12–13.
17. Kligler, *Epidemiology and Control of Malaria*, 13.

tists began to shift in the nineteenth century from seeing malaria as an environmental problem to one that was partially manmade (particularly by the "native" population). This shift was part of a larger critique of indigenous practices and lifestyles.[18] Both British and Zionist scientists espoused this perceptual shift. They supplemented this perception about the human causes of malaria with a broader image of the general Arab population as neglectful, indifferent, and lazy. These scientists identified watering holes and old, leaky irrigation ditches, both linked to what was considered the *fellah*'s primitive agricultural methods, as prime places for *Anopheles* breeding. The "neglect and carelessness" of Palestine's residents became mirrored in the terrain of the "wasted and forsaken" land.

Other malaria surveys provide the historian with additional clues about the ways people lived in Palestine during the Mandate period and the ways in which the survey's authors perceived these residents. From malaria surveys of the Beisan area in the Jezreel Valley, for instance, we learn that it was one of the oldest continually inhabited places in the world. The village lay at the juncture where the Jalud River intersects the Jordan Valley (the *Ghor*). The report stated that in 1923, Beisan had about 1,600 people consisting of a Muslim majority, Christians, and a small number of Jews. High mortality from malaria was evident. One survey for Samaria village reported that "not one child out of ten ever reaches maturity. 'Fever' carries them off before they are ten years of age."[19] The authors remarked that because of its location with easy crossing to Transjordan, the town of Beisan continued to serve during the Mandate, as it had done for centuries, as a trading post for northeastern Palestine. Residents traded products such as tobacco, cutlery, and cloth. Leather craftsmanship and small-scale farming provided the staple occupations of the village. The report's authors believed that the movement of people across the Jordan added to malaria incidence and infection.

The Beisan report—written by bacteriologist and medical entomologist P. A. Buxton, field director of MRU no. 1, P. S. Carley, and sanitary engineer of MRU no. 1, J. J. Mieldazis—attempted to capture the effect of malaria on the socioeconomic history of the area. It described in detail the socioeconomic conditions of Beisan and the surrounding villages, including the water supply, topography, and residential patterns. So, for

18. Harrison, "'Hot Beds of Disease,'" 13, and idem, *Climates and Constitutions.*

19. P. A. Buxton, P. S. Carley, and J. J. Mieldazis, "Malaria Survey of Beisan, 1923," 15–17, ISA M1503/58/1.

instance, bedouin tribes inhabited the area in wadis on both sides of the village. The bedouin presence was noted because British and Zionist health officials (and the engineers they hired) believed the tribes to be primary reservoirs for malarial infection.

Such a history conveyed a particular narrative about the beginning of the malaria problem in Beisan, Islam, and about the indigenous population. By constructing this account, the sanitary engineers claimed that they were the catalysts for changing a long tradition of fatalism and neglect to a history of productivity and energy.

According to the historical timeline of the Beisan survey, extensive canals kept swamps from forming during Roman and Byzantine control. A battle between Heraclius and the Muslims in A.D. 636 ruined the canals, causing swamps to form. Yet the authors blamed the origins of the malaria problem on the coming of Islam:

> From the foregoing it can be reasoned therefore that there was no malaria in Beisan district until the capture by Islam in 636. There is no evidence at hand to show that any attempt at a definite canalization and drainage scheme was ever tried out until the advent of British occupation almost 1,300 years later.[20]

All Ottoman efforts at drainage were ignored in this description, making the British the harbinger of progress. Western, colonial ideas about Islam, a related, conceptual stagnation of the Orient, and Arab neglect of their environment were clearly designated as contributing factors to the spread of malaria in this report.

Data about women, residents' marital status, and marriage practices in malaria surveys exposed a particular view of the Palestinian Arab population in the area. The engineers commented that polygamy existed but that most residents practiced monogamy with four to eight children per family. Most women were said to marry between the ages of thirteen to fifteen. In addition to concerns about modesty, a supposed continual state of pregnancy for married women complicated malariologists' splenic examinations in their attempt to assess the malaria situation of the area. Pregnancy complicated the splenic exam because one could not accurately measure the enlarged spleen of a pregnant belly. Doctors conducted spleen examinations with the person lying on his/her back with knees up and the lower part of the body undressed.[21]

20. Buxton, Carley, and Mieldazis, "Malaria Survey of Beisan, 1923," 1–4.
21. Buxton, Carley, and Mieldazis, "Malaria Survey of Beisan, 1923," 13.

Information on family size may have given the malariologists some idea of the influence of malaria on fecundity; however, most of the contemporary research does not note a direct correlation between polygamy and malaria incidence. It seems that malaria experts during the Mandate included this information more out of a colonial description of *fellahin* customs and in line with the descriptive epidemiological writing of the period. As such, malaria reports provided ethnographic material related not only to malaria transmission but also about scientists' ideas of the indigenous population and about general communal practices before and during the modern development project of Palestine. I now turn to the case studies of the Jezreel Valley (Nuris and Nahalal) and Huleh Valley projects to show how subsequent, dramatic technological, demographic, and environmental transformations played out.[22]

Environmental Factors and Human Consequences: Two Pathological Areas and Their Reclamation

Among other initiatives, Zionist land reclamation agencies carried out two major drainage interventions in the "pathological" areas of the Jezreel and Huleh valleys. Issues of land purchase, Jewish settlement, immigration, and land transformation, introduced in the previous chapter, coalesce in the stories of the Jezreel and Huleh valley drainage projects. That fusion posits not only the extreme complexity of malaria eradication in Palestine but suggests that the notions of environment and ecology include the mutually constitutive elements of physiographical, social, and political processes.[23]

22. Other drained areas in the Jezreel Valley include Merchavia and Tel Adashim. For additional information on major schemes such as Kabbara, Khulde, Beisan, and Kishon as well as their connections to land purchase debates, see MRU, "Malaria Control Demonstrations," 20–21 and tables 1, 8; and subsequent MRU reports and British Health Department of Palestine annual reports, all in ISA; Ayalon, "Yibush habitzot Kabbara," 233–239; ESCO Foundation for Palestine, Inc., *Palestine: A Study of Jewish, Arab and British Policies*, vol. 1 (New Haven: Yale University Press, 1947), 310. See also Geremy Forman and Alexandre Kedar, "Colonialism, Colonization and Land Law in Mandate Palestine: The Zor al-Zarqa and Barrat Qisarya Lad Disputes in Historical Perspective," *Theoretical Inquiries in Law* 4, no. 2 (July 2003): 491–540; Bunton, "Land Law and the 'Development' of Palestine"; "Pratim 'al 'avodot hayibush shel KKL" [Details of drainage work by KKL in Kishon], April 20, 1936, CZA A246/442; 'Avodot hayibush 'al adamat KKL [Drainage work on KKL land] for chart of different schemes (Yehuda, Sharon, Jezreel Valley, Emek Zevulun, and Emek HaYarden), including dunams reclaimed and cost, July 10, 1936, CZA KKL5/8816.

23. Philips convincingly makes this argument in *Civilizing Natures*, 30, 40.

Case One—Jezreel Valley: Nahalal and Nuris

When a person entered the Jezreel Valley in the 1910s and 1920s he or she saw a vast terrain filled with Jewish settlements, Arab villages, springs, and swamps. The Jezreel Valley was a unique locale where irrigation canals, springs, wadis, and watering holes provided ample places for *Anopheles* breeding. The valley was a strip of land lying between two ranges of hills. Many springs at the foot of the hills spread and flooded low-lying areas, giving rise to extensive swamps.[24]

There were two types of swamps in the Jezreel Valley: seasonal swamps and permanent swamps. The seasonal swamps covered large areas of territory in the winter but dried up at the end of the rainy season. The swamps were relatively easy to dry by drainage channels and planting eucalyptus trees, which soak up water at a fast rate. This drainage system did not curb the problem of flooding, which caused losses in harvest especially in rainy seasons. Planting cotton for several years finally alleviated the situation.[25] Permanent swamps existed but did not form a large portion of the swamps in the Jezreel. The permanent swamps did, however, provide a favorable breeding ground for the *Anopheles* and therefore caused great concern from a public health perspective.

Visualizing these swamp areas before drainage helps the historian understand the topographical transformation that occurred in the aftermath of their reclamation; the destruction of these swamps laid the foundation for molding a new technological, "developed" topography of the land by changing wetlands into areas for cultivation and settlement. The names of the marshes are presented here as evidence of the topographical features that no longer exist.

Ten major swamp areas spanned the entire region of the valley: Wadi Zarrain, Ein Mayeti (which was not drained by KKL), Ein Jalud, Ein Taboun and a stream, Wadi Boket, Jalud River between its source and Tel Joseph, Ein Rasal and a stream, the Rihaniya springs and streams, Shatta swamp, Ein Sakhna, and Wadi Sakhni.[26] The following swamps areas formed a complete circle around the Jewish colony of Nahalal: Ein Semuniya and neighboring springs, Wadi Mitvah, Ein Sheikh, Ein Beida,

---

24. Elan Zaharoni, "Bitzot she baEmek," 88.

25. Zaharoni, "Bitzot she baEmek," 88.

26. "Drainage Operations of Various Jewish Agencies," 6. Also, JA, "Antimalaria and Drainage Work by Jewish Bodies," 9–10.

FIGURE 4.1. Jezreel Valley: buffalo in Background. Courtesy of the Central Zionist Archives.

and Ein Medora and neighboring springs.[27] At Nahalal (Malul in Arabic), the Ein Beida and Ein Sheikh springs caused the greatest problems for Jewish malaria morbidity as they contributed to the formation of a large swamp at the south end of the Jewish settlement. Nahalal was built in a circle with housing for the settlers located on the inside of the circle while livestock dwellings were located in the circle's periphery. This arrangement was meant to divert mosquitoes from biting men to biting animals.[28]

The marsh areas of Nuris were mostly caused by uncontrolled springs and by the Jalud River and its tributaries, Wadi Zarrain, Wadi Taboun, and Wadi Sachni.[29] Large amounts of vegetation accompanied these swamps. The vegetation reflected the arable nature of the soil but also hid some of the areas with blocked springs and caused water to stagnate.

27. "Drainage Operations of Various Jewish Agencies," as part of British Health Department of Palestine MRU annual report, Haifa 1923, 9; also JA, "Antimalaria and Drainage Work by Jewish Bodies," 10.

28. Saliternik, *History of Malaria Control*, 13.

29. Although the distinction between swamps (natural) and marshes (manmade) is sometimes made in malaria surveys and correspondence, in many cases the terms are used interchangeably. I have tried to replicate the exact terms used in each respective report and thus, reflect the surveys' fluctuations.

Considerable demographic and topographical changes occurred in this valley (Marj ibn 'Amir in Arabic or Emek in Hebrew[30]) during the Mandate period. The roots of those changes lay in the promulgation of the Ottoman Land Code of 1858 described in chapter 3.[31] The fateful story of the Jezreel Valley began when the Ottoman government sold Marj ibn 'Amir in 1872 to the Sursock family of Beirut. The Zionists began to show interest in buying the Jezreel Valley in 1891, but the Palestine Land Development Company (PLDC), a Zionist land purchasing agency, only made its first purchases in 1910. The PLDC acquired land for the Jewish National Fund (JNF), private individuals, and private colonization companies.[32] Yehoshua Hankin of the KKL transacted the final settlement of purchase in 1921 for those parts under discussion here. Hankin originally worked for the PLDC and then became the main land speculator for both agencies.[33]

Ideological and practical considerations compelled the KKL to buy this large tract of land from the Sursock family. Urgency for large colonization, quick purchase, and rapid self-sufficiency by means of general agriculture made the Jezreel Valley the focus of the Zionist Organization's land purchasing plans. The Zionist Organization thought the Jezreel Valley to be more desirable, for instance, than even the coastal region where smaller parcels of land were available for purchase. Buying parcels in the coastal region involved more complicated purchasing negotiations. The coastal region was also the center of orange cultivation, a type of agriculture that would not yield profit for several years, thus potentially impeding the Labor Zionist idea of self-reliant settlement.[34] Moreover, at the time of the purchase of the Jezreel Valley, "uncertain financial conditions" plagued the Zionist Organization. This instability made purchasing a tract of land in the coastal region a very risky endeavor

30. According to Barbara Smith, the Jewish Agency did not distinguish between the Valley of Esdraelon and the Jezreel Valley. My documents confirm that statement but since I examine the specific cases of the Nuris and Nahalal swamp drainage programs, both of the Jezreel Valley, I have used that term. Smith, *Roots of Separatism in Palestine*, 98.

31. Shafir, *Land, Labor and the Origins of the Israeli-Palestinian Conflict*, 32–36.

32. Smith, *Roots of Separatism in Palestine*, 89; Khalidi, *Palestinian Identity*, 113. For a short history of Zionist purchases in the Jezreel Valley before the Mandate, see Shilony, *Ideology and Settlement*, 193–204.

33. J. Ettinger, "Mavo shel hatza'at tochnit le'avodot KKL beEretz Israel beshnati'im habaot," June 16, 1921, 1, CZA KKL3/53; Arthur Ruppin's account, "Buying the Emek" (1929) reprinted in *Three Decades in Palestine*, 182–190.

34. Ruppin, "Buying the Jezreel Valley," 188.

since settlers would be dependent upon Zionist funds in the first years of settlement. Furthermore, the Palestinian Arab population opposed the transaction. They expressed their grievances in an article in *Al-Carmel*, an Arabic newspaper based in Haifa. Arab resistance compelled Arthur Ruppin of the KKL to push for a quick transfer of the property because "news from newspapers, like that mentioned, are suitable to strengthen the opposition against the purchases and it is an old experience that in such a case the best step is the creation of a fait accompli." The article was translated into Hebrew and published in *Doar HaYom*, illustrating Zionist interest and concern with Arab responses to land issues.[35]

Another practical reason for purchasing the Jezreel entailed internal competition with other Jewish land purchasing agencies (PICA, for instance) and a general increase in land speculation.[36] Both Ettinger and Ruppin argued that failing to take advantage of the chance to buy the tract would have dire consequences for the KKL: most notably, lowering the institution's prestige and calling into question the legitimacy of the entire agency. Buying the Emek, Ruppin and Ettinger explained, was an attempt to "save the honor of the KKL and its value" and to make the KKL the most important Zionist land purchasing agency of the time. Others opposed the purchase, like Nehemia De Lieme, managing director of the KKL. In addition to contesting the price, he and fellow Zionist Executive Reorganization Commission members felt that preference should be given to urban acquisitions since the danger of delaying purchase of those lands was exceptionally great and would lead to higher prices.[37] The Reorganization Commission's bias for urban areas clashed with the practical and ideological mission of the KKL—to provide land for "wide national settlement" that would enable agricultural workers to "strike roots in the land."[38] Towards that end, the Jezreel Valley purchase included 225,000 dunams of land.[39]

35. Arthur Ruppin, *Abschrift Juedischen Nationalfonds* (The Hague: 1921). On resistance in al-Fula before the Mandate, see Khalidi, *Palestinian Identity*, 108–109, 188.

36. Lehn and Davis, *Jewish National Fund*, 51.

37. Other members included Julius Simon and Robert Szold. De Lieme objected to the failure to ask approval of the KKL board of directors for the purchase. Lehn and Davis, *Jewish National Fund*, 51.

38. J. Ettinger, "Mavo shel hatza'at tochnit le'avodot KKL beEretz Israel beshnatim habaot" [Introduction of the suggestion for a program of KKL works in Palestine during the coming years], June 16, 1921, 1–2, CZA KKL3/53.

39. Kimmerling, *Zionism and Territory*, 39.

Considerations about draining the Jezreel Valley, as well as other swampy areas, also fit the KKL's central objectives of large agricultural settlement and Hebrew labor. The Central Executive Minutes of the KKL, for example, explicitly stated that sanitary concerns figured only as a secondary consideration in executing drainage measures.[40] Participants at the Zionist Conference in London in 1920 therefore overruled Nehemia De Lieme's position. At the Twelfth Zionist Congress in Carlsbad in 1921, authorization was given to proceed with the purchase. As a consequence of this opposition, De Lieme eventually resigned from both the Executive of the World Zionist Organization and from his position in the KKL.[41]

Two governmental actions influenced the timing of the Jezreel purchase: the Land Transfer Ordinance of 1920 (clause 8) and the reopening of the state land registries. Clause 6 of the Land Transfer Ordinance insisted on the immediate development of a land purchase, hence encouraging quick swamp reclamation projects in the Jezreel Valley.[42] The need to provide agricultural work for newly arrived immigrants (1920–1921) who were working in governmental public works, in particular, also played a part in acquisition deliberations. Buying the Jezreel "had special importance" for widening the scope of Jewish land to enable the transfer of *haluzim* (Zionist pioneers) from government work to Zionist-sponsored work.[43] Agricultural labor expansion consequently facilitated the development of several different markets for Zionist intensive cultivation. Land within the valley also served as a land reserve for expected immigrants. Finally, the Keren HaYesod's establishment (Palestine Foundation Fund) in 1920 significantly aided the timing and financial ability of the KKL to purchase large tracts of land in the 1920s.[44]

Strategically, the purchase of the Jezreel Valley contributed to the N-shaped pattern of Jewish settlement (north along the coast, southeast along the Valley of Jezreel axis and north along the shores of Lake Tiberias). The N-shaped pattern of Jewish settlement formed, according to Rashid Khalidi, the "demographic backbone of the *Yishuv*."[45] Zionist

40. "Metoch din veheshbon shel hanhala mirkazit shel KKL" (1921–1923), 3.
41. Lehn and Davis, *Jewish National Fund*, 48, 52.
42. Smith, *Roots of Separatism in Palestine*, 91–92, 96.
43. Ettinger, "Mavo shel hatza'at tochnit le'avodot KKL," 2.
44. Letter from Ruppin and Ettinger of KKL to Nehemia De Lieme, "Rechishat hakarkaot beEmek Yizra'el" [Purchase of lands in the Jezreel Valley], March 27, 1921, 2, CZA KKL3/53; Smith, *Roots of Separatism in Palestine*, 89.
45. Khalidi, *Palestinian Identity*, 98.

land purchasing agencies commonly purchased large tracts of land like the Jezreel during this decade. These agencies tended to buy the tract and then expand its boundaries as much as possible, creating a "territorial continuum" that the land purchasers felt would protect Jewish settlements. It also led, as Baruch Kimmerling notes, to the external image of Zionist settlement as a "powerful, homogenous, political unit."[46] Finally, in the case of the Jezreel Valley, the price was right—the KKL feared that delaying the transaction would risk eventual purchase altogether. The KKL paid six yearly payments for the Jezreel Valley in its final purchasing agreement with the Sursock family.

Strategy and cost were not the only considerations for the KKL in purchasing the Jezreel. The KKL considered the valley to be sparsely populated, having approximately 5,000 settled Arabs in the area. As explained in the previous chapter, this sparse population was due in part to the traditionally high endemicity of malaria in the valleys. The KKL thought that buying the Jezreel Valley would provide the least amount of local Arab resistance and do the least amount of harm to their interests.

Contemporary research by scholar Yoram Bar-Gal, however, has challenged the KKL's claims made during the Mandate period about the sparse demography of the Jezreel Valley. Bar-Gal has argued that determining the number of Arab inhabitants in the Jezreel Valley depended on what was considered the actual Jezreel (in geographic terms) and what was considered the margins of the Jezreel Valley. Bar-Gal has concluded that there were Arab villages in the Emek/Marj ibn 'Amir and in the low hills and that it was not "completely empty." Whatever the exact boundaries of the Jezreel Valley, and who decides them today or in the past, Jewish settlement in the purchased tract increased the density of the area tenfold.[47]

This drastic increase in population density proved problematic for the bedouin who, before such settlement, preferred to settle in low-density areas in order to use them for grazing and agriculture. Bedouin preference for low-density areas (relative to other areas in Palestine) was based on the workings of Ottoman rule where inhabitants of low-density areas were usually the least affected by the governor's decrees and could subvert payment of taxes.[48] After Jewish purchase of the Jezreel Valley,

46. Kimmerling, *Zionism and Territory*, 40.

47. Yoram Bar-Gal, "Emek Yizrael vebitzotav-teshuva leteguvot" [The Jezreel Valley and its swamps—reply to responses], *Cathedra* 3 (1983) 185–195, quotation on 194.

48. Falah, "Pre-State Jewish Colonization," 295.

arrangements were made to allow the *fellahin* (Arab peasant farmers) and bedouin to continue cultivating the land. The KKL primarily recognized this provision as a way to avoid confrontation with the Mandatory government rather than as an act of benevolence toward the *fellahin*. Arthur Ruppin, in his report of the KKL in 1921, made it clear that leaving a reserve for *fellahin* may not necessarily be a possible (or likely) provision in future land transactions.[49] According to the amended Land Transfer Ordinance of 1921, purchasers of land would have to put aside sufficient land for the *fellahin*. This provision became an integral part of subsequent land agreements that included swamp drainage plans during the course of the Mandate period. Under the Protection of Cultivators Ordinance of 1929, purchasers could replace this provision with the option of offering financial compensation to the tenants, a stipulation that only further exacerbated the problem of Arab peasant landlessness resulting from Zionist purchase.[50]

Once in KKL possession, the whole Jezreel Valley was considered the prized possession of the Jewish people. It was the first large tract of land owned by a Zionist national agency for land purchase (as opposed to private agencies). [51] It was also the first large tract of land bought after a change in KKL policy beginning in 1920 that called for an emphasis on purchasing huge tracts.[52] Yehoshua Hankin, the procurer of the purchase, deemed the valley the best land in Palestine at the time. He noted that the only thing that prevented close Jewish settlement was its large swamps.

The first two places the Jews settled in the Emek/Marj ibn ʿAmir during the Mandate period were Nahalal (Malul) in the west and Nuris (Ein Harod) in the east. An article in the *New Palestine* about the choice of name for Nahalal illustrated the imagination and naming of place that used the language of potentiality, privileging the valley's promising future rather than what it deemed as its present state. This rhetoric consequently erased the local history of the area; it dehistoricized the Jezreel in order for an exclusive Jewish history to be introduced as its "authentic" geographical story:

49. Ruppin, *Abscrift Juedischen Nationalfonds.*

50. Smith, *Roots of Separatism in Palestine*, 92–93; Stein, *Land Question in Palestine*, 80–172.

51. Private agencies in Palestine were funded by private individuals, usually for profit, without explicitly nationalist purposes and were not managed by Zionist institutions. Zionist "national" agencies were funded by membership fees of the movement and donations given to the World Zionist Organization or Jewish Agency. Monies were distributed to the nationalist leadership in Palestine to be implemented in ways that promoted the establishment of a Jewish national home in Palestine.

52. See Lehn and Davis, *Jewish National Fund*, 48–51.

FIGURE 4.2. Jezreel Valley: swamp drainage, 1920. Courtesy of the Keren Keyemet L'Israel Photo Archive.

Mahalul was the name of the tract when the Twenty [Jewish settlers] went up. So the Arabs had called it time out of mind and so it was referred to in the Talmud. But the Moshav preferred the historic Biblical name—Nahalal—...for "Nahalal" apparently meant "a well watered pasture"; and that expressed their idea of what their village should be. If Nahalal had no past history, they would give it a place in the history of the future. The best was "yet to be."[53]

Drainage of the Nuris and Nahalal areas took place soon after purchase between late 1921 and 1924, with 1923 being the year of most significant work. For its part, the Histadrut (the Jewish Federation of Labor) wanted to become involved in draining the valley as part of their public works activities, and very possibly as an employment measure. In response to this request, the KKL approved a joint contract on May 15–16, 1922, for the Histadrut and Ran'ania Company for drainage.[54] Minutes of a KKL

53. Samuel Dayan, "Village in Nahalal," *New Palestine*, January 21, 1927, 77, PRO CO 733/135/11.
54. "Drainage Operations of Various Jewish Agencies," 6; letter from Office of Public Works of Histadrut to Executive of KKL, April 16, 1922, CZA KKL3/56; letter from Schocken on memo of meeting, May 16, 1922, CZA KKL3/56.

Central Executive meeting for the years 1921–1923 noted that the agency expended 34,960 Egyptian liras for the Nahalal and Nuris projects out of a total of 49,096 Egyptian liras spent on all its reclamation projects from 1921 to April 1923. An Egyptian lira equaled 100 piasters, or just over 1 pound sterling.[55] The figure for reclamation of all lands made up 13.4 percent of all KKL expenditures for that same time period. It seems that the KKL figure included building roads and a water supply system.

The project was extensive with regard to both land area and labor. The Central Executive minutes stated that 19,000 dunams of land were drained—16,000 in Nuris and 3350 in Nahalal.[56] At the beginning of the work in November 1921, there already were about four hundred men working on ploughing, drainage, irrigation, and road building. Pinchas Rutenberg, the founder of the Palestine Hydro-Electric Company, had an obvious interest in harnessing the various water resources in northern Palestine. He was therefore not surprisingly also involved as an engineer in the drainage of Nahalal and Nuris.[57]

Besides ridding the area of swamps, significant changes in the landscape resulted from common mechanical interventions set out in the drainage plan, such as setting subsoil cement drains, irrigation and water supply pipes and channels, as well as redirecting the courses of the springs. As opposed to using open drains (practiced during this time in other contexts), engineers preferred using subsoil drains in Palestine because of its hot climate.[58]

As part of implementing drainage plans, Jewish workers also planted trees to mark and strengthen land borders between Jewish colonies and Arab villages. These trees were mostly eucalyptus trees since they helped drain swamps. As opposed to drainage, which changed both the under-

55. During this time 1 pound sterling equalled approximately 5 U.S. dollars. The currency of Palestine until 1927 was the Egyptian pound (EL). For more details on currency changes in Palestine, see Smith, *Roots of Separatism in Palestine*, xiv.

56. "Metoch din veheshbon shel hanhala mirkazit leKKL" (1921–1923), 3, 5–6. See also Gideon Biger and A. Kartin, "Habitzot shel Emek Yizra'el-mitos or metziut?" [The Swamps of the Jezreel Valley—myth or reality?], *Cathedra* 30 (1983): 181, for drainage expenses in English pounds.

57. Letter from Ettinger to Kessel, September 9, 1922, CZA KKL3/57b; letter from Ettinger of KKL to general manager of the Palestine Railways, November 24, 1921, CZA KKL3/55; "Metoch din veheshbon shel hanhala mirkazit leKKL" (1921–1923), 3.

58. For case of Khulde, see Technical Description of the Scheme for the Drainage of the Swamp near Khulde, CZA KKL3/99 aleph; letter to director, Zionist Commission, Jerusalem, from British Health Department of Palestine, November 4, 1921, CZA KKL3/99 aleph.

ground and surface of the land, planting trees transformed the exterior landscape of this area, demarcating inclusionary and exclusionary frontiers. Workers often built roads alongside drainage areas in order to increase the value of the land and to facilitate the easy movement of future agricultural goods for the urban market. In the case of Nuris, these operations were said to open up trade in Haifa and allow the area to become an "important centre [where] there will be a considerable traffic of passengers and goods to and from the settlements."[59] As such, it was meant to show the Mandatory government the extent of Jewish devotion to the development of Palestine.[60]

Engineers considered other methods, all practiced in U.S. reclamation projects, for malaria eradication in Palestine. They considered such plants as alga chara foetida as a way to prevent the growth of mosquito larvae. Chara grew on the surface of pools in some areas of Palestine. The Americans used it in Cuba, and Mr. Briarcliffe of the Palestine Health Department recommended it to the KKL. In 1922, the Palestine Department of Agriculture and the British Health Department of Palestine carried out investigations to determine whether alga chara foetida actually prevented the development of mosquito larvae. Conclusions proved negative. Similarly, scientists tested certain fish in Palestine, such as tilapia and cyprinoden, for their ability to eat mosquito larvae. These tests proved that cyprinoden ingested *Anopheline* larvae. Zionist health officials also contemplated using bats to eat mosquitoes, as had been done in San Antonio, Texas, by Dr. Charles Campbell. They seriously considered bats as an alternative method of malaria control although it is unclear if they ever actually used them. These examples show, however, that the Zionists utilized international methods of malaria eradication, thus exposing their engagement in transnational networks of scientific exchange during this period.[61]

To many Zionists, the ability to cross the drained land with one's feet or a cart after drainage testified to the tangible effects of such work, as well as marking clear, bordered ownership in a way hitherto impossible.

59. Letter from Ettinger of KKL to general manager of the Palestine Railways.

60. Letter from Ettinger, Council for Matters of KKL, to Official Agency of KKL, December 19, 1921, CZA KKL3/55.

61. See Antimalarial Service, British Health Department of Palestine, annual report, 1922, part C, ISA M1669/130/18; letter form Ettinger to Official Agency of KKL, December 19, 1921, CZA KKL3/55; and letter from Briarcliffe to director of KKL, May 30, 1922, CZA KKL3/19-4. For botanical networks in India, see Philip, *Civilizing Natures*, 53.

A member of the League of Nations Malaria Commission of 1925 reported, "The whole aspect of the valley has been changed...; the plantations of eucalyptus trees already begin to give a new character to the landscape;...what five years ago was little better than a wilderness is being transformed into a smiling countryside."[62]

Such transformation, however, affected other residents of the area like the bedouin. Drainage of Marj ibn ʿAmir/Emek significantly affected bedouin tribes, particularly the Kaʿbiyyah and Saʿayidah, since it reduced their pasture land and transmuted their settlement preference pattern for low-population density. In the case of the Kaʿbiyyah, Jewish colonization impinged upon the bedouins' grazing area, eventually forcing them to relocate to alternative leased land (which did not include their original grazing area) and ultimately introduced private land while disbanding tribal groups.[63]

Healing the land through swamp drainage was intended not only to facilitate the healing of the Jewish people but also to rejuvenate the land. In many cases, it succeeded. At the beginning of drainage, about 9.2 percent of the population in Nahalal had experienced bouts of malaria, but by 1923, only 0.9 percent exhibited signs of the disease with no new cases.[64] This substantial decrease in the prevalence of malaria during and after drainage led Dr. Deutsch in 1924 to remark, "It is possible to say that man has overcome nature here and even has overcome death."[65]

Whereas Deutsch inferred immortality inherent in the connection between the land and the Jewish people, Dr. Hexter of the Jewish Agency believed in human control over death, albeit with limitations. The great improvement by 1932 of health conditions in the Emek/Marj ibn ʿAmir compelled Hexter to write to Henrietta Szold: "The record of the Emek is simply stupendous and indicates that health is a purchasable commodity and that literally the people can determine their own death rate."[66] For some Zionist health and land activists, health, like land and agricultural products, became something that could be bought, controlled or exchanged.

---

62. League of Nations Health Organization, Malaria Commission, *Reports on the Tour*, 23.

63. Falah, "Pre-State Jewish Colonization," 305–306; Metzer, *Divided Economy of Mandatory Palestine*, 202, for Arab peasant land alienation and loss of grazing rights.

64. "Metoch din veheshbon shel hanhala mirkazit leKKL" (1924–1925), 2.

65. As quoted in Zaharoni, "Bitzot she baEmek," 91.

66. Letter from Dr. Maurice Hexter of Jewish Agency to Henrietta Szold of Vaad Leumi (and of Hadassah), Jerusalem, May 13, 1932, CZA J1/4355; see also Robin A. Kearns and Wilbert M. Gesler, conclusion, *Putting Health into Place*, 290, about the contemporary need to buy health rather than become healthy.

Case Two: Huleh Valley

The Huleh Valley, located in the northernmost section of the Jordan Valley, was the site of a complicated story of land purchase, malaria, and swamp drainage during and after the Mandate period. Before its drainage, the Huleh lake was one of the oldest lakes in history. It was the southernmost wetland in the Levant, the "limnological lungs" of the River Jordan, and one of two large, freshwater lakes in the Middle East (the second is in southeastern Turkey). Scientists today consider the Huleh lake as one of the first permanent places of human settlement.[67]

Although complete drainage of the Huleh did not begin in full force until after the Mandate period, Zionist plans and negotiations to acquire the concession along with antimalaria activities did occur during British rule. The issues of Zionist land transactions and settlement along with the high endemicity of malaria in the area make the Huleh case an important one to examine.

Like those of the Jezreel Valley, malaria plans and reports to drain the Huleh tell us about the landscape of the area prior to drainage, including the Arab villages therein.[68] We learn, for instance, that, according to the scientific reports, before drainage 5,000 mostly Arab inhabitants lived in the Huleh area. The main Arab villages were Khalisa and Salihiye, while Jewish settlements in the area included Yesod HaMaʿala, Tel Hai, Kfar Giladi, and Metulla.[69] Residents engaged in agriculture and cattle breeding as well as weaving papyrus mats and fishing.[70] At the time of the original grant, the Huleh had nineteen Arab villages with about three to four thousand people, mostly of the Ghawarna tribe.[71] The majority, if not all, of the inhabitants had suffered from recurrent bouts of malaria fever. Their economic subsistence was based on making products of papyrus, tending water buffalo and cultivating rice, corn sugar-cane, cotton,

---

67. Dimentman, Bromley, and Por, *Lake Hula*, 1, 5.

68. For a survey of scientific research done in the Huleh before, during, and after the Mandate, see Dimentman, Bromley, and Por, *Lake Hula*, 8–17.

69. "Rapport Cantor adresse à la Commission sanitaire," 1.

70. Rendel, Palmer, and Tritton, "Huleh Basin: Report, Preliminary Scheme and Estimates for Reclamation of the Huleh Lake and Marshes and Drainage and Irrigation of the Basin" (July 1936), 5, ISA M4431/02/3/22. EMICA was a subsidiary of PICA established after the 1929 disturbances. Rendel, Palmer, and Tritton were British engineers consulting on the project. See Mica Levana, "Yibush haHuleh" [Draining the Huleh], *Ecologia vesviva* 1, no. 4 (August 1994): 213.

71. Warwick Tyler, "The Huleh Lands Issue in Mandatory Palestine, 1920–1934," *Middle East Studies* 27, no. 3 (July 1991): 343–373.

and sorghum while an expansion of cotton production in the Huleh region took place during the American Civil War.[72]

Of the 5,000 buffalo in Palestine in 1930, most of them were found around the Huleh lake.[73] Migratory birds and a variety of fish made up the fauna of the area.[74] Eel were also known to be abundant in the coastal marshes of Palestine.[75] In one drainage report (1919), Louis Cantor, a sanitary engineer employed by the Zionists during the Mandate period, described to the Sanitary Commission of Hadassah the exact location of Salihiye village. It was, he noted, "at the point where Nahar Aboulahaf (River Aboulahaf) and Kheviat Salhiye cross."[76] In addition to understanding exactly where these villages were located, such programmatic descriptions inform the historian about the waterways that existed before drainage. These descriptions show the great transformations drainage imposed on the Palestine landscape.

Malaria reports also tell us about the seasonal conditions of the *fellah* and the number of laborers in the area. According to M. Franck, in 1919, 218 *fellahin* worked in the Huleh region during all seasons, while 7,024 laborers worked only in the summer. Low water levels of the Jordan made the summer the hardest season for the *fellahin*, causing difficulty for *fellahin* to operate some of the small mills on Huleh Lake's shores and, consequently, to cultivate the soil.[77]

Whereas during the Mandate period Zionist officials considered Huleh lands among the most fertile and valuable, the MRU considered the Huleh swamp itself as "the worst and the most extensive in Palestine," which could only be eliminated by "radical drainage."[78] The indigenous population instead saw utility in the marshes—as watering places for buffalo, areas for birds' to build their nests, and as a site for pelicans and swans. They therefore named it al-Sheri'at al-Kebira, which means large watering

---

72. Dimentman, Bromley, and Por, *Lake Hula*, 109–110.

73. Dimentman, Bromley, and Por, *Lake Hula*, 109–110; Tyler, "Huleh Lands Issue."

74. Rendel, Palmer, and Tritton, "Huleh Basin," 5.

75. League of Nations Health Organization Malaria Commission, *Reports on the Tour*, 27.

76. Louis Cantor, "Rapport Cantor addresse à la Commission Sanitaire," 5.

77. Extrait d'une lettre de M. Franck, 5 January 1919, CZA S25/595; and Cantor, "Rapport Cantor addresse à la Commission Sanitaire," 1–4.

78. MRU, "Malaria Control Demonstrations," part 1, 4–5; and Cantor, "Rapport Cantor addresse à la Commission Sanitaire," 5. The Cantor report also quotes John MacGregor's impressions of the Huleh made in 1870. John MacGregor, *Rob Roy on the Jordan* (London, 1870), 197, CZA S25/595. John MacGregor carried out the first modern mapping of the Huleh lake and marsh. Dimentman, Bromley, and Por, *Lake Hula*, 9.

place. Furthermore, different activities in the swamps, like papyrus harvesting, had a significant influence on the distribution and composition of the flora in the area.[79]

These two distinct Zionist/Arab visions about the same place, the Huleh swamp, illustrate different ideas about the relationship between man and nature and the ecological utility and potential of the land. The Huleh case provides a good example—both in terms of its similarities and differences—of what Kavita Philip has described as differential "rational systems of knowledge about nature, under-girded by unequal structures of economic, political and social practice."[80] As she has explained, when indigenes attributed a subjectivity to natural objects (like nicknaming the swamp), colonial scientists took it to mean that they failed to detach themselves from their surroundings, thereby hindering an ability to act upon and transform the landscape. In contrast, the non-indigenous way of approaching nature was to objectify it, separating out nature from culture by applying scientific measurement and estimating commercial profitability. Europeans established boundaries in order to transition into modernity and industrial production.[81]

The Zionist approach to nature, science, and modernity generally followed Philip's description above, but it strayed in certain important ways. Labor Zionists attributed both subjectivity and objectivity to natural objects. They added a nationalist twist by merging nature and culture through ideologies like *kibush hakarka*, *kibush ha'avoda* in order to effect change. They saw nationalist renewal in natural objects. At the same time, Zionist scientists and settlers believed that their manipulation of nature through scientific measurement (and the drainage work that resulted) in fact established part of their attachment to the land. They also considered commercial profitability in their drainage plans. As discussed in chapter 1, and in minor distinction to Philip's interpretation, the Zionists did not believe the Arab population held a strong attachment to the land; they used the existence of swamps as proof of a supposed Arab indifference and detachment. For Zionists, Arab attribution of subjectivity to natural objects had less to do with a failure to transform the landscape and more to do with ignorance, laziness, and lack of will and resources.

79. Dr. Masterman, in his speech delivered to the Royal Geographical Society on January 14, 1918, in MacGregor, "Comments on Rob Roy," 5, CZA S25/595; also, Dimentman, Bromley, and Por, *Lake Hula*, 110.

80. Philip, *Civilizing Natures*, 98.

81. Philip, *Civilizing Natures*, 36.

FIGURE 4.3. Huleh swamp with lily pads. Courtesy of the Keren Keyemet L'Israel Photo Archive.

To be sure, both systems, indigenous and non-indigenous, attributed value to the land—albeit very different—but there was a power gap between each side's scientific and fiscal authority during this period. As we shall see in chapter 7, the colonial/Zionist one ultimately trumped the Arab one, but not without resistance and negotiation. The contested terrain of the Huleh literally and figuratively shows us how the particular, interconnected configurations between science, economics, land politics, and national ideology effected specific transformations in Palestine's landscape.

Before it was drained, the lake and the permanent Huleh swamp covered a third of the Huleh Valley, which had a total area of 177 square kilometers. Sanitary engineers that studied and mapped the Huleh during the Mandate period divided it into two parts, a southern and northern part. The southern part had a total area of fifty-three square kilometers. The Huleh basin was situated at the northeastern tip of this southern section. The swamp area was located in the northern part, whose total area was 124 square kilometers. The Zionists, from the end of the nineteenth century onward, planned to drain the northern part.[82]

During Ottoman rule, the area of the Huleh Valley was located along two administrative subdistricts: Acre and Beirut.[83] Already at the beginning of the twentieth century, the Turks were interested in leasing the Huleh lands to the Jews because of their financial and scientific capability to undertake antimalaria measures and agricultural cultivation in that area.

Wider European colonial activity in the field of malaria control certainly influenced the Ottoman decision to undertake drainage projects. The Ottoman government sponsored a series of experiments for the Huleh (1887, 1904–1905) in order to decrease the number and level of the swamp area to allow for better water flow. In 1917 they hired the German team of Muhlens, Zoller and Weidner to draw up other reclamations schemes for the Huleh. Zoller submitted a report on the Huleh only in 1924. Although it was not put into practice, the team's mission and recommendations provided an important basis and reference for subsequent surveys conducted in the area during the Mandate period. Sanitary engineer Cantor mentioned the Turko-German venture in his January 1919 report to Hadassah. His references contradicted British and Zionist claims made during the Mandate period that the Ottomans did not undertake any efforts to drain swamps or to improve the malaria situation.[84]

Baron Edmund Rothschild and his reclamation company, ICA (Jewish Colonization Agency), considered buying the Huleh and undertaking

82. For maps of the Huleh Valley and concession area, see Tyler, *State Lands*, 83, 114.

83. In 1923, a delegation from Huleh district arrived in Beirut with a petition from residents of twenty-six villages in the Huleh, protesting its potential incorporation into Palestine. Dispatch Reg. no. E4510/231/65 from Consul General Satow, April 23, 1923; and letter from Satow of Beirut to principal secretary of state for foreign affairs, April 23, 1923, PRO FO 371/9000.

84. Shimon Rubinstein, "The Huleh Concession: The Rising and Falling of Zionist Attempts to Attain the Huleh," *At the Height of Expectation: Land Policy of the Zionist Commission in 1918*, appendix 18 (Jerusalem, 1992), 293, Yad Ben Tzvi Library; Rendel, Palmer, and Tritton, "Huleh Basin," 6–7; Goldstein, "Tochnit zioni," 162; Y. Karmon, *Emek haHuleh hatzfoni* (Jerusalem, 1956).

drainage at the end of the nineteenth century. Rothschild thought that the Huleh would provide land on which to settle Jews during the first and second waves of Zionist immigration.[85] Negotiations between the Ottoman government and the ICA took place between 1901 and 1909, but ICA intentions to buy rather than lease the land derailed the process. Scholar Yaakov Goldstein has stated that the Zionist movement's heightened interest in buying and draining the Huleh can be seen as early as 1907; it was a direct result of the Eighth Zionist Congress, which stressed practical ways to encourage the Zionist national endeavor.

In addition to the aforementioned parties, American Zionists were interested in malaria and swamp drainage programs in the Huleh area. Their strong influence in these and other health affairs of the *Yishuv* can be seen both before and during the Mandate. Driven by a wider American Zionist concern for economic reconstruction in Palestine after World War I, for instance, the Zionist Society of Engineers and Agriculturists submitted a detailed program in 1919, called the National Merom Project after Lake Merom, the main lake of the Huleh region. The National Merom Project was a program to drain the Huleh swamps and prepare the area for settlement. This project was the first plan that a Jewish group brought forth for the Huleh that dealt with all aspects of the swamp drainage problem: agricultural, hydrological, electrical, and industrial. The society saw this endeavor, especially in terms of the energy provision of its plans, as a way to quickly acquire national assets "in order to give our national aspirations a safe base and strong foothold to proceed in the successive developments on national principles exclusively." The plan never came up for serious discussion among the central departments of the Zionist Organization of America, nor was it ever practically implemented. Yet recommendations similar to those found in this program were made in subsequent plans drawn up during the Mandate period.[86]

85. Goldstein, "Tochnit zioni," 162.
86. Criticism of American Zionism as not "imbued with the true Zionist spirit" but only having a "platonic interest" is described in a letter from Kligler to Robert Szold of Palestine Endowment Funds, September 29, 1940, 1, Hadassah Archives RG 1/100/2; Goldstein, "Tochnit zioni," 161, 166–168. For the debate about the role of Hadassah in the life of the *Yishuv*, see the memorandum between Dr. Katznelson and Yassky about Anti-Malaria Control, April 17, 1940, 8–9, Hadassah Archives, Hadassah Archives RG1/100/2. For examples of additional American involvement in Zionist health affairs in Palestine, see Sandy Sufian and Shifra Shvarts, "Legacy of the Past: An Introduction to the History of Palestinian Arab Health Care during the British Mandate of Palestine, 1920–1948," in *Separate and Cooperate, Cooperate and Separate: The Disengagement of the Palestinian Health Care System from*

The Zionists' final acquisition of the Huleh concession was a drawn out process involving property, economic, agricultural, and political concerns. The complex process of the Huleh transfer was in part because it was considered *miri* land. As such, the Mandatory government was obliged to reconfirm a concession of the area that the Ottoman authorities had granted to Syrian Arabs in 1914.

The concession area itself covered almost a third of the Huleh Valley. There were no sources of water in the concession area other than the springs in the bed of the lake and in the marshes. Irrigation for the area therefore required water from outside of the concession.[87] The area started from the south at the old Jisr Benat Ya'qub Bridge (which was then relocated further south in 1934 as part of the drainage project) and comprised lands northward including the lake area, the marsh at the north of the lake, and a cultivated area (which was eventually reclaimed by the local Palestinian Arab population along the north of the marshes). Scientists conducted the survey in 1935 when tensions were rising between Arabs, Jews, and the British in Palestine. The report reflected this political friction. In its acknowledgement, the authors noted the effects of contentious Arab/Jewish relations upon implementation of drainage survey work: "At the time when the Arabs at Huleh were in an unsettled state and inclined to interfere with our surveyors, the Police Department provided escorts and enabled us to carry on with the work with very slight delay."[88] Surveyors expected the Huleh basin land to the north of the concession area to increase in value and agricultural possibility once drainage had taken place.[89]

Although the original concession was certified under the Syro-Ottoman Agricultural Company, the Huleh Valley and the concession area were split between British and French jurisdiction after World War I. In 1920, the company applied for and was granted recognition of the original concession. Tyler explains, however, that under the British military administration recognition of concessions were only granted to those

---

Israel and Its Emergence as an Independent System, ed. Tamara Barnea and Dr. Rafiq Husseini (Westport, CT: Praeger, 2002), 9–30; Shifra Shvarts and Ted Brown, "Kupat Holim, Dr. Isaac Max Rubinow and the American Zionist Medical Unit's Experiment to Establish Health Care Services in Palestine, 1918–1923," *Bulletin of the History of Medicine* 72, no.1 (Spring 1998): 28–46.

87. Tyler, "Huleh Land Issue," 346.
88. Rendel, Palmer, and Tritton, "Huleh Basin," 14, 46.
89. Tyler "Huleh Land Issue," 346.

that had been in full operation before actual British occupation. Zionist land agencies raised doubts about the validity of the company's approval since the original concessionaires had not undertaken reclamation work in the Huleh. Even after British sanction and repeated extensions, the Beiruti concessionaires did not fulfill their obligations. Instead, they wanted to sell it. British approval of the concession to the original concessionaires thus frustrated Zionist land agencies, which had, by the early 1920s, developed a strong interest in attaining the Huleh Valley. They regularly took issue with the concessionaires' continual failure to develop the land.

Negotiations for the transfer of the Huleh Concession between the original concessionaires and the Zionist Organization (ZO) ensued during the early years of the Mandate period.[90] These efforts were quite sporadic and were influenced by hesitations concerning business and agricultural interests.[91] At first, the ZO hesitated to move forward with concession negotiations without a survey of the costs of reclaiming the area, especially when compounded by the "serious doubt entertained in responsible quarters in Palestine as to the value of the Huleh concession and possible even as to its validity." It should be noted that drainage was not mandatory outside of the concession area for the concessionaires. The Malaria Survey Section conducted a study in 1924 to detail a scheme that suggested malaria control methods for the concession area as well as the area north of the concession. By 1925, a Jewish survey of the area had been completed that estimated the cost of drainage at 250,000 LP after a commitment by the Keren HaYesod of 100,000 LP. In 1926, Hankin then tried to undertake secret negotiations regarding the Huleh concession. The ZO ultimately found these negotiation terms unacceptable on financial grounds. The offer did not delineate compensation to be made by Selim Bey Salam, head of the Syro-Ottoman Agricultural Company, for ripari rights (claim of loss and value of land adjoining the lake by residents),

90. Short summary of preliminary remarks by Dr. Ruppin, JDC Archives, AR21/32/294, microfilm 40. For summaries of the 1924–1925 Huleh negotiations, see "Huleh Concession" (summary), JDC Archives, AR21/32/294, microfilm 40; "Huleh Concession: Short Summary and Report by McDonald and McCorquodala, Consulting Engineers," JDC Archives, AR21/32/294, microfilm 40; summary of statement by Mr. Ibish (agricultural expert to concessionaires), JDC Archives, AR21/32/294, microfilm 40.

91. On the debate between Selim Bey Salam and the Zionist Organization about the Huleh concession, see handwritten letter from Mr. Triphon, representative of PICA, at Metulla, to Kisch, Metulla, June 18, 1926, CZA S25/595; letter from Kisch to political secretary of the Zionist Organization, June 29, 1926, CZA S25/595; and letter from Kisch to Triphon, June 29, 1926, CZA S25/595.

grazing rights, rights of cattle movement, rights for cutting papyrus, and fishing rights. This would have left the Zionist Organization responsible for such compensation to the residents. The compensation costs, in addition to the concession and drainage costs, were an amount the ZO could not afford, having recently acquired the Jezreel Valley.[92]

The hesitation of Pinchas Rutenberg, owner of the electrical power concession for Palestine, also posed a source of opposition to the quick transfer of the concession. Rutenberg, a Zionist entrepreneur, worried that drainage work on the Huleh would change the level of the Jordan River at the same geographical point where his electricity concession became operative. He feared that drainage would eliminate the Huleh lake as a storage reservoir to draw on for electrical energy and wanted to secure electrical energy and also submit provisions for the Huleh's drainage and irrigation. The ZO took Rutenberg's views seriously in concession negotiations since the organization did not want to cause a conflict of interest between two significant Zionist endeavors, the Palestine Electric Corporation and exploitation of the Huleh area.[93]

Issues of hiring non-Jewish labor in the Huleh drainage work posed another point of contention for Zionist officials. Some experts suggested using Arab labor in order to lower labor costs. Henriques, one of the experts of the Joint Palestine Survey Commission in 1928 who reported on the pos-

---

92. Letter from L. Stein to secretary of the Palestine Zionist Executive, February 4, 1925, CZA S25/595; memo on the Huleh Concession, note on meeting at Hotel Cecil, January 22, 1925, with Dr. Weizmann, Mr. Stein, Mr. Salam, and Mr. Salam, Jr., in attendance, January 1, 1925, CZA S25/595; Huleh Concession, note of interview at Great Russell Street with Mr. Shoucair and Mr. Stein in attendance, March 12, 1925, CZA S25/595; extract from cable from Mr. Kaplansky to Keren HaYesod Foundation, June 30, 1925, CZA S25/595. For proposed provision of the land reserve, see translation from telegram in Polish, no date, CZA S25/595.

93. Confidential letter to Sir John Evelyn Shuckburgh, Government House Jerusalem from unsigned, April 21, 1925, CZA S25/595; letter from L. Stein of ZO to secretary of the Palestine Zionist Executive, March 11, 1925, CZA S25/595; confidential letter to Feiwel from unsigned [but likely Kisch], June 29, 1925, CZA S25/595; letter to Kisch from Stein, October 8, 1925, CZA S25/595; letter from PEC to PICA, on anti-malarial work in the neighborhood of Jisr el-Mejamieh, July 17, 1936, Haifa, CZA J15/5589; and Rendel, Palmer, and Tritton, "Huley Basin," 44–45. See also Harry Sachar's hesitations about losing money in the venture in letter from Kisch to L. J. Stein, Esq., copy sent to Harry Sacher and Dr. Ruppin, marked confidential, July 14, 1924, CZA S25/595; P. Carley, controller of the MSS, summary of the activities of the Malaria Survey Section of the Government of Palestine during 1924, January 2, 1925, 2, ISA M1670/130/35. Barbara Smith notes that the Mandatory government continued to keep the concession alive under Salam so as to prevent the PEC from gaining the concession. Although my documents do not explicitly convey this view, it is a possibility, but I think it was probably one consideration among many. Smith, Roots of Separatism in Palestine, 125.

sibilities for drainage of the Huleh, estimated in his survey that it would only cost 14 percent more to execute the work by Jewish labor rather than Arab labor. He was involved in other colonial drainage projects outside of Palestine. Henriques argued that if Arab labor were to be used to carry out drainage work, Labor Zionist leaders would have to allow Mr. Salam to provide the labor force. Henriques suggested that leaders could be convinced to use non-Jewish labor if they were presented with the argument of promoting national interests through preserving loss of Jewish life.[94]

As part of their financial concerns concerning the concession, Zionist land agencies were particularly curious about potential agricultural yield and income from the area. In a Zionist survey of the area before signing the concession, scientists included figures on barley production and its financial yield in order to show the value in the investment.[95]

Along with the fish industry, papyrus use was a major economic consideration for these agencies. Before drainage, the Huleh Valley was known for its abundance of papyrus. In fact, before drainage, the Huleh contained the largest quantity of papyrus in the world.[96] Residents in such villages as Mallaha and Buweiziya used papyrus to make mats, baskets, and huts.[97] The ZO was interested in commercially exploiting the Huleh's papyrus supply upon the concession's transfer to the Keren HaYesod (the financial institution of the ZO). It wanted to introduce Jewish basket makers into the area. As Mr. Singer of the Zionist Organization's Financial and Economic Committee stated: "It will be interesting for us to know not only what the government receipts are but who is making use of the papyrus, the quantities cut, the purpose for which it is used and the methods of utilization." The EMICA preliminary report (EMICA was a merger of the Palestine Emergency Fund and the ICA) stated in 1936 that because

94. Negotiations were then broken again by Salam on February 17, 1926. Letter from Kisch to political secretary, February 9, 1926, CZA S25/595; letter from Singer to KKL, January 21, 1926, CZA S25/595; letter from Kisch to Zionist Executive, Palestine Zionist Executive, and Board of Directors of KKL, April 25, 1926, CZA S25/595.

95. Note to Kisch [informal] about barley production, no date, CZA S25/595.

96. MacGregor, in his *Rob Roy on the Jordan*, first issued in 1869, described his visit to the Huleh region and its abundance of papyrus. In his fascination with the plant, he wrote that "the papyrus plant is called 'babir' by the Arabs of the Huleh...In Arabic its name is Berdi, and in Hebrew Gome, a word found four times in the Bible." MacGregor, *Rob Roy on the Jordan*, 264–265. MacGregor notes the dwelling of Zweer in the Huleh marshes. "Comments on Rob Roy," 3, CZA. S25/595.

97. For comparison, see *merdjas* (marshlands) and papyrus production in colonial Morocco. Swearingen, *Moroccan Mirages*, 50.

of the questionable nature of the Huleh soil in the marsh and the lake, it was uncertain whether development of the concession area (based on the original concession and its transfer to the PLDC) was worth undertaking as a commercial endeavor. Development of the concession area was likely to include the commercial exploitation of the papyrus industry.[98] The secretary of the Palestine Zionist Executive wrote to the Mandatory government director of lands to find out the uses for papyrus and to inquire about the government's income from selling the plant. British officials were not always swiftly forthcoming when presented with Zionist inquiries about the Huleh. Colonel Kisch commented, for instance, on the refusal of one British official, Mr. Badcock, to show Mr. Henriques a report on the Huleh area. Kisch thought Badcock's refusal had to do with his preference for the Huleh lands to remain in the hands of Arab cultivators. Thus, it seems that political leanings frustrated the collection of material for what was supposed to be a "scientific" survey about potential returns on this investment. Ultimately, what information was given to Zionist officials showed that the government's income from papyrus was quite small. The ZO grew curious about how revenue from papyrus escaped taxation and, upon concession transfer, if it was legally necessary to compensate the Huleh Arab residents who were involved in the papyrus enterprise. They reiterated this query to the Trade and Industry Department in 1926.[99]

Despite concerns, clearing the papyrus and reeds in the lake became part of the larger drainage scheme by 1934. Engineers used explosives to excavate drains so that the peat soil would dry, enabling easier burning of the papyrus. Engineers considered burning papyrus the cheapest method for excavating the area, and it was also deemed beneficial to the soil because burning freed the mineral nutrients stored in the plant.[100]

98. Singer asks Thischby to hold out on implementation of the basket maker plan so as not to raise opposition by the residents to Huleh drainage. Letter from Singer to Thischby, Esq. Trade and Industry Department, March 17, 1926, CZA S25/595; letter from Singer of Zionist Organization Financial and Economic Committee to Kisch, September 24, 1925, CZA S25/595; Storrs, director of lands, "State Domains Lake Huleh Papyrus Rights," Official Gazette, December 15, 1924, CZA S25/595. For the case of India and profits, see Philip, *Civilizing Natures*, 123.

99. Letter from Storrs, director of lands, to secretary of Palestine Zionist Executive, October 20, 1925, CZA S25/595; letter from Singer to Kisch, October 29, 1925, CZA S25/595; letter from Henriques to Kisch describing his visit to the Huleh, no date, CZA S25/595; letter from Kisch to Singer, January 7, 1926, CZA S25/595.

100. Rendel, Palmer, and Tritton, "Huleh Basin," 10–11, 32–33.

FIGURE 4.4. *Fellah* cutting papyrus reeds in the Huleh Valley, February 11, 1926. Courtesy of the Keren Keyemet L'Israel Photo Archive.

The Zionist agencies adapted burning from an Arab custom of clearing portions of the papyrus area when the water level was low. The *fellahin* cut the papyrus at the end of the summer and burned it along with the upper layer of peat so that a layer of ash covered the soil (which was known as a way to prepare peat soil for cultivation). When the level of the water fell after the rainy season, the *fellahin* cultivated the land. Even before the beginning of the actual Huleh drainage scheme, the craft of cutting papyrus (taken from the Arab residents of Mallaha) was adopted as part of a preliminary survey of the area. The adoption of Mallaha papyrus cutting showed that the Zionist scientists utilized local knowledge and integrated it into "modern" reclamation projects. Such integration provides another example (outside of colonial South India) of what Philip calls "mixed technoscientific modernity." Technoscientific modernity, where indigenous practices were adopted in and adapted for colonial scientific and agricultural projects, was emblematic of imperial science.[101]

In order to survey the bed of Lake Huleh and the Huleh marshes, Zionist engineers wanted to make straight lines across these bodies of

101. Philip, *Civilizing Natures*, 79, 98.

water at regular intervals and take soundings. Making straight lines would allow these scientists to produce cross sections of the land, which they then could study. Such concern with regularity and order—correcting what was considered the wild landscape—was typical of tropical scientific practices during this time.[102] Yet the marshes' abundance of papyrus posed problems for implementation of this orderly plan. So Dr. Mer of the Rosh Pina Malaria Research Station recruited papyrus cutters to manually cut straight lanes through the marsh. Significantly, drainage interventions eventually caused the loss of the papyrus commodity from the landscape.

Zionist land agencies considered acquisition of the Huleh concession important not only to secure economic interests, like the papyrus industry, but for Zionist settlement and politics, especially in relation to the most northern Jewish settlements. As Colonel Kisch remarked in December 1925: "The recent critical situation in the North has made apparent the general political importance of our securing the Huleh area for Jewish colonization and development . . . Politically it would of course be disastrous for us to carry out any such evacuation [in the North, i.e., Metulla] therefore we have to take steps to strengthen our position in the North."[103] The "recent critical situation" that Kisch noted likely referred to the shape of Palestine's northern border. Zionist land agencies deemed the border's configuration vital for the Zionist endeavor at this time for two reasons: up until 1923, the designation of the final northern boundary of Mandatory Palestine was contested and might not have, under the Sykes-Picot and Occupied Enemy Territory Administration proposals, included the Huleh Valley and some Jewish settlements in the area. The latter proposal, for instance, would have excluded the Jewish colonies of Metulla, Tel Hai, and Kfar Giladi. These colonies posed the second reason for the northern border's importance and the attainment of the Huleh concession. Because they were defended against Arab attacks in 1920 in the Upper Galilee, where the noted Joseph Trumpeldor died, the Zionists sought the inclusion of this area in Mandatory territory in order to afford continued defense of these colonies and increased settlement.[104] The 1923 agreement in fact included these areas in Mandate Palestine. Still, explicit political urgency on behalf of the Zionists to buy the Huleh concession

102. For a similar Cartesian approach in forest clearing in India, see Philip, *Civilizing Natures*, 51.

103. Letter from Kisch to Singer, December 7, 1925, CZA S25/595.

104. Tyler, *State Lands*, 86–87.

was counterbalanced by the recognition of political difficulties in actually draining the area.

The Joint Survey Commission of the Agricultural Colonization Advisory Commission, led by the American experts, also acknowledged the political ramifications of selling the concession. In addition to noting the expected improved health and wealth of Palestine after drainage, the report recognized the complexity of drainage due to the fact that Jewish settlements existed near the margins of the concession while Arab tribal chiefs owned a large part of the Huleh area. Drainage involved "so many questions relating to tribal rights and the acquisition of titles to land." The commission wrote that "the Zionist policy in farm settlement has gone too far in exalting muscular efforts and in belittling or restricting the opportunity for management." The commission's reference to "muscular efforts" most likely referred to the sentiment of self-sacrifice among many Zionist settlers. The commission felt that no other agency but the Mandatory government could simultaneously resolve these issues successfully and undertake the drainage scheme. Moreover, because it valued the Huleh's economic and agricultural potential, the commission recommended that the Mandatory government sponsor reclamation rather than approve the sale of the concession. Given the Huleh's potential as the "largest, most valuable body of agricultural land in Palestine," drainage in the area had large importance for the future of Palestine. Elwood Mead of the Committee of Reclamation of the U.S. Department of the Interior; Frank Adams, of the Irrigation Investigation and Practice of University of California; and A. R. Knowles of the Horticulture Agricultural Experiment Station of Haiti signed the recommendation formulated by the commission. The commission's members and input showed the extent to which international drainage forums were active in Palestine.[105] Ultimately, however, the Mandatory government did not heed the American recommendation, and the transfer of the concession was awarded to the Zionists.

Zionist interest in the Huleh fluctuated until its final purchase in 1934. These hesitations were reflected in malaria survey reports of the area. In 1919, Mr. Franck, a PICA engineer, approximated the cost of the entire Huleh drainage plan to be five million francs. This figure included building agricultural roads, buildings, materials, livestock, clearing of the marsh, and its drainage. Then, in 1923 a change of heart occurred, as the MRU

105. Joint Survey Commission Agricultural Colonial Advisory Commission, December 1927, PRO CO 733/156/3.

reported that draining the Huleh was a "hopeless" endeavor—it was "impossible to do anything against mosquito breeding and mechanical protection and quinine are the only protective palliatives available."[106] Once Zionist interest in the concession peaked again, the engineers suddenly saw drainage of the Huleh as a possible undertaking. Dr. Werber, for instance, performed additional investigations for the PLDC from December 1934 to May 1935.[107] The descriptions and recommendations of his reports reflected not only scientific issues at hand but also pointed to the political status of gaining the concession and the possibility of drainage. Clearly, report recommendations closely depended upon land speculation interests.

In October 1934, the Mandatory government transferred the Huleh concession from the Syro-Ottoman Agricultural Company to the PLDC at 185,000 LP; double the price as had been offered in the 1925–1926 negotiations. Upon the transfer, the Zionists acquired 56,940 dunams of land: 21,453 dunams of marsh, 16,919 dunams of lake, and 18,568 dunams of actual land. According to land policies at the time of the transfer (1934), British provisions required land to be reserved for Arab cultivators as part of the sales negotiation. The Mandatory government requested that 13,000 dunams (30 dunams per family) be set aside for extensive cultivation (land for 285 Arab families) while 10 dunams per family be set aside for intensive cultivation (land for 435 families). This final settlement differed from the one suggested the year before when the secretary of state in Palestine informed Weizmann that at least 40 dunams per family reserve would be required. At that time, the government estimated that 2,000 Arab families would have to be provided for within the whole Huleh area (concession area plus adjacent land), which totaled 80,000 dunams out of 160,000 dunams of land. A clause to protect the rights of the Arab cultivators was attached to the purchase contract of the concession.[108]

The Mandatory government required a reserve provision in the Huleh transfer, as opposed to the Jezreel Valley transfer, because of the problem of Arab landlessness that largely emerged from Jewish land purchases and

106. MRU, "Malaria Control Demonstrations," part 1, 14.

107. Extrait d'une lettre de M. Franck, 5 January 1919 [the text says 1910 but is surely a typographical error since Cantor did not submit his report until January 3, 1919], CZA S25/595. For Henriques's comments, see letter from Henriques to Kisch, no date, CZA S25/595; Treidel, "Technical Problems in Soil Drainage," 8. Joseph Triedel (1876–1929) had served as a scientific advisor in the German colonies and did cartographical work for the Commission on the Exploration of Palestine (established in 1903), directed by Otto Warburg. See Penslar, "Zionism, Colonialism and Technocracy," 15–151.

108. See comments, folio 1, Mr. Williams from Downie, April 5, 1934, PRO CO 733/263/8.

which peaked in the 1930s.[109] Stipulations in both the Huleh and Jezreel Valley drainage agreements reflected a recognition of this demographic shift and political problem. A Jewish Agency for Palestine memorandum submitted to the Palestine Royal Commission in 1936 took issue with the terms of the land reserve provision. It disagreed that Jews needed to

> allocate about one-third of the drained area to the native inhabitants without any contribution by them to the concession or for the drainage. This procedure is unique in the annals of reclamation. If it is the duty of the Government to protect the cultivators, it is equally its duty to participate in the cost of drainage and irrigation, at least in respect of that part of the land which will fall to the Arabs.[110]

Here the Jewish Agency for Palestine referred to the Protection of Cultivator's Ordinance originally promulgated by the Mandatory government in 1929 as a response to Jewish land sales and heightened political tensions. According to Stein, the ordinance confirmed the use of monetary compensation to terminate tenants' rights to work purchased land. At first it also removed the stipulation of reserving land for a majority of Arab tenant farmers. Amendments between 1929 and 1933, however, widened the definition of a tenant and allowed land to be used for compensation.[111] The Jewish Agency for Palestine contention in the quotation above referred to the land reserve issue and touched upon the ongoing debate between the *Yishuv* and the Mandatory government regarding the extent of government financial aid in Jewish development projects. Interestingly, however, the JA's assertion implied the importance of equal Arab participation in drainage as a general step toward "healing the land" (but *not* the people, which was reserved for Jews alone; Jews were considered a coherent nation).

Arab opposition to the Huleh transfer in particular was based on the physical displacement of the Arab population as the culmination of years of Zionist land purchase and the subsequent physical manipulation of the

109. For landlessness due to intercommunal land transfers, see Metzer, *Divided Economy of Mandatory Palestine*, 93–94. Metzer and others note that whereas large landowners (like those of the Jezreel Valley and the Huleh) used the proceeds from their land sales to reinvest in other, profitable ventures, *fellahin* who owned land often sold it to pay off debts (90).

110. See Stein, *Land Question in Palestine*, chapters 4 and 5, for further discussion; JA, "Antimalaria and Drainage Work by Jewish Bodies: Memorandum Submitted to the Palestine Royal Commission," December 1936, 21, CZA Library.

111. Stein, *Land Question in Palestine*, 52–53.

land. Arab leaders threatened to severely criticize the Mandatory government if there was a failure to provide fully for Arab cultivators in the Huleh agreement. Awni Bey warned High Commissioner Wauchope that he would lead such opposition. These threats impelled Wauchope to make sure to implement protective measures. Andrews, the commissioner on special duty, investigated Arab claims in the Huleh to make sure infractions did not take place. He acknowledged water rights as potential factors in Arab opposition although he felt that undertaking comprehensive antimalaria work in the entire Huleh basin was considered the only way to attempt the complete sanitization of the Huleh. In a 1935 report, "Huleh Reclamation Project," engineer E. J. Buckton noted the political concerns of carrying out any scheme, stating that "the necessary control of water supplies seems impossible until the unsatisfactory system of water rights in Palestine has been straightened out."[112]

Articles in Arabic newspapers called for village leaders to make clear provisions for the protection of inhabitants' rights, mentioning letters written by Palestinian Arab leaders to uphold the special terms of the concession. A series of articles in *Filastin* on November 6, 7, and 8, 1934, pointed out manipulations of the concession's text that facilitated the transfer of the concession to the Jews. Such manipulations, the authors claimed, were based on the government's disregard of the statement that all shareholders of land should be an "Ottoman descendent coming from a father with Ottoman residence and citizenship." An article that appeared on November 8, "Imtiyaz al-Huleh wa huquq al-daʿwa" (The Huleh Concession and Rights of Appeal), argued that Ottoman citizenship of yesteryear was the Palestinian citizenship of the present. It also stated that although the Mandatory government easily granted Palestinian citizenship to Jews, the Jews still did not have Ottoman or even Palestinian ancestors. Such conditions therefore did not comply with the latter portion of the provision. The article voiced the importance of facing the government with demands to cancel the implementation of the concession sale and challenge Britain's "manipulation of Arab land and

---

112. Extract from letter of Wauchope to Secretary of state, April 14, 1934, PRO CO 733/263/8; and letter from Andrews, commissioner on special duty, to secretary, Palestine Royal Commission, January 21, 1937, PRO CO 733/345/6; Tyler, "Huleh Land Issue," 346; E. J. Buckton, "Huleh Reclamation Project: Note by Visiting Partner of Messrs. Rendel, Palmer, and Tritton to Dr. Maurice Hexter of EMICA Association," April 20, 1935, 2, PRO CO 733/271/7; Rendel, Palmer, and Tritton, "Huleh Basin," 10, 11, 39. Hexter was also part of the International Health Board of the Rockefeller Foundation.

neglect in its destruction."[113] That same article also took issue with the Mandatory government for promising the Jews the right to settle in Palestine. Jewish settlement, it noted, was causing the displacement of *fellahin* and fisherman from the Huleh and was ultimately "kicking the Arabs from their country and tearing them from their land."[114] Two days earlier, the article "Mashru'a al-Huleh 'arabi lil-'arab" (Arab Huleh Scheme for the Arabs) appeared. That article mentioned that Ya'qoub Farraj, vice-president of the Arab Executive Committee (1934) and later Jerusalem Christian representative to the Arab Higher Committee (1936), wrote to the high commissioner about certain violations of the concession sellers in fulfilling the concession terms that protected inhabitants' rights. Awni Abdul Hadi, founder and president of the Palestinian Istiqlal (Independence) party, also expressed outrage about this matter. By pointing out the background of the concession and its transfer to British jurisdiction, the article clearly reflected Palestinian Arab frustration of Zionist manipulations of such texts. It stated "this problem is a plain fact and cannot be manipulated by Mandatory politics or given satanic Zionist explanations."[115] An article appearing a day earlier, on November 7, 1934, described the original owners and negotiations of the Huleh concession as well as details of the Tiberias concession citing "Jewish greed" in buying land and attaining different concessions with the inauguration of the Mandate. As a result, the author argued, the country would then turn into a "Jewish kingdom as Jewish as England is English" with the parallel destruction of the Palestinian people.[116] In essence, Arabic newspaper editors, authors, and nationalist leaders saw the transfer of the Huleh concession to the Jews and the Huleh's drainage as another step—in addition to overall Jewish land purchase and settlement—in the ultimate displacement of the Palestinian Arabs from their land.

Such opposition put politics in the forefront of considerations for scientific surveys and reclamation. Arab resistance caused the government

113. "Imtiyaz al-Huleh wa huquq al-da'wa" [The Huleh concession and rights of appeal], *Filastin*, November 8, 1934, front page. See also residence requirement of Land Transfer Ordinance 1920, as discussed in Stein, *Land Question in Palestine*, 48.

114. "Imtiyaz al-Huleh wa huquq al-da'wa," front page. Articles also appeared in *al-Difaa* on the same days. These articles are among the small number of Arabic documents about malaria and swamp drainage during the Mandate period.

115. "Mashru'a al-Huleh 'arabi lil-'arab" [Arab Huleh scheme for the Arabs], *Filastin*, November 6, 1934, front page.

116. "Watan bil-mazad: imtiyaz al-Huleh" [Nation at auction!: The Huleh concession], *Filastin*, November 7, 1934, front page.

to tread cautiously in making plans to perform a hydrographic survey of the area extending north of the Huleh concession area. The Mandatory government hesitated in directly affiliating itself with a Jewish body (EMICA) that had already started to undertake such work in the concession area in March–April 1935. Potential opposition to partnership with Jewish organizations, however, often depended upon exact location. The government did finally request that the same consulting engineers for EMICA perform a similar survey for another area north of the concession, and EMICA agreed.[117]

Like all other drainage interventions, significant manipulations planned for the landscape were evident in the EMICA scheme for the Huleh.[118] The EMICA preliminary survey report (1936) noted that Zionist land agencies would set aside 15, 772 dunams of irrigated and drained land as reserve for the Arab residents at the concessionaires' expense in the final agreement. After official transfer of the concession, the parties would demarcate boundaries on the ground.[119] The plan included straightening, widening, and deepening the Jordan River down to Jisr Banat Ya'qub (which led to Damascus) even though scientists recognized that there was no available scope for widening the river within the concession limits. Even at this time, scientists acknowledged shrinkage of the peat soil as an obstacle to the proposed embankment of the Jordan River and to drainage in general. They considered the particular peat soil of the Huleh unique to Palestine. Its impact upon drainage schemes was still deemed more favorable than the soil of the Kabbara swamp, where the peat "proved exceedingly difficult and expensive to drain."[120] Mechanical manipulations consisted of cutting and enlarging a channel in the Jordan to carry

117. See confidential telegram, no. 155, May 18, 1935, 5, PRO CO 733/271/7; J. H. Hall, chief secretary to Dr. Hexter of EMICA Association, May 21, 1935, PRO CO 733/271/7; letter from Wauchope to Parkinson, March 30, 1935, 5, PRO CO 733/271/7; follow-up by Wauchope to Sir Philip Cunliffe-Lister, HMG principal secretary for state for the colonies, marked confidential, May 29, 1935, PRO CO 733/271/7. See also letter from high commissioner of Palestine, Wauchope, to Mr. Parkinson, March 30, 1935, 3–4, PRO CO 733/271/7.

118. Israel Kligler, however, was ultimately given credit for formulating the plan for the sanitation of the Huleh. I did not find the details of his plan. Letter from Abramovitsch, Palestine Research Department, to Rose Bradley, Palestine Economic Corporation, August 7, 1948, Hadassah Archives, RG 1, box 100, folder 2.

119. Rendel, Palmer, and Tritton, "Huleh Basin," 39.

120. Commentary on the peat soil issue can be found on page fourteen of the Rendel et al. report. Existence of this peat soil of the Huleh caused spontaneous combustion in contemporary Israel, thus compelling the Israeli government to reswamp the area. See Rendel, Palmer, and Tritton, "Huleh Basin," 9, 14, 19, 22.

off any possible flood. Other measures included the construction of cultivation feeder channels and tunnels from the Upper Jordan through the northern Huleh area, the installation of primary and secondary drains, the diversion of several wadis (i.e., Bureight) in the northern area of the basin, the building of a barrage in the southern area of the basin, and the construction of roads and bridges. Scientists recommended building an unembanked channel through Huleh Lake as well.[121] Scientists and engineers made these plans with the understanding and goal that if malaria was eradicated in the Huleh, the area would be "healthy and pleasant for European settlers." It would be an important asset in the "development of the Jewish National Home."[122] These experts estimated the cost of drainage and of the tax, in addition to the sale of the concession at 20 LP per dunam, or a total of almost one million lira. The Mandatory government was to pay about 222,600 LP of it, and the Palestine Land Development Company was to pay 710,400 LP. The cost for the Huleh scheme was considerable in comparison to other drainage expenditures; it was ten times the estimated cost for drainage of the Jezreel Valley in the early 1920s. Moreover, settlement expenditures were considered separately. Showing explicit comparisons between Palestine and other colonial drainage projects, the EMICA report noted that Arab labor was four times as expensive as Indian labor. Similarly, the Huleh project was compared in its extent and significance to the Pontine marsh drainage project in Italy and to the Tennessee Valley Project in the United States.[123] After complete drainage, with the area free of malaria, land experts estimated that approximately 50,000 families would be able to settle in the Huleh area. Full drainage of the area, however, was delayed because of political events: the Arab Revolt (1936–1939), proposals for the partition of Palestine, and World War II. Despite the delay in drainage, Jewish settlements continued to be established in the area in the 1940s.[124]

In the first years after statehood, the Israeli government organized a committee to discuss the Huleh matter: the continued presence of malaria and lack of drainage there. The committee rejected the EMICA plan—which entailed digging one channel—but accepted a plan drawn up by an

---

121. Buckton, "Huleh Reclamation Project," 2; Rendel, Palmer, and Tritton, "Huleh Basin," 21, 26. See also Treidel, "Technical Problems of Soil Drainage," 7.

122. JTA bulletin, "The Huleh Concession," January 11, 1935, 7, PRO CO 733/271/7.

123. Rendel, Palmer, and Tritton, "Huleh Basin," 27, 30, 38, 52.

124. The war in 1948 caused the removal of Arab villagers from the Huleh Valley. Rabinowitz and Khawalde, "Demilitarized, then Dispossessed," 516.

engineer of the Jewish Agency for Palestine, Kobelnov. Kobelnov prescribed two channels that would meet at the lowest point of the lake and
then run southward in one channel.[125] Drainage of the Huleh finally took
place in the 1950s and was completed in 1958. The Israeli government, the
JA, the KKL, and the Keren HaYesod funded the drainage work. It cost
eight million LP.[126] As stated in the introduction, the alleged reasons for
the drainage vary, but the consequences were serious.

Conclusion

As this and previous chapters have shown, Zionist swamp drainage projects effected several transformations of Palestine's landscape—physical
and discursive, technological and ecological—in an effort to promote Zionist ideological, practical, and political agendas.[127] As both development
and public health efforts, these drainage projects were an essential part of
the Jewish nation-building project in Palestine and played an important
role in the Zionist aim to transform Palestine and heal themselves. By
understanding the epidemiological, economic, ecological, and physical conditions before drainage, we can gain an appreciation of just how far-reach-
ing the technological work of drainage was from a health and landscape
perspective. To be sure, antimalaria measures in the Mandate period significantly reduced the incidence of malaria from its pre-Mandate endemic
profile. Scientists replaced prophylactic and general drainage measures
with plans that relied heavily upon scientific measurement. These measures implemented new malariological discoveries resulting from shifts
in etiological theories and required large technological and labor investments. The malaria and drainage surveyors that prescribed these methods
also offered ethnographic material of particular areas, reconstructing the
histories of swamp areas and their surrounding villages in ways that revealed their views about the land and its inhabitants.

The cases of the Jezreel Valley and the Huleh Valley show that sani-
tary/health concerns and political/settlement interests merged in the site

125. Levana, "Yibush haHuleh," 213. Levana gives the technical reasons why the EMICA
plan was rejected and the Kobelnov plan was accepted.

126. Levana "Yibush haHuleh," 213.

127. Biger describes British involvement and policy in the physical transformation of
Palestine in *An Empire in the Holy Land*, but gives scant attention (162–163) to the topic
of swamp drainage.

of the swamp. In addition to controlling malaria, Zionist land agencies used drainage interventions to create more living and agricultural area for Jewish settlers.

Another component to the transformation of the land's topography was the use of scientific measurement and mapping. As the next chapter will demonstrate, for Zionist health professionals, medical and scientific experimentation accompanied by medical cartography furnished the tools for a "rational" transformation of Palestine.

Perceptual Landscape

Scientific Experimentation, Colonial Medicine, and the Medicalization of Palestine

A Medicalized Land

Zionist health officials in the Mandate period believed that only by knowing the land (*yedi'at Ha'aretz*) and knowing the enemy (*yedi'at ha'oyev*, the mosquito) could effective malaria control and nationalist transformation take place.[1] In their explicit connections with nationalist agendas, Zionist malariologists' discoveries and experiments were social and political events that add a new layer to the sociocultural history of Mandate Palestine.

For Zionist medical bureaucrats involved in malaria control, a "rational" transformation of Palestine meant the extensive use of science and medicine in drainage interventions and a shift in the nature and scope of drainage projects in the Mandate period as compared to the pre-Mandate period. Although science was a tool for achieving Zionist goals during the pre-Mandate and Mandate periods, Zionist malariologists during the Mandate contrasted their "rational" transformation with earlier Zionist drainage works,[2] which they saw as not using scientific data widely or

---

1. The pamphlet, *Conversation on Malaria*, explicitly states this necessity. Hadassah National Library, no. 3 (Jerusalem, 1921), 7, CZA Library; Fleck, *Genesis*, 76; Pratt, *Imperial Eyes*, chapter 2.

2. Penslar, *Zionism and Technocracy*, 154.

systematically and which they considered haphazard and hasty in their execution. Their call for a "rational" approach to drainage schemes distinguished Zionist methods of land use from an alleged misuse of the land by the Palestinian Arab population. Contemporary scholarship has shown that Arab land use had its own rationality: Arab farmers adjusted agricultural methods and land institutions to changing external conditions. A lack of opportunities or money, not irrational behavior, accounted for periods when adjustments were made slowly.[3]

This shift to stressing scientific methods meshed with a larger colonial trend in the interwar period in which the British Colonial Office linked imperial development to the biological and ecological sciences. U.S. and Australian officials also stressed scientific methods in the face of soil erosion crises during that time, calling on science for long-term management of nature. Colonial, U.S., and Australian officials framed long-term agendas such as pathogen eradication as serving or "developing the nation," following earlier discourses of "science for development."[4] A similar trend occurred with the Zionist case in Palestine.

The *Haskalah* movement and the culture of colonialism influenced the Zionist emphasis on science and rationality in their antimalaria campaigns and in settlement policy at large. Zionism redirected the ideals of reason and progress to serve nationalist ends.[5] As Gershon Shafir and Derek Penslar have demonstrated, practical means were incorporated into Zionist labor ideology; scientific knowledge informed the way *Kibush ha'avoda* (Conquest of Labor) and *Kibush hakarka* (Conquest of the Land) were carried out. In other words, science was a critical element in the struggle for Jewish self-transformation, in physical labor and in the reclamation of national land.[6]

A shift in the Mandate period to the expansion and coordination of scientific professionalism in malaria projects contributed to the amplification of the role of medical science in *Kibush ha'avoda* and *Kibush hakarka*

3. Metzer, *Divided Economy of Mandatory Palestine,* 96.

4. Libby Robin, "Ecology: A Science of Empire?" in *Ecology and Empire: Environmental History of Settler Societies*, ed. Tom Griffiths and Libby Robin (Seattle: University of Washington Press, 1997), 67, 70–71.

5. Penslar, *Zionism and Technocracy*, 66–79; Mayer, *Jewish Identity in the Modern World*, 61; Nicholas Dirks, "Introduction: Colonialism and Culture," in *Colonialism and Culture*, ed. Nicholas Dirks (Ann Arbor: University of Michigan Press, 1992), 6.

6. Shafir and Penslar deal with settlement and agricultural engineering rather than medicine. Penslar, *Zionism and Technocracy*, 132; Shafir, *Land, Labor and the Origins of the Israeli-Palestinian Conflict*.

during that time. The KKL, for instance, took pride in this precise scientific work, asserting itself to be

> [the] first organization that approached solving the healing (*havra'a*) of wide tracts of land in both a scientific and practical way by using specific programs and modern techniques. The technical problems connected to draining the swamps were new in Palestine, even though Petach Tikva and Hadera were established in the areas of dangerous swamps. But they [the settlers of Petach Tikva and Hadera] did not care for the complete healing [of the land] in a logical way, which is what caused many deaths/victims. There was no organization then that was able to take responsibility for this type of work and for its investment.[7]

The KKL did not always uphold its own standards, but what is instructive about this quotation is how it illustrates the shift in thinking about land policy and drainage. The "heroic" acts in first- and second-generation *Aliya halutz* (pioneer) mythology were replaced in the Mandate period with what was considered to be the more legitimate, scientific, modern, and "rational" acts of the land-purchasing agency. Science, rationalism, and modernism became the hallmarks of drainage projects in the Mandate period. Scientific rationality was regarded as the main determinant of the success of the endeavor; land agencies and malariologists blamed other, less scientific approaches for death and destruction. Thus, beginning with the Mandate period, Zionist officials (land agents, scientists, and politicians) considered the scientific manner in which one drained land as the most vital factor in healing or correcting the land.[8] A bond between the land and the Jewish people could form only if nature was conquered by scientific means.

The use of medical cartography was one significant, scientific way of reconceptualizing the land. In their attempt to make Palestine a malaria-free locale, Zionist medical professionals and their British and U.S. colleagues used maps of swamps as well as tables and diagrams of malaria rates to construct a medicalized landscape of Palestine.[9] Swamp maps and

---

7. "Metoch din veheshbon shel hanhala mirkazit leKKL" (1924–1925), 1.

8. See Moulin, "Tropical without the Tropics," in Arnold, *Warm Climates*, 172, for the Algerian case.

9. Kligler, *Epidemiology and Control of Malaria*, and all British Health Department of Palestine annual reports at ISA; Kearns and Gesler, *Putting Health into Place*, ix–x.

malaria tables and diagrams made medicine and disease the new conceptual lens through which scientists observed, experienced, and reclaimed
the landscape of Palestine.

The act of medicalizing the landscape involved a two-part process
whose constituents most often worked simultaneously. The first consisted
of gathering and analyzing specimens from the land and its inhabitants
(humans and mosquitoes), and the second part involved consolidating
and organizing this information in maps, diagrams, or tables in the Hippocratic tradition of colonial medicine and medical geography. According to Michael Osborne, the Hippocratic tradition emphasized the description of "local meteorological, dietetic, and geographic features and
their correlation with the health, habits, race, and character of a region's
inhabitants."[10] Through monthly and yearly surveying of the ecological and biological environments, Zionist malaria professionals constantly
constructed, reconstructed, and "re-objectified the space of Palestine"; a
circuit of knowledge was formed about the mosquito, malaria, the indigenous population and the *Yishuv*.[11] Once they completed drainage plans
and observed resultant fluctuations in malaria incidence, Zionist malariologists and engineers produced maps and tables that showed new rates
of malaria, new topographies, and new areas of control. Doctors in Jewish
settlements and British medical officers made monthly and yearly malaria
reports and then submitted them to the British Health Department of
Palestine for the government's annual report. Just as Jews were undergoing a process of nationalist transformation, Palestine was a land that was
conceptually and physically being remade.[12]

The Zionist faith in the use of technology in antimalaria measures was
consistent with colonial medical services in Palestine and other contexts.[13]
"Correcting" nature's defects entailed an environmental manipulation that

10. Michael Osborne, "Resurrecting Hippocrates: Hygienic Sciences and the French Scientific Expeditions to Egypt, Morea and Algeria," in Arnold, *Warm Climates*, 82.

11. A. Young, "Mode of Production of Medical Knowledge," *Medical Anthropology*
(Spring 1978): 99–100; all British Health Department of Palestine annual reports, ISA; and
Dan Rabinowitz, *Overlooking Nazareth: The Ethnography of Exclusion in the Galilee* (Cam-
bridge: Cambridge University Press, 1997), 16.

12. Geores, "Surviving on Metaphor," 36; Kearns and Gesler, conclusion, *Putting Health
into Place*, 290.

13. Randall Packard, "Visions of Postwar Health and Development and Their Impact on
Public Health Interventions in the Developing World," in *International Development and the
Social Sciences: Essays on the History and Politics of Knowledge*, ed. Fred Cooper and Randall
Packard (Berkeley: University of California Press, 1997), 95, 101.

tried to make the land more fit for human habitation but less fit for mosquitoes.[14] Though British and Zionist doctors and engineers, either separately or collaboratively, undertook mapping and scientific experimentation, the Zionist medicalization process is distinct in its connection to Zionist national agendas. Having no political or military power, the Jews could not claim Palestine, as the British did, by military conquest or "right of protectorate."[15] They attempted to do so by the intellectual, practical, and financial means available to them through land purchase, settlement, labor, and scientific conquest.

Zionist malariologists translated medicalization of the land into swamp drainage projects and practical settlement for the establishment of the Jewish national home.[16] They insisted that the larger Zionist movement draw upon extant scientific knowledge about tropical medicine in order to prevent loss of life, money, and labor. Jewish medical professionals explicitly recognized their first priority as dealing with malaria for sanitation in settlements; once sanitation was under control, medical professionals would pay more attention to other health issues in the Jewish colonies.[17] Zionist doctors, malariologists, and sanitary engineers raised "the flag of medicine and health" in the advancement of the national project.[18] Scientists involved in medicalizing the land took what they imagined as desolate, primitive, untamed land and rendered it ready for "rational" use, meaning the most effective exploitation of the land.

As land passed from Arab to Jewish hands (through land purchase), a simultaneous cognitive transfer from indigenous understandings of the land to medical, scientific understandings of it took place. Zionist health professionals' scientific analyses made the land discursively modern, ready for Jewish settlement and development. By implication, this discursive move further reduced Arab rights to the land. In other words, Zionist doctors and sanitary engineers, perhaps unconsciously or perhaps consciously, implanted an "exclusive national identity" onto the landscape of Palestine by erasing the indigenous perceptions of the land and applying scientific

14. See Harrison, *Mosquitoes, Malaria and Man*, 2; Vaughn, *Curing Their Ills*, 25.

15. Shafir, *Land, Labor and the Origins of the Israeli-Palestinian Conflict*, 17, 23.

16. See Penslar, *Zionism and Technocracy*, 154; Levine, "Discourse of Development in Mandate Palestine," 97; letter from Henriques, engineer, to unknown, copied to Agricultural Experiment Station and Col. Kisch, April 7, 1926, CZA S25/595.

17. Report on the second meeting of the Health Council, April 13, 1926, Jerusalem, CZA S25/740.

18. "Magamatenu" [Our aim], *HaRefuah* 1, no. 1 (March–May/June 1920): 4, CZA Library.

principles and measurements about swamps, insects, and inhabitant morbidities. They then literally and practically eliminated the existing landscape by implementing drainage interventions. The motivations behind this process, couched here in a discourse of health, were akin to what Swedenburg has observed for the Zionist erasure of Palestinian Arab sites, wherein "every parcel of land had to be plowed with Jewish history."[19] Making maps, conducting experiments, and surveying the land were a part of making that history. Land agencies and health professionals used scientific land and malaria surveys to plan drainage schemes that, in effect, made the medicalization of the land a part of the process of nationalizing it.

The application of medical and sanitary science to the landscape in surveys and maps was a symbolic possession continually reenacted through scientific control. The overlaying of medical terms onto the landscape resembled the nationalist ritual of taking Zionist nature hikes (*moledet* worship, that is, worship of the motherland/homeland) common during this time. Meron Benvenisti, Yael Zerubavel, and others have studied the cultural effects of these hikes as an "actualization of a link to the land."[20] In their studies of the epidemiology of malaria, Zionist scientists and doctors made efforts to know the land (*yedi'at Ha'aretz*) by direct contact with its very minutiae.

Organizing a Rational Transformation: Professional Agencies and Collaboration

The Zionist project to control malaria mobilized a substantial amount of expertise and money. The leaders of the *Yishuv* devoted a good deal of its financial, scientific, and labor investments to the scientific gathering of data and the implementation of antimalaria control measures. The cost

19. Swedenburg, *Memories of Revolt*, 7–10, 60–61. During the Mandate period, afforestation and swamp drainage were both part of land reclamation activities of the Zionist movement in Palestine. Afforestation was very likely initiated not only for beautification but also because of the cutting down of forests for fuel during World War I in Palestine. See League of Nations Health Organization, Malaria Commission, *Reports on the Tour*, 22.

20. Meron Benvenisti, *Conflicts and Contradictions* (New York: Villard Books, 1986), 23, as quoted in Swedenburg, *Memories of Revolt*, 56; Zerubavel, *Recovered Roots*, 28, 90; and O. Ben-David, "*Tiyul* as an Act of Consecration of Space," in *Grasping Land: Space and Place in Contemporary Israeli Discourse and Experience*, ed. Eyal Ben-Ari and Yoram Bilu (Albany: SUNY Press, 1997), 129.

of control operations in 1923 in Ein Harod, for instance, was 24,106 PL. For Hadera, the cost was 38,095 PL. These per capita costs were similar to those in the United States.[21]

Scientific research and study of malaria before the Mandate period was not centralized or organized, yet no Zionist official questioned its importance. The Pasteur Institute and the Straus Health Bureau, both in Jerusalem, dealt with the general sanitary issues of the *Yishuv*. Individual Jewish doctors such as Hillel Yofe, Arthur Brunn, L. Goldberg, and L. Puchovsky disbursed malaria treatment and determined malaria incidence for the Jewish community. Their scientific experiments included the analysis of types of malaria plasmodia and their development, methods of checking blood for the parasite, the determination of parasite incidence in patients, and studying the effectiveness of quinine prophylaxis.[22]

Researchers who conducted similar investigations for the Palestinian Christian and Muslim populations before the Mandate period included Professor Muhlens, director of the Tropical Diseases Institute in Hamburg, Germany, Dr. E. W. G. Masterman of the English Mission Hospital, and Dr. Cropper, also of English origin.[23] Muhlens resided in Palestine from August 1912 until January 1913, heading a malaria expedition sponsored by the German government. Muhlens developed a "thick drop" technique for microscopic examination of blood that increased the accuracy of diagnosis and decreased the time it took to get results. He also did scientific work on syphilis. Drs. Masterman and Cropper conducted epidemiological studies on malaria in Jerusalem in 1912, where they found cisterns to be the cause of urban malarial transmission. Muhlens and Masterman also collaborated with each other during that year. Muhlens, Masterman, and Palestinian Arab Dr. Tawfiq Canaan lectured about malaria in German and English to scientific audiences. Canaan was a central medical figure among the Arab population before and during the Mandate period. Dr. Cropper was the first to study malaria in Palestine in light of the new scientific knowledge of malaria etiology. Cropper found that

21. For other areas in 1923, see MRU, "Malaria Control Demonstrations," 21–22, and table 8.

22. Dr. Brunn, "Hadiagnosa shel Malaria" [Diagnosis of malaria], *Zichronot HaDavarim* 2–3 (May 1913): 89, CZA Library.

23. Fleck, *Genesis*, 71; Yekutiel, "Masterman, Muhlens and Malaria," 4, 12, and also for discoveries regarding the high incidence of tropical malaria in Jerusalem and the importation of infection from the "Ghor" region of Palestine (9); Karakara, "Ma'arechet habriut hamandatorit vehavoluntarit ve'aravi Eretz Israel: 1918–1948," 7–32.

malaria in Palestine was endemic, occurring mainly, although not entirely, in children. He also found that blackwater fever occurred chiefly in immigrant Jews. Blackwater fever results from the prolonged destruction of red blood cells and the consequent release of haematin, which, in malignant malaria, amasses in the spleen, liver, bone marrow, and sometimes the brain, causing these organs and urine to turn black.[24] Dr. Brunn, whose work was supported by the Nathan Straus Health Center and who primarily did research in the Jewish colony of Hadera, worked with Muhlens on malaria matters in Jerusalem during this period. Brunn and Goldberg conducted the cooperative work of recording spleen rates among children in the Jerusalem district and found that they varied between 40 and 83 percent in 1912.[25] In a conference held at the end of 1912, foreign consuls, local physicians, and these doctors decided to collaborate with international and local medical figures of all religions on malaria and other health matters. They resolved that Muhlens's laboratory would be converted into the International Health Bureau. World War I, however, would abort this endeavor.[26] Still, these scientists' reports provided the foundation for further scientific experimentation conducted in the Mandate period.

As part of an emerging scientific and medical relationship between the U.S. government, American Jews, and the *Yishuv* that began during World War I, Justice Brandeis of the U.S. Supreme Court sponsored the earliest organized Jewish malaria initiative. He proposed that a malaria survey be conducted for the *Yishuv*.[27] The American Zionist Medical Unit, soon to become the Hadassah Medical Organization (HMO), funded the initiative in 1919. In February 1921, Hadassah sponsored a malaria conference in Jerusalem in order to bring together antimalaria specialists and discuss a plan of action against the disease.

With the official establishment of the British Mandate over Palestine in 1922, malaria work became more centralized and coordinated through several malaria agencies. These agencies shared scientific data, making

24. Bynum and Porter, *Companion Encyclopedia*, 386; League of Nations Health Organization, Malaria Commission, *Reports on the* Tour, 7, 11.

25. League of Nations Health Organization Malaria Commission, *Reports on the Tour*, 8.

26. Yekutiel, "Masterman, Muhlens and Malaria," 13.

27. Sandy Sufian and Shifra Shvarts, "Mission of Mercy and the Ship That Came Too Late," in *Proceedings of the European Association for Jewish Studies Conference Toledo 1998: Jewish Studies at the Turn of the Twentieth Century, Volume 2: Judaism from the Renaissance to Modern Times*, ed. Judit Targarona Borras and Angel Saenz-Badillos (New York: E. J. Brill, 1999), 389–398.

some distinctions between Zionist data and work and those of the British Mandatory government and the Malaria Survey Section of the Rockefeller Foundation harder to discern. (Swamp and pool elimination projects outside the jurisdiction of these agencies also included leaders of Jewish agricultural settlements and Palestinian Arabs.) Further, the agencies were scientific research centers, institutions vital to the nation-building effort. The call to establish research centers was originally made in a proposal for the foundation of the Hygiene Research Institute. The proposal described Palestine as a *terra ignota* (unknown or unexplored land), a land of "half culture" in need of "research observatory stations with European science"; the centers were necessary for Jews as part of building a national identity.[28] The proposal described Palestine as unknown, unexplored, and inhabited by uncivilized beings without a history, a land ready for "discovery" through scientific, "rational" observation.

Medical research was supplemented by agricultural research conducted by other Zionist scientific agencies such as the Agricultural Experiment Station, the Hebrew University, and the Daniel Sieff Research Institute. The number of Zionist malaria organizations and agricultural centers during the Mandate facilitated the proliferation of scientific research about the disease's manifestation in Palestine, which, as David Arnold has observed, became "essential to the emerging character of colonial medical science."[29] Malariologists focused on local conditions in Palestine, a fact that supports Richard Grove's assertion that there was no "monolithic colonial scientific mentality" during this time. Connections between local experts and administrators were often more consistent and regular than those between scientists in the colonies and in the metropole.[30]

Specialization and proliferation of malaria research also depended upon scientific knowledge of the disease. Once the basic etiology of malaria was known, scientists could concentrate on variations and particularities in different settings. Until the 1920s, doctors made most diagnoses of malaria by clinical means without laboratory proof.[31] Physicians' and scientists' increased and consistent use of the microscope for diagnostic purposes during the Mandate period helped further the growth of scientific knowledge.

28. Letter for Hygiene Research Institute (Berlin [no date but most likely 1920s]), CZA L1/69; Penslar, *Zionism and Technocracy*, 62.

29. Arnold, *Colonizing the Body*, 19, 37.

30. Grove, *Ecology, Climate and Empire*, 2.

31. Kohn, Weiss, and Flatau, "History of Malaria," 178.

Conducting experiments on the land and on the mosquito made up the first part of the medicalization of the land. Studies of soil conditions, discharges of rivers, springs, flooding trends, and evaporation measurements formed a portion of each malaria survey ritual. Scientists took these measurements in order to know the details of each respective area in order to design specific plans to eliminate swamps. Such data allowed the engineers to determine what part of the drainage work specifically needed repair and extension and/or what topographical features would enable or impede a certain drainage plan.[32] In addition, drainage schemes had to take into consideration that certain species of *Anopheles* mosquitoes behaved differently in varying environments.[33] According to David Arnold, local modifications made medicine a comparative rather than exclusively European exercise in the colonies.[34] The act of gathering details about the land was particularly significant in Mandate Palestine because, in contrast to other malaria control projects around the world, engineers generally applied all available and affordable antimalaria methods rather than relying upon one particular model.[35]

Malaria Research Unit

The main Zionist institution for antimalaria work during the early Mandate period was the Malaria Research Unit (MRU), founded on September 15, 1922, after a bad year of malaria epidemics. The MRU took over Hadassah's malaria control efforts initiated by Brandeis before the Mandate period. On that day, Hadassah and the Zionist Organization officially transferred its malaria and sanitary organization to the British Health Department of Palestine, while the American Jewish Joint Distribution Committee (JDC, a Jewish umbrella organization based in New York that provided relief work for the benefit of Jewish communities around the world) supplied the funds to establish the unit. Funding from the JDC is not surprising since it had also funded the American Zionist Medical Unit (the precursor to the Hadassah Medical Organization) effort to Palestine

32. Buxton, Carley, and Mieldazis, "Malaria Survey of Beisan, 1923," 12. ISA M1503/1/58.
33. League of Nations Health Organization Malaria Commission, *Reports on the Tour*, 52, 55.
34. Arnold, *Colonizing the Body*, 40.
35. League of Nations Health Organization Malaria Commission, *Reports on the Tour*, 25.

after World War I.[36] The JDC carried two-thirds of the financial burden, while the Mandatory government contributed one-third of the cost of the unit. Leaders of the JDC, especially Bernard Flexner, head of the JDC's health and relief work, conceived of the MRU as expanding the scope of what he thought was much-needed malaria work in Palestine.

In exchange for his guarantee to carry a large portion of the financial burden of the MRU, Flexner took the opportunity to submit a proposal to the Mandatory government to raise a large capital sum that would be devoted to drainage projects on government land that was not given out to concession. He proposed the establishment of a nonprofit company with a board of directors consisting of representatives of donors and the Mandatory government. The fund would serve as a revolving fund until the fulfillment of its purpose. The company would obtain a concession of *jiftlik* marshland and drain it. After the area had been drained, the government commission would fix a minimal value and offer the land for sale or lease in "accordance with government policy to non-profit-making companies for agricultural development, in the first instance to Jewish agencies established for the public good." Flexner's proposal was an indirect way to use drainage as a means to gain more government land at low cost for the Zionist project, and it addressed the fulfillment of article 6 of the Mandate charter. Judah Magnes and Israel Kligler feared the proposal, stating that development of Arab lands through drainage before purchase would actually be harmful to "Jewish progress," because the price of the land would either become too high or it would never be sold.[37] The Mandatory government apparently never accepted the proposal.

Despite Flexner's failed attempt at gaining additional lands, the birth of the MRU was a significant step in the Zionist history of medicine of Palestine. The establishment of the MRU, with its JDC sponsorship, illustrated the status of the United States as a quasi-metropole for the *Yishuv* in terms of financial investment, scientific training, and technological importation for malaria and health endeavors.[38] The medical exchange

---

36. Letter from Szold to Briercliffe, September 15, 1922, and July 15, 1922, CZA J113/554.

37. Letter from Kligler to Flexner, June 3, 1924, JDC file 279; "Proceedings of the Ninth Meeting of the Antimalarial Advisory Commission," May 22, 1924, 2, 4, JDC file 280. Bernard Flexner was the brother of Abraham Flexner of the AMA Flexner Report for Medical Education (1910), and of Simon Flexner of the Rockefeller Foundation.

38. Fred Cooper and Ann Stoler, "Between Metropole and Colony: Rethinking a Research Agenda," in *Technologies of Empire: Colonial Cultures in a Bourgeois World*, ed. Fred Cooper and Ann Stoler (Berkeley: University of Berkeley Press), 1–56; Roy MacLeod,

between the United States and the *Yishuv* before and during the Mandate period foreshadowed a growing relationship between these two parties. Medical exchange was also a way for non-Zionist American Jews to get involved in Jewish settlement in Palestine, a fact often overlooked in the literature on American-*Yishuv* relations. The scholarly literature has generally focused on links established during and after World War II.[39] The Mandatory government administered the MRU from the British Health Department of Palestine because some Zionist health bureaucrats believed that the "campaign against malaria and other infectious diseases is a government function."[40] In reality, however, prominent Zionist medical figures in the *Yishuv* ran the MRU.

Despite their focus on Jewish settlements, Zionist antimalaria agencies consulted with the Mandatory government in eradicating the disease. This cooperation was possible because of the special political and scientific relationship between the British and Zionists during this period, especially with regard to public works. The British found in the Zionist endeavor an opportunity to build certain types of infrastructure, promote development within Palestine, and use Zionist scientific and medical information without having to dedicate large sums of British money to do so.[41] In fact, the only way that the government could afford to rule the country was to promote the fulfillment of Zionist development aims.[42] Understanding this reality, the Zionists pressed for increased government involvement in development schemes, especially when it came to financial support. After British rule was established, annual reports of the British Health Department of Palestine and special surveys of the MRU and of the Malaria Survey Section of the Rockefeller Foundation provided the main sources of information used by the Zionists and Mandatory government on malaria incidence and control in Palestine.

"On Visiting the 'Moving Metropolis': Reflections on the Architecture of Imperial Science," *Historical Records of Australian Science* 5 (1982): 1–16.

39. For examples, see, Menachem Kaufman, *An Ambiguous Partnership: Non-Zionists and Zionists in America, 1939–1948* (Jerusalem: Magnes Press, 1991); and Peter Grose, *Israel in the Mind of America* (New York: Knopf Press, 1983). For exceptions, see Sufian and Shvarts, "Mission of Mercy"; Michael Brown, *The Israeli-American Connection: Its Roots in the Yishuv, 1914–1945* (Detroit: Wayne State University Press, 1996); Ruth Kark, *American Consuls in the Holy Land, 1832–1914* (Jerusalem: Magnes Press, 1994).

40. Kligler, "Report of the Activities of the Malaria Research Unit" (1922), 1, ISA M1670/ 130/33a.

41. Scott Atran, "The Surrogate Colonization of Palestine, 1917–1939," *American Ethnologist* (1989): 719–744.

42. Levine, "Discourse of Development in Mandate Palestine," 104.

Dr. Israel Kligler, an American who had formerly been employed by the Rockefeller Foundation and served as chief bacteriologist of the Hadassah Medical Organization, was appointed director of the MRU. In addition to his work with the MRU, Kligler served as professor of bacteriology and hygiene at Hebrew University beginning in 1926. Kligler became the chief Zionist specialist on malaria and sanitation during the Mandate period. Other leading figures in the MRU were Shapiro (field medical officer), Lieberman (field engineer), and Weitzman (sanitarian engineer). Most subinspectors were Jews.[43] The Vaad HaBriut (Health Council of the Palestine Zionist Executive) significantly influenced decisions and directions of the MRU; the MRU was treated as a Zionist entity.[44]

The MRU held the unique status of a Jewish scientific research and development agency *within* the Mandatory government.[45] In addition to its Jewish staff, the scope and nature of MRU's work explicitly sought to promote and facilitate Zionist settlement; in effect, it was an instrument of the Zionist movement. Whereas the government concentrated its initial antimalaria efforts on urban malaria, the MRU chiefly dealt with the sanitation of Jewish settlements and "Arab vicinities" that threatened Jewish health.[46] Kligler described the establishment of the MRU:

> We have extended our areas of operation to include adjacent Arab villages. This has given us useful information regarding the relative prevalence of malaria among Jews and Arabs in given areas. But more important still it will undoubtedly facilitate control in the Jewish section and benefit all concerned.[47]

The MRU served as the official liaison of Jewish organizations for government purposes and was charged with supervising compliance with the Antimalaria Ordinance and with other "government regulations in connection with drainage, reclamation, and other antimalarial schemes

43. Memo: Staff of the MRU, January–December 1923, ISA M1670/130/33a/6420.

44. See JA, "Antimalaria and Drainage Work by Jewish Bodies," 16–18.

45. Letter from Kligler to Flexner, March 23, 1923; and extract of letter sent from Bernard Flexner, chair, Committee on Medical Affairs, to Billikopf, Cohn, Fishel, Liebmann, Rosenau, Stein, Wolk, Mrs. Cohn, and Miss Flexner, May 1, 1923, both in JDC Archives file 279.

46. Kligler, "Report of the Activities of the Malaria Research Unit," 1; League of Nations Health Organization Malaria Commission, *Reports on the Tour*, 41.

47. Kligler as quoted in letter from Flexner, chairman, Committee on Medical Affairs, JDC, to various members of that committee, May 1, 1923, JDC Archives, file 279; also same letter from Kligler himself, March 23, 1923, JDC Archives, file 279.

proposed by Jewish organizations in Palestine."[48] The position of the
MRU as a Zionist agency that was a department of the Mandatory government was unique. In many other ways, however, the arrangement reflected the obligation of the British administration set forth in the Balfour Declaration, then integrated into the Palestine Mandate, to facilitate
the development of the Jewish national home while also safeguarding the
civil and religious rights of the indigenous population. Its status was consistent with the particular policy of Herbert Samuel, the high commissioner of Palestine (1920–1925), who promoted and relied upon Jewish
capital investment for Palestine's development.[49] Under this setup (and
consequently consistent with the general organization of health services in
Palestine), the government maintained primary responsibility for malaria
control for the Palestinian Arab population, who had no separate or private malaria agency.

In the late 1920s, the government gradually began to take over antimalaria work that previously had been held by the MRU. This transfer signaled the ultimate disintegration of the MRU as a distinct agency,
yet it did not end Jewish malaria work.[50] The Vaad Leumi, the Jewish
Agency, and the Zionist Commission in London in late 1931 sought to
overturn the final decision to disband the MRU in late 1931. In addition
to worrying about losing their privileged place within antimalaria work in
Palestine, they claimed that the government alone could not manage or
finance rural work as efficiently and productively as the MRU. They were
also concerned that Dr. Shapiro and other subinspectors would be fired.
They pressured the government to retain Dr. Shapiro and the other Jewish inspectors, arguing that Jewish presence was essential for the proper
fulfillment of antimalarial measures.[51] Furthermore, these Zionist agencies pressed for the supervision of Jewish settlement areas by Jewish medical officers.[52] In some cases, they succeeded.

48. Document signed by Kligler and Briercliffe on behalf of Health Department, "Proposals for the Continuation of the Malaria Research Unit" (March 23, 1924), 1 JDC Archives,
file 279.

49. Smith, *Roots of Separatism in Palestine*, 13, 20.

50. Communiqué regarding JDC/government cooperation in sanitary/malaria affairs,
CZA J113/555.

51. Letter from Szold to Brodetsky, November 12, 1931, CZA J1/1676; letter from
Brodetsky to under secretary of state, Colonial Office, December 23, 1931, CZA J1/1676;
"Hamachon lechekirat malaria," *HaAretz*, CZA J1/1676. See also CZA S25/379.

52. Letter from Kligler to Kisch, February 16, 1927, CZA S25/701.

The breakup of the MRU was caused by other factors as well. Controversy about the dissolution of the MRU was probably also a debate about Zionist bureaucratic turf. Limited documentation available on the topic suggests that the MRU's disintegration was partly a result of internal Zionist dynamics in which the Vaad HaBriut tried to gain full control over Jewish sanitary work.[53] The government also attributed the MRU's dissolution to the nature of the agency's work; as malaria incidence decreased, the need for the extensive services the MRU provided also diminished. By 1928, the MRU staff had been reduced to a controller and five subinspectors. By 1930, most of the remaining employees had been absorbed by the government. The economy caused the MRU's final collapse, especially after the O'Donnell Commission recommended cutbacks for the British Health Department of Palestine.[54] By the end of its tenure as an agency within the health department, the MRU had expended 40,000 LP in malaria control for Jewish settlement.[55]

While in operation, the MRU was made up of field and laboratory sections. Scientists in the field section implemented control operations, while those in the laboratory section performed blood tests and general laboratory work relevant to malaria research. In contrast to some medical contexts where laboratory work was never tested in the field, these sections worked in tandem. The arrangement of the MRU resembled the general colonial medical situation in which colonies were places to test scientific findings made in the laboratories of the metropole. It differed from agencies in other colonies, however, in that it had both laboratory and field operations under the same roof in Haifa. The coordination of the MRU's laboratory and field sections and the extent of its scientific work led Professor Swellengrebel, member of the League of Nations Malaria Commission of 1925, to remark that the unit gave antimalaria efforts "more the character of a huge scientific experiment (such as might have been carried out in the year 1900, to test Ross's theory) than of a practical antimalarial campaign founded on an economic basis." The MRU was an example of the kind of work recommended by the League of Nations Malaria Commission. Members of the commission included well-known tropical medicine figures Professor Nocht (director of the Institute of Tropical Diseases, Hamburg), Dr. C. Chagas (director-general of the Health

53. Letter from Kligler to Kisch, February 16, 1927, CZA S25/701.
54. 'Inyanei Briut (Matters of Health), 1931, CZA J1/2607.
55. British Health Department of Palestine, annual report, 1930, 35, ISA M4475/06/1.

FIGURE 5.1. Members of the International Commission on Malaria inspecting a sick child, 1925. Courtesy of the Central Zionist Archives.

Service of Brazil), Dr. L. Raynaud (inspector-general of the Public Health Service of Algeria), Lt. Col. S. R. Christophers (Central Research Institute, Kassauli, India), and Dr. C. M. Wenyon (director of the Welcome Bureau of Scientific Research, London). The commission's report stated that it "is unanimously of the opinion that the study of malaria must be continuously pursued in the laboratory and the field."[56]

The commission's mission was to determine what measures were appropriate for countries where costs to undertake malaria eradication measures were limited. By comparing the MRU's malaria work to the elaborate laboratory work of Ronald Ross (the discoverer of the malaria vector), another member of the commission, Professor Swellengrebel (Institute of Tropical Medicine, Amsterdam) contrasted the vast scientific field work of the MRU with the more modest antimalaria measures done elsewhere. He saw the MRU as an agency situated within the tropical medicine community. Swellengrebel's remarks were made in the context of comments about the MRU's receipt of substantial funding as opposed to the smaller

56. League of Nations Health Organization, Malaria Commission, *Principles and Methods*, 8–10.

funds at the disposal of the government of Palestine. He was therefore hesitant to recommend applying MRU schemes to other places studied in the League of Nations Malaria Commission, like Romania or Bulgaria, given the financial restrictions of those projects.[57]

As a central actor in the medicalization of Palestine, the MRU conducted four main types of scientific activities for the purpose of combating malaria. First, it established control demonstrations where inspectors determined the types and quantity of mosquito species in various regions of the country; experimented with various larvacide, petroleum mixtures, and mosquito repellants; determined the topographical and climatological characteristics of different areas of Palestine; and assessed the socioeconomic/sanitary conditions of the populations that included conducting malaria screening tests.[58] These preliminary studies followed the basic standards for colonial malaria work, including discerning the most prevalent type of malaria in the country, the incidence of disease, the extent of infection, the probable sources of mosquito breeding, entomological surveys, and seasonal prevalence.[59]

The scientists' ultimate purpose for control demonstrations was to calculate the cheapest and most effective method of malaria eradication for Jewish settlements in Palestine. From 1921 to 1923, the MRU performed this work in nine areas in both the Jewish and neighboring Arab areas.[60] After reducing malaria incidence in these areas, the MRU expanded their efforts to all Jewish, rural, malarious areas of Palestine.

In addition to using scientific findings to sustain existing Zionist settlements, the MRU (and its affiliated land agencies) used the results of these control demonstrations and preliminary surveys to evaluate new locales for Jewish colonies. Land agencies did not adopt this practice but slowly conceded to its value. In fact, until the late 1920s, Kligler criticized the KKL for their failure to coordinate settlement operations with preliminary scientific work. Kligler repeatedly argued that such coordination was

57. Swellengrebel performed malaria work in Indonesia and was active in "species sanitation." Report by Professor N. H. Swellengrebel, part 2, League of Nations Health Organization Malaria Commission, *Reports on the Tour*, 40; Bradley, "Specificity and Verticality," 7.

58. Kligler, "Report of the Activities of the Malaria Research Unit" 2. For similar work in British India, see Arnold, *Colonizing the Body*, 34.

59. HMG, *Campaign against Malaria in Palestine*, 2–3; "Some Problems Awaiting Solution," 191–193, ISA M4390/01/3/332.

60. Settlements included in the Control Area, attachment to Kligler, "Report on the Activities of the Malaria Research Unit" (1922), ISA M1670/130/33a.

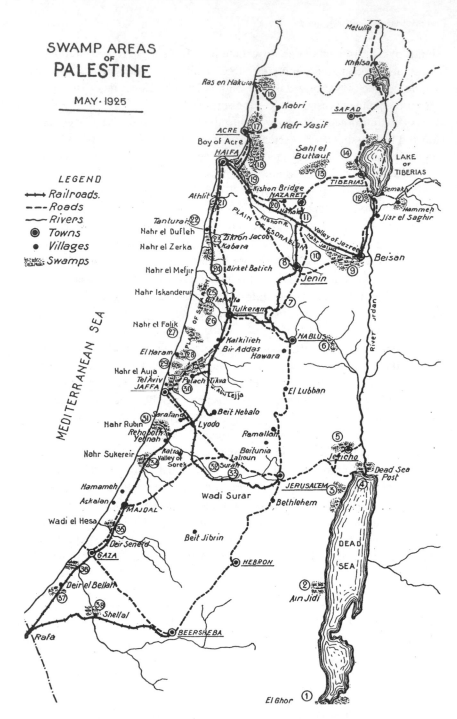

FIGURE 5.2. "Swamp Areas of Palestine, May, 1925." From League of Nations Health Organization, Malaria Commission, *Reports on the Tour of Investigation in Palestine in 1925* (Geneva, 1925), 20–21.

the only way to develop self-supporting Jewish settlements, a key goal in labor Zionism.[61] In May 1924, for instance, Kligler wrote to Colonel Kisch of the Palestine Zionist Executive about his frustration:

> For the last two or three years I have been urging the Colonization Department that they notify us at least a few months in advance with new settlements they contemplate so that the preparatory study regarding malaria conditions, etc., might be undertaken and the necessary measures adopted prior to the arrival of new settlers ... Last year ... a group was allowed to settle on Yazur land without our knowledge ... Now the engineers have rendered their report. From it [it] is clear that drainage this year is out of the question because it will take at least a year to accumulate the necessary hydrographic and other data ... The absurdity of this situation ought to demonstrate the desirability of studying a new place before any steps are taken to colonize it.[62]

Kligler felt that scientific surveying of the land before settlement was so important that he included it as a provision in a set of proposals concerning the continuation of the MRU. One of the provisions included the MRU's responsibility to become involved at the very beginning of a new settlement rather than, as done in 1924, after it had already been founded and people had contracted malaria.[63] He felt strongly about putting the provision into the MRU proposal because he believed that a lack of preliminary surveying would create unnecessary damage to the settlers' health and would cause a loss of effective manpower. Significant economic loss would arouse disfavor among the Joint Distribution Committee, the American Jewish financier of MRU projects. Kligler and the JDC saw the KKL's quick settlement policy as wasting resources. They stressed efficiency in settlement and drainage interventions, efficiency that could be gained from the use of precise scientific knowledge. The economic depression of the *Yishuv* in the 1920s and into the 1930s also forced an emphasis on efficiency and businesslike methods. Larger colonization trends of the time also used the language and practice of efficiency.[64] Colonel Kisch of the Palestine Zionist Executive replied to Kligler's claims: "It

---

61. Confidential letter from Kligler to Kisch, June 29, 1924, CZA S25/740.

62. Letter from Kligler to Kisch, May 27, 1924, CZA S25/740; "The Malaria Situation in 1927" (April 3, 1927), 27, CZA J1/1631.

63. "Proposals for the Continuation of the Malaria Research Unit," 2.

64. Penslar, *Zionism and Technocracy*, 154.

may be that Dr. Kligler does not understand that it will sometimes be necessary for us to acquire land which is affected by malarial conditions. It would, however, appear that in cases where this is inevitable, steps should be taken to free the land of malaria before colonization is undertaken."[65]

Even earlier, Zionist medical professionals had raised the issue of settling before malaria control had taken place. Before colonization of Nahalal, for instance, Zionist officials asked Hillel Yofe to determine the possibilities of habitation with regard to health. Calling the area "very bad," Yofe advised Zionist land agencies and settlers not to settle in the area before drainage could take place. He recommended that settlers stay temporarily on the hills nearby and go to the area daily to arrange their future settlement. The settlers did not take heed of Yofe's advice and settled there anyway.[66]

Kligler constantly battled with settlement officials (especially with those of the KKL) about this change in colonization policy. At the beginning of the Mandate, Kligler harshly criticized the KKL for ignoring his public health advice and failing to enforce the policy of land reclamation before settlement in a strict manner. In his *Epidemiology and Control of Malaria in Palestine* (1930), Kligler noted that with the progress in eradication of malaria, he hoped that colonization agencies would "come to the further wisdom of realizing that the land must be prepared before, and not after, the settlers arrive."[67]

It seems that the KKL understood Kligler's argument, at least in theory. Citing *geulat ha'aretz* (redeeming the land) as the primary goal of the KKL, the agency enumerated health concerns as a precondition in its executive's minutes. This recognition reflected an agreement the KKL had concluded with Kligler and the Vaad HaBriut in 1927 to clear all sanitary criterion before buying or settling the land. In this agreement, supplying water, swamp drainage, terracing, and provision of roads all came under "health concerns" that had to be resolved before settlement could be officially inaugurated.[68] The KKL declared in the 1927 minutes of its head council: "The settlements that were founded in places where causes of disease and death [that] were controlled beforehand, changed for sure, to

65. Kisch to members of the Palestine Zionist Executive, November 4, 1924; copy sent to Van Vriesland, Pick, Ruppin, Sprinzak, and Kaplansky, CZA S25/740.

66. Shmuel Din, "Yibush habitzot beNahalal," 93, CZA Yellow Folder, Swamp Drainage.

67. Kligler, *Epidemiology and Control of Malaria in Palestine*, x.

68. "Metoch din veheshbon shel hanhala hamircazit leKKL" [From the Central Executive minutes of the KKL], Jerusalem, 1922–1927, I, CZA A246/441.

the most healthy places in Eretz Israel." In a memorandum to the Palestine Royal Commission, the KKL recognized its policy shift in settlements where reclamation efforts were implemented "first along with and subsequently even before settlement."[69]

By 1945, the KKL wanted to publicize its change of heart. The agency highlighted malaria eradication in its propaganda films, photos, and publicity material. In a letter that expressed the KKL's displeasure at not being given credit for its role in the transformation of Palestine, the agency tried to bring attention to antimalaria activity as an exceptionally important aspect of its work. The KKL wrote to Abraham Katznelson of the Antimalaria Committee of the Palestine Zionist Executive:

> Reclamation (*havra'a*) work of the land that is handed over for new settlement and antimalaria work is one of the basic branches of work of the KKL during all the years of new settlement in Palestine, and the KKL is interested that the public appropriately knows of these activities.[70]

In a letter to the *Palestine Post*, Abraham Katznelson tried to assuage the KKL's discontent by recognizing its role in making the "most significant contribution to the antimalaria campaign in Palestine" and how it "continues to carry out antimalaria work and maintains a highly experienced staff of inspectors and other antimalaria workers for this purpose."[71] Yet even with the eventual adoption of Kligler's principle, many factors complicated colonization and successful malaria control among immigrants: poor housing facilities, lack of mosquito netting, poor eating habits, and confusion over the importance of consistent quinine prophylaxis (thereby obstructing some antimalaria measures).[72]

The struggle between Kligler and the KKL over surveying the land before settlement reflected the practical urgency for immediate Zionist settlement and indicated the gradual, increasing importance attributed to specialized medical knowledge in the conquest of the land. It also points to issues surrounding professional boundaries and tensions between the

---

69. "Din veheshbon meat halishka harishit shel KKL, 1922–1927" [Minutes by the Head Council of the KKL], 1, CZA A246/441; JA, "Antimalaria and Drainage Work by Jewish Bodies," 13.

70. Memo from KKL to Dr. Katznelson of Antimalaria Committee, July 3, 1945, CZA J1/2284.

71. Letter from Katznelson to editor of *Palestine Post*, July 8, 1945, CZA J1/2287.

72. Ruppin, "Sanitation of Palestine," 2, CZA.

agendas of scientists and other agencies. Kligler's frustration revealed the strong belief among Zionist medical professionals in scientific and medical analysis of the land as the only way to "rationally" prepare it for Jewish settlement. For the malariologists of Palestine, a medical rationale made antimalaria measures "rational" tools in effecting that change. By February, 1928, Kligler's rationale won out. Land agencies officially coordinated their settlement policies and plans with antimalaria agencies. From then on, nationalization followed the medicalization of the land.[73]

The next stage of medicalizing the land was the actual execution of several swamp drainage projects. For the most part, the MRU carried out these projects with the financial help and labor of Jewish land-buying and settlement companies. MRU officials suggested methods of control and implementation procedures. They also drew up engineering schemes for both major and minor drainage projects.[74] In most instances, the MRU carried out more substantial and costly operations, allowing Jewish colonies and Arab villagers to carry out minor schemes.[75] For those development companies or new colonies interested in drainage but without engineering personnel, the MRU would survey the land, prepare the scheme, supply estimates of cost, and supervise the work with the help of government medical and district officers.[76] They dried man-made swamps and stagnant pools around water-powered mills and supply channels. Significantly, Arabs and Jews worked together in drying manmade swamps and stagnant pools. Most medical and district officers were Arab and were employed by the government.[77]

In addition to fulfilling tasks, the MRU served as a scientific research organization that conducted experiments on mosquitoes and people to determine the etiology and epidemiology of malaria and the infectivity,

73. "Recommendations for the Regulation of Relations between the Colonization Agencies of the Zionist Organization and the Secretariat for Medical Affairs of the Palestine Zionist Executive," submitted by the subcommittee to the Vaad HaBriut and adapted by it at its meeting, February 8, 1928, CZA J1/7734; "Report of the Committee for Establishing Relations among Settlement Agencies and the Secretariat for Health Affairs of the PZEs" (Hebrew), February 9, 1928, CZA S25/6675; Shafir, *Land, Labor and the Origins of the Israeli-Palestinian Conflict*, 192.

74. "Proceedings of the Seventh Meeting of the Antimalaria Advisory Commission," 15, 17.

75. MRU, "Malaria Control Demonstrations," 22–23.

76. "Proposals for the Continuation of the Malaria Research Unit," 2.

77. MRU, "Malaria Control Demonstrations," 23.

prevalence, and breeding habits of certain *Anopheles* species in Palestine. Scientists all over the colonized world pursued these same directions of research.[78] The MRU studied the effectivity of various control methods as well as different types of clinical treatment, including the therapeutic value of quinine derivatives, particularly for resistant benign tertian malaria.[79]

Central Malaria Council

With the disintegration of the MRU, Jewish industrial institutions (like Nesher, the Dead Sea works, the Palestine Electric Corporation works, etc.) integrated individual Jewish medical officers in order to help with antimalaria work when they needed it. The KKL hired malaria supervisors in their reclamation projects in the Emek Zevulun and Emek Hefer areas. With time, Jewish health and land reclamation institutions—Kupat Holim, KKL, Vaad Leumi, Hadassah, and Hebrew University—formed the Central Malaria Council. The council's mission was to coordinate all Jewish antimalaria activities.[80] The Antimalaria Control Service, the Central Malaria Control Committee, and the Joint Hygiene Service attempted to recentralize in the 1940s, after the Arab Revolt. Dr. Katznelson of the Palestine Zionist Executive and Dr. Yassky of Hadassah Medical Organization saw centralization as a way to "obviate differences of opinion regarding the organizational setup of malaria control activity" among Zionist antimalaria activities.[81] Besides coordinating Zionist malaria eradication efforts, the Central Malaria Council acted as a liaison between the various antimalaria agencies of the *Yishuv* and government malaria control services. Drs. Katznelson, Meir, and Yassky made up the Central Malaria Control Committee, while Dr. Saliternik took leadership in April 1945.

78. Bynum, "'Reasons for Contentment,'" 22.

79. Kligler, "Report on the Activities of the Malaria Research Unit," 4; Saliternik, "Hamalaria veshlavei," 356–365, Ein Kerem Medical Library, WC765 JP2 40.

80. Saliternik, "Hamalaria veshlavei," 360.

81. Memorandum on meeting between Dr. A. Katznelson and Dr. Yassky regarding antimalaria control, April 17, 1940, Hadassah Archives, RG1/100/2; Antimalaria Campaign, Hadassah Medical Organization Malaria Control to Dr. Yassky, June 1, 1945, Hadassah Archives, RG 1/100/2; Central Agency for Public Hygiene Services of Department of Health of Vaad Leumi, Hadassah Medical Organization and Kupat Holim, Central Antimalaria Council, May 30, 1945, CZA J1/2284; Activities of Department of Health of Vaad Leumi in emergency period, 6, CZA J/7718.

Rosh Pina Malaria Research Station

By the time the MRU had completely disbanded, the Rosh Pina Malaria Research Station had been established by Hebrew University (1927) under the direction of Dr. Gideon Mer. The station soon became internationally recognized for its malaria research conducted mostly in the Huleh area fourteen kilometers away.

In *A Trip through the Upper Galilee*, Dorothy Bar-Adon described the Rosh Pina laboratory, which was also Mer's home. Rows of clay bowls containing mosquito eggs for breeding and research lined the laboratory. Dr. Mer drove his "portable clinic," a gray Ford, during his inspection rounds, while looking at his gigantic diagram of a mosquito, trying to "keep an eye on the swamps."[82] Mer was most likely trying to identify which particular species of *Anopheles* inhabited the area so he could decide what specific methods would prevent its breeding. This technique, called "species sanitation," privileged the specificity of the local over a general theoretical model of malaria eradication. Malariologists worldwide emphasized the discernment of local conditions as the best way to control malaria. The League of Nations Malaria Commission, for instance, came to this conclusion:

> In every country and very largely in every area, there must be preliminary examination to ascertain what method is best suited to the local conditions. At present, it cannot be said that for malaria control there is a method of choice superior to all others.[83]

Specificity was central to the medicalization of the land and the nationalist meaning of the malaria project for Zionist malariologists because it supposedly allowed malariologists (and settlers, once they were educated about these characteristics) to attach to the land and facilitated nationalist transformation. To fulfill *yedi'at hakarka* (knowing the land) and *yedi'at ha'oyev* (knowing the enemy), the station's researchers (including Mer, Reitler, Saliternik, Yekutiel, and Yoeli, among others) studied a variety of scientific questions. They studied the endemic nature of malaria in the

---

82. Dorothy Bar-Adon, *A Trip through the Upper Galilee* (Tel Aviv: Lion the Printer, 1941/42), 6–22.

83. League of Nations Health Organization Malaria Commission, *Principles and Methods*, 17.

Huleh and upper Galilee and examined the bionomics (flight traits) of the *Anopheles* in the Huleh. They also conducted a study on the development cycle of one malarial parasite, researched the pathology of blackwater fever, and organized inquiries into hyperendemic regions that lacked antimalaria efforts. These scientists also experimented with DDT spraying as a method for malaria control.[84] During his tenure as director of the Rosh Pina Malaria Research Station (1929–1961) and in his enthusiastic pursuit of scientific knowledge, Dr. Mer even conducted a study on the experimental transmission of quartan malaria by infecting his pregnant wife and himself in order to see if malaria could be transmitted to the embryo.[85] He soon found out that it could. Mer also developed a system of determining the age of the *Anopheles*. Ahmed Aloub, Mer's assistant, helped with the dissection and identification of mosquitoes.[86]

In addition to experimenting on the mosquito, the station's staff performed medical experiments on residents in the area to determine the effects of different pharmaceutical drugs and to document changes in the nervous systems and eyesight of malaria patients.[87] In one experiment, the station's staff exposed healthy nonparasite-carrying volunteers to mosquitoes carrying quartan malaria in order to add to their knowledge about quartana parasites and their transmission.[88] Through such experimentation, the human body and the mosquito became part of the project to medically document the land and the body in Palestine.

Malaria Survey Section

Another private malaria agency attached to the Mandatory government was the Malaria Survey Section (MSS). Initiated in 1922, the Rockefeller

84. Saliternik, "Hamalaria veshlavei," 359; C. L. Greenblatt, "Historical Trends in the Anti-Malarial Campaign in Palestine," *Israel Journal of Medical Sciences* 14, no. 56 (May 1978): 508–517, quotation on 513.

85. Experimenting on oneself as a medical researcher was relatively common. Davidovitch notes that there are several examples of researchers who infected themselves with relapsing fever and leishmaniasis during this time. Personal correspondence with Nadav Davidovitch, September 2006.

86. Rivka Ashbel, *As Much As We Could Do: The Contributions Made by the Hebrew University of Jerusalem and Jewish Doctors and Scientists from Palestine during and after the Second World War* (Jerusalem: Magnes Press, 1989), 75, 81, 87.

87. Ashbel, *As Much As We Could Do*, 83.

88. Ashbel, *As Much As We Could Do*, 85. Ashbel relates these malaria experiments uncritically, celebrating the scientific positivism of the doctor's account.

Foundation funded the MSS, whose express goal was to undertake anti-malaria work during the Mandate. This initiative was one of many public health schemes that the Rockefeller Foundation supported around the world at this time. Beginning in the 1920s, the Rockefeller Foundation became involved in malaria campaigns in Asia, Latin America, and the Middle East.[89] The MRU, the MSS, and the Mandatory government collaborated on antimalaria projects. In contrast to the MRU, however, the Malaria Survey Section examined malarial incidence and prepared schemes for Palestine's population at large. By following the general orientation of the Rockefeller Foundation's health initiatives around the world, the MSS emphasized "scientific medicine" (linking a disease to a microbe) over reducing socioeconomic inequities as a way to reduce mortality.[90] The agency therefore performed tasks similar to those of the MRU: it helped devise malaria control operations, including both major and minor irrigation and drainage schemes; determined the prevalence of malaria in Palestine; supervised drainage projects; made field surveys of areas ridden with mosquito breeding; carried out flight-distance experiments on mosquitoes; and constructed maps and plans for antimalaria measures.[91] Dr. P. S. Carley served as the MSS controller, and J. J. Mieldazis was its engineer.

89. Paul Weindling, "Social Medicine at the League of Nations," in *International Health Organizations and Movements, 1918–1939*, ed. Paul Weindling (Cambridge: Cambridge University Press, 1995), 135–136; Watts, *Epidemics and History*, 272; also *Medical Anthropology* (special volume on malaria and development, guest editor, Randall Packard) 17, no. 3 (1997); E. Richard Brown, "Public Health in Imperialism: Early Rockefeller Programs at Home and Abroad," in *Cultural Crisis of Modern Medicine*, ed. John Ehrenreich (New York: Monthly Review Press), 252–270, especially 258–259, 266.

90. A. E. Birn, "Eradication, Control or Neither? Hookworm vs. Malaria Strategies and Rockefeller Public Health in Mexico," *Parassitologia* 40, nos. 1–2 (June 1998): 139; Nancy Gallagher, *Egypt's Other Wars: Epidemics and the Politics of Public Health* (Syracuse: Syracuse University Press, 1990), 11–28.

91. See MSS work in Haifa and Acre, Beisan, Jericho, Sanour, Dead Sea, Huleh, and Mejdal, "Metoch din veheshbon shel halishka hareshit shel KKL el hacongress hazioni ha-17 be Basel" [From minutes of the KKL Head Council to the Seventeenth Zionist Congress in Basel], 1929–1931 CZA A246/441; "Proceedings of the Seventh Meeting of the Antimalarial Advisory Commission," 2; letter from Heron to general director, Rockefeller Foundation, and attached annual report, 1923, Malaria Survey Section, April 28, 1924, ISA M1503/1/58 (54); "Malaria in Samaria Village; Tantura."; "Survey of Sanour Swamp"; "Investigation of Supposed Roman Sump Hole Drainage at Sanour"; "Malaria in Beisan Gendarmerie Camp," all of ISA M1503/1/58; "Medical Research in Colonies, Protectorates and Mandated Territories, 1930," 104, ISA M4390/1/3/332.

Antimalarial Advisory Commission

The Antimalarial Advisory Commission served as the umbrella committee for all malaria agencies in Palestine. High Commissioner Herbert Samuel founded the commission in 1920. Colonel Heron, the director of the British Health Department of Palestine, presided over the commission, which consisted of official and unofficial members W. Tyrell, wing commander and medical officer of Palestine Command; Dr. Hillel Yofe; J. F. Rowlands, director of public works; A. F. Nathan, director of agriculture; R. Briercliffe, deputy director of health; Dr. Tucktuck, medical officer of Nablus; Flight Lt. J. A. Musgrave, command sanitary officer; Dr. Carley of the MSS; J. J. Mieldazis, engineer for the MSS; W. K. Bigger, medical officer of Samaria district; L. Cantor, sanitary engineer for the government; Dr. Kligler, controller of the MRU; Dr. Shapiro, field medical officer of the MRU; and M. Lieberman, field engineer of the MRU. The commission provided a forum on the malaria problem in the country.[92] It met every three months in order to review work completed and discuss and advise on new projects in all sectors of Palestine.

Zionist Malariology as Colonial Medicine

Architects of antimalaria campaigns in Palestine used colonial "investigative modalities," the sciences of medicine, parasitology, geography, and cartography to reshape Palestine's landscape.[93] By documenting the land and its conditions, malaria and drainage surveys transformed scientific knowledge into "textual forms" that were then used by the Zionists and the British to settle and rule Palestine. The drive underlying the enormous accumulation and analysis of data was to gain a precise understanding of Palestine's material world, its ecology and inhabitants, in order to combat malaria. Malaria surveys and the statistics within them were one component of an enormous collection of medical statistics produced and used by both the Mandatory government and the Vaad HaBriut. Medical statistics are found in every health department report. The Vaad HaBriut had a bureau of social hygiene and medical statistics.[94]

92. "Proceedings of the Eighth Meeting of the Antimalarial Advisory Commission," 1.
93. Goubert, *Conquest of Water*, 109; Cohn, *Colonialism and Its Forms of Knowledge*, 5–8.
94. Letter from Dr. Ratner to Vaad HaBriut, June 24, 1927, CZA J1/3005.

Zionist malariologists and doctors fit into the colonial medical world not only by virtue of their techniques (gathering specimens, drawing maps, and conducting surveys), but also in their training and dialogue with other colonial malariologists.[95] Engineers working on Zionist drainage activities also worked in other areas of the world. Jacob Zwanger, an engineer who performed similar work in Argentina, supervised the Jezreel Valley project in Nahalal and Nuris. Zwanger's participation exhibited the extent to which Zionist drainage activities were part of an international effort toward malaria eradication[96] that took place within the context of scientific exchanges with international health organizations during the interwar period, when these were also occurring in forestry, medicine, and other disciplines.[97] As David Lowenthal has observed, scientists and colonial officials promoted the association of learned societies as part of a "coordinative" phase of colonial science emerging in the dominions of the British Empire.

Many prominent Zionist malariologists received their training in leading schools for malariology and tropical medicine. For example, both Hillel Yofe and Gideon Mer, who along with Israel Kligler and Zvi Saliternik were considered among the chief Zionist malariologists of Palestine during the Mandate period, received their training in Italy with renowned malariologist Giovanni Batista Grassi. Grassi was famous for illustrating the life cycle of the malaria parasite (1898).[98] Mer eventually completed postgraduate work in tropical medicine in Amsterdam and Paris.[99] Yaakov Yofe and Fritz Yekutiel, other known malaria workers in the later decades of the Mandate, attained their degrees in tropical diseases from Liverpool University.

95. Douglas Melvin Haynes, "Social Status and Imperial Service: Tropical Medicine and the British Medical Profession in the Nineteenth Century," in Arnold, *Warm Climates*, 214–215.

96. Letter to KKL from unknown, May 29, 1922, 4; letter from KKL to Jacob Zwanger, May 26, 1922, CZA KKL3/57b.

97. Weindling, "Introduction: Constructing International Health between the Wars," in *International Health Organizations and Movements, 1918–1939*, ed. Paul Weindling (Cambridge: Cambridge University Press, 1995)..." 3; MacLeod, "On Visiting the 'Moving Metropolis," 11; Lowenthal, "Empires and Ecologies," 229.

98. Greenblatt, "Historical Trends," 510; Ashbel, *As Much As We Could Do*, 74; Roy Porter, *The Greatest Benefit to Mankind: A Medical History of Humanity* (New York: W. W. Norton & Company, 1997), 470. Aviva Zuckerman was also an important malariologist but she did not come to Palestine until the late 1940s, at the end of the Mandate. I therefore do not discuss her work here. Greenblatt, "Historical Trends," 514–515.

99. Greenblatt, "Historical Trends," 512.

For their part, Palestinian Arab doctors were generally not specialists but did work in malaria treatment and prevention. Dr. Canaan worked before the Mandate in malaria control, among other things. Palestinian Arab medical officers and antimalaria subinspectors working for the Mandatory government on malaria control were not, on the whole, formally trained as malariologists. Two Palestinian Arab doctors, Krikorian and Dabbagh, worked in the British Health Department of Palestine laboratories on malaria and other infectious diseases. An associate of these two doctors, Dr. Berberian, was part of the Department of Parasitology at the American University in Beirut. Berberian published a manual of *Anopheline* mosquitoes in Syria and Lebanon in 1944 to be used in a malariology course for government doctors. He also wrote several articles on malaria for the journal of the Palestine Arab Medical Association. One article, published in July 1946, was "The Species of Anopheline Mosquitoes found in Syria and Lebanon: Their Habits, Distribution and Eradication."[100] In this article, Berberian referred to the work of Kligler, Mer, and Saliternik. Indeed, exchange between Arab doctors in Egypt, Syria, Lebanon, and Palestine was frequent. This occurred as well between Jewish doctors and those in the Arab world in the early Mandate period.

Zionist malariologists contributed to the global literature on malaria at this time.[101] Kligler, Saliternik, Mer, Shapiro, and Weitzman, for instance, published their scientific research in esteemed journals like the *Royal Society for Tropical Medicine and Hygiene*, the *American Journal of Tropical Medicine*, *American Journal of Hygiene*, and the *Journal of Preventive Medicine*. Gideon Mer shared his work with Russian scientists. Dr. Yofe published a total of thirty-seven articles, most on malaria.[102]

---

100. Dr. Berberian, "The Species of Anopheline Mosquitoes found in Syria and Lebanon: Their Habits, Distribution and Eradication" (written in English), *Al-Majala al-Tabiya al-'Arabiya al-Filastiniya* [Journal of the Palestine Arab Medical Association], 1, no. 5 (July 1946): 120–146.

101. "Types of Malaria Transmitting Anopheles in Rural Palestine," 2, ISA M1670/130/33a; Sandy Sufian, "Colonial Malariology, Medical Borders and the Sharing of Scientific Knowledge in Mandatory Palestine," *Science in Context* (Fall 2006): 381–400.

102. I. Kligler, "Quinine Prophylaxis and Latent Malaria Infection," *Trans. Royal Society of Tropical Medicine and Hygiene* 27 (1923): 259, in MRU annual report, 1923, 24, ISA M1670/130/33a/6420; I. Kligler, "Malaria Control Demonstrations in Palestine," *American Journal of Tropical Medicine* 4 (1924): 139–174; idem, "The Control of Malaria in Palestine by Anti-anopheline Measures," *Journal of Preventive Medicine* 1 (1927): 149–183, idem, Shapiro, and Weitzman, "Malaria in Rural Settlements in Palestine," 280–316, idem,

Their research, books, articles pamphlets, and maps produced substantial medical literature on the subject of malaria that was used both within
the medical community of Palestine and read by international scientists
in malaria control.

In this literature on malaria, Zionist medical professionals often referred to or compared their experiments and conclusions with other studies in their own annual reports. They cited the work of Celli, Suvas, C. M.
Wenyon (director of the Welcome Bureau of Scientific Research, London), LePrince, or Lewis Hackett (director of Rockefeller Foundation
work in Italy and epidemiologist of Europe). They discussed the results of
work conducted in territories such as Macedonia, Sicily, Jamaica, Greece,
and Panama.[103] They reprinted articles about malaria research elsewhere
in Hebrew medical journals.[104] Through these publications, Zionist malariologists engaged in colonial medical debates on malaria, like the one
about the effectiveness of prophylactic quinine, a debate that had taken
place even before Kligler's arrival to Palestine between Hillel Yofe
and other doctors of the Hebrew Medical Society in 1912.[105] Other debates that Kligler and his colleagues addressed included the seasonal

"Flight of Anopheles Mosquitoes," *Trans. Royal Society of Tropical Medicine and Hygiene*
18 (1924): 199–206, idem, "The Epidemiology of Malaria in Palestine: A Contribution to the
Epidemiology of Malaria," *American Journal of Hygiene* 6 (1926): 431–449; I. Kligler and
R. Reitler, "Studies on Malaria in an Uncontrolled Hyperendemic Area (Hule, Palestine). I.
The Seasonal Prevalence, Distribution and Intensity of Malaria in an Untreated Indigenous
Population," *Journal of Preventive Medicine* 2 (1928): 415–432; G. Mer, "Notes on the Bio-
nomics of Anopheles Elutus," *Bulletin of Entomological Research* 22 (1931): 137–145, idem,
"The Determination of the Age of Anopheles by Differences in the Size of the Common
Oviduct," *Bulletin of Entomological Research* 23 (1932): 563–566; Shapiro and Saliternik,
Malaria beEretz Israel: Hasefer shimushi [Malaria in Palestine: A practical book] (Jerusalem:
Hasefer Press, 1930). Saliternik's other work is published after the Mandate. Nissim Levy,
"Zichron Yaakov," 93; Ashbel, *As Much As We Could Do*, 84–85.

103. MRU annual report, 1923, 18, ISA M1670/130/33a/6420; "Proceedings of the Seventh
Meeting of the Antimalarial Advisory Commission," 5, 7–8; Hillel Yofe, "The Problem of
Quinine Prophylaxis," *HaRefuah* 1, nos. 4–5 (April 1925): 1–2, CZA Library; "Proceedings
of the Ninth Meeting of the Antimalarial Advisory Commission," 7.

104. Dr. H. Necheles, "A Contribution to the Physiology of Anopheles," Hamburg University. *HaRefuah* 1, nos. 4–5 (April 1925): 2–3, CZA Library.

105. Hadassah Newsletter, section on Malaria Research, 5, no. 6 (April 1925): 3, Ed.
Ms. Edward Jacobs, CZA number 1270; "Hoda'ah leaguda hamedicinit ha'ivrit she beYafo"
Zichronot HaDavarim 2–3 (May 1913): 54. 1913, CZA Library; "Vichuchim [on Puchovsky's]
'Hashe'lah al davar heripui shel kadachat," twelfth scholarly meeting, November 31, 1912,
Zichronot HaDavarim 2–3 (May 1913): 92–93, CZA Library; letter from Heron to Dr. Cohn,
Malaria Research Unit, February 5, 1923, JDC Archives, file 279; letter from Kisch to Salkind,
acting director of HMO, June 7, 1923, CZA S25/584; "Proceedings of Antimalarial Advisory
Commission," May 24, 1923, 6, JDC Archives, file 280.

occurrence of malaria and its parasites and the utility of blood and spleen exams for measuring prevalence.[106] In the 1940s, Kligler and Abraham Katznelson of the Vaad HaBriut debated the usefulness of DDT in malaria eradication. They also argued over the use of atabrine, a new drug considered more effective than quinine. HMO helped import drugs and sprays from the United States to Palestine on behalf of the Jewish population in the latter part of the Mandate period. In the last decade of the Mandate period, information on the effectiveness and availability of atabrine became highly important because many of the battles of World War II would take place in the tropics. A *Washington Post* article quoting Willam Lavarre, tropical explorer and head of the Commerce Department's American Republic Unit, was sent to the Hadassah Medical Organization in Palestine, from where he wrote, "Unless something is done, and done quick, more men will be killed by malaria than by bullets."[107]

Correspondence between Zionist malariologists and their U.S. counterparts frequently compared malaria in Palestine and America and made use of each other's findings. In 1942, the U.S. Department of Agriculture adopted modifications of Kligler's antimalaria control measures. The department also circulated Kligler's report on malaria control work in 1941 in Palestine "to [U.S.] specialists in this field for their information."[108] Likewise, the *Yishuv*'s leadership regularly considered or adopted U.S. methods of antimalaria control. For example, Hadassah and the Vaad Leumi discussed the idea of airplane spraying of DDT, a method used by the U.S. Public Health Service in the 1940s to combat malaria. They even hired a Jewish pilot for spraying DDT in Palestine,[109] hoping to add "another leaf to the exciting book of making Palestine—the classical land of epidemic and endemic diseases—a completely healthy place to live in." After Israeli statehood was established, the government used DDT sparingly. It never became the method of choice to eradicate malaria.[110]

106. Sandy Sufian, "Colonial Malariology, Medical Borders and the Sharing of Scientific Knowledge in Mandatory Palestine," *Science in Context* (Fall 2006): 381–400.

107. Jack Bearwood, "U.S. Will Seize Nazi Patents to Beat Malaria," *Washington Post*, May 3, 1942, Hadassah Archives, RG1/100/2.

108. Letter from P. N. Annand, chief of bureau, to Mrs. A. P. Schoolman, chairman, Palestine Committee, Women's Zionist Organization of America, August 31, 1942, Hadassah Archives, RG1/100/ 2.

109. Memo on malaria control by the airplane spray method, Hadassah Archives, RG1/100/2.

110. Yassky, director of HMO to schoolman, chairman, Palestine Committee of Hadassah, July 4, 1946, Hadassah Archives, RG1/100/2; Kitron, "Malaria, Agriculture and Development," 302.

Zionist malariologists used their expertise for military purposes.[111] During World War II, Mer, Yekutiel, Yoeli, and Yaakov Yofe were recruited into the British Army. They directed malaria field laboratories in Palestine and other colonial and occupied areas such as Egypt, Transjordan, Iraq, Iran, India, and Burma. By performing experiments on mosquitoes, residents, and soil and by undertaking drainage projects, they aided the success of military maneuvers against the Axis by preventing high malaria fatality rates among British soldiers.[112]

Zionist agencies actively participated in and contributed to colonial malaria tours such as the League of Nations Health Organization Malaria Commission. Besides comparing different malaria problems in various region, League of Nations' tours set out to resolve disputes between the two schools of thought on malaria control, as described by scholars Randall Packard and Peter Brown. The biomedical group, influenced by U.S. and British scientists, saw malaria primarily as a problem of vector control, while the social-uplift group saw the problem of malaria as connected to socioeconomic conditions such as housing and agricultural methods. The League sided with the social-uplift model in recognizing the great limitations to total elimination of the mosquito with contemporary scientific methods.[113]

The League of Nations' authority over Palestine as a mandated territory gave it the opportunity to assess its health status; the commission's visit was part of its larger effort to assess malaria control in Europe between the years 1924 and 1926. The commission's scope spanned from areas of Spain all the way to the Soviet Union.[114] The League of Nations Malaria Commission visited Palestine in May 1925, to assess its antimalaria efforts and suggest methods for improvement. The tour was a

111. Biger, "Ideology and the Landscape of British Palestine, 1918–1929," 194; also Dobson, *Contours of Death and Disease*, 6.

112. Amadouny, "The Campaign against Malaria in Transjordan," 469; Greenblatt, "Historical Trends," 513; "Danger of Malaria Epidemic," *Palestine Post*, June 25, 1945, 4, CZA J1/2284. For information on other Jewish malaria workers in Palestine (Dr. Oscar Theodor, Dr. Saul Adler, Dr. David Birnbaum, Alexander Russell, Aharon Shulav, etc.), see Ashbel, *As Much As We Could Do*, 118–137.

113. Randall Packard and Peter Brown, "Rethinking Health, Development and Malaria: Historicizing a Cultural Model in International Health," *Medical Anthropology*, special volume on Malaria and Development, 17, no. 3 (1997): 184–185.

114. Weindling, "Introduction: Constructing International Health," 2, 8, 11, idem, "Social Medicine at the League of Nations," 135, 142; Packard, "Visions of Postwar Health and Development," 95; League of Nations Health Organization Malaria Commission, *Principles and Methods*, 6.

collective inquiry and comparative evaluation of Palestine, Syria, and Turkish Asia Minor.[115] On their way to Syria from Palestine, some of the members of the commission were involved in a fatal car accident that curtailed the trip, making Palestine the only country visited during the tour.

The Vaad HaBriut prepared a memorandum for the commission on medical and antimalarial work by Jewish institutions and sponsored an honorary dinner for members of the League of Nations Commission.[116] Similarly, in their pamphlet *Amelioration Works in the Valley of Jezreel, Keren Keyemeth L'Israel*, the KKL took the opportunity to submit to the commission detailed information about its reclamation efforts, especially with regard to its prized possession, Nahalal and Nuris.[117]

In his report on the tour in Palestine, Professor Swellengrebel, one of the members of the commission, noted that given the scarcity of successful antimalarial schemes, the Palestine case was useful for legitimizing modern scientific knowledge of the epidemiology of malaria in its practical application. Doctors and engineers in Palestine had used scientific findings of their own to design practical drainage measures. To Swellengrebel, the complementarity of these two spheres—the laboratory and the field— proved the practicality of epidemiology. Coordination of laboratory work and field work was something the commission highly recommended for all countries. Palestine's unique combination of high capital investment and exceptional brain power made the lessons learned and methods used there hard to apply in other malarial areas with lesser means.[118] Nonetheless, Swellengrebel concluded that the experience gleaned from Palestine was an "invaluable addition to practical malariology and the men who

115. League of Nations Health Organization, Malaria Commission, *Reports on the Tour*, 6.

116. Jewish Medical and Antimalaria Work in Palestine, memorandum submitted by the Vaad HaBriut of the Palestine Zionist Executive to the Malaria Commission of the League of Nations, Jerusalem, May 1925, CZA J1/3116; speech by Dr. Salkind at dinner given to Malaria Commission of League of Nations, May 14, 1925, CZA S25/647; report of public health in Palestine (June 17, 1925), 8–9, CZA S25/740; letter from Heron to Kisch, May 6, 1925, CZA S25/647.

117. "Metoch din veheshbon shel hanhala hamirkazit leKKL" [Out of minutes of Central Executive of KKL] (Jerusalem 1924–1925), 1, CZA A246/441; Jewish Medical and Antimalaria Work in Palestine, memorandum submitted by the Vaad HaBriut of the Palestine Zionist Executive to the Malaria Commission of the League of Nations, Jerusalem, May 1925, 18, CZA J1/3116.

118. League of Nations Health Organization Malaria Commission, *Reports on the Tour*, 35, 40, 66.

carried it out can be regarded as benefactors not only to the Palestine population but to the world as a whole."[119]

This favorable assessment of Palestine's malariology and malariologists was exactly what Zionist medical professionals had hoped for. Although Zionist doctors realized that they could not "compete with the great European or American countries," in their work and exchange, they sought to be accepted as legitimate actors and contributors to the greater world of science.[120] Sometimes they were. The *Journal of the American Medical Association* acknowledged Kligler's contribution in an editorial on malaria control demonstrations in Palestine. For the Jewish scientists, this validated the [fledgling] Zionist national movement and enhanced its credibility in the rest of the world. In addition to alleviating Palestine's malaria condition and strengthening the Zionist nationalist project, the MRU saw its mission as contributing to widening antimalaria research worldwide. Even at the establishment of the MRU, the agency wanted its findings to "not only be of help to us but will also perhaps be of use to others engaged in antimalaria work."[121] Noting the international acceptance of Zionist malariology, Kligler wrote,

> The work on the behavior of infected *anopheles* and their long-range dispersion during the fall has thrown light on a number of obscure aspects of malaria epidemics. The international significance of this work is attested by the fact that the [Rosh Pina] station is one of a number designated by the Health Section of the League of Nations for research on these problems.[122]

The Mandatory government also distributed its public health reports (including the MRU malaria surveys within them) to the Department of Health of India, the International Public Hygiene Office in Paris, the directors of the Liverpool School of Tropical Medicine, and the London School of Tropical Medicine. The director of public works in Palestine, for

119. League of Nations Health Organization Malaria Commission, *Reports on the Tour*, 66.

120. Zondek, "Contemporary Medicine and Jewish Physicians," 9. His article was read at the opening of the Scientific Convention of the Hebrew Medical Organization of Palestine; and I. Kligler, "Antimalarial Fight in Palestine Wins Recognition," *Hadassah Newsletter* 5, no. 2 (November 1924): 6, CZA Library.

121. Kligler, "Report of the Activities of the Malaria Research Unit," 3.

122. Kligler, "Fighting Malaria in Palestine" (April 16, 1931), Hadassah Archives, RG 1/100/2; Kligler, "Report of the Activities of the Malaria Research Unit," 11–13.

instance, compared Kligler's findings that questioned the efficacy of quinine prophylaxis with similar findings in Jamaica.[123] The Zionist malaria
project, alone and in its association with British antimalaria efforts, made
its mark in the worldwide attempt to control malaria.

Zionist Particularity

Reconstructing Palestine from an ancient, pathological land "of epidemic
and endemic diseases" to a vibrant, modern, and completely healthy place
was the essence of the Zionist project. Though Zionist antimalaria figures
were contributing to the global project of malaria eradication, their interest was particular to the Zionist national project. References to waves of
Jewish immigration, Jewish settlement conditions, demographic trends,
and Zionist organizations were all made in Zionist literature on malaria.
They noted, for instance, that scientific experiments thought to endanger
the "well-being of our [Jewish] human material" would not be done in
Palestine; instead, findings from other places would be adapted to "serve
the peculiar needs of the Zionist Organization."[124]

The Zionist leadership acknowledged the importance of scientific work
by inscribing Kligler's name in the Golden Book of the KKL. The Golden
Book was a registry of outstanding persons or projects connected with
KKL activities. To register a name in the Golden Book, an application
had to be made, usually with a donation. The decision to include a name
or project in the Golden Book was made by a board of directors. In a letter to Abraham Katznelson, Granovsky recognized Kligler's contribution
and the importance of antimalaria work for the success of sustaining existing Zionist settlements and establishing new ones: "I hope that Professor
Kligler will continue to be our loyal consultant in questions of malaria
in general and especially in regions of new settlement."[125] Dr. Meir sent
a letter to Kligler congratulating him on this honor. He thanked Kligler
for his efforts to achieve *havra'a shlema shel ha'aretz*, full healing of the
land.[126] Upon his death, Kligler's malaria work in Palestine was compared
to that of other prominent malaria figures: "His achievement in conquer-

---

123. See 1925 Public Health Department report considered by Colonial Advisory Medical
and Sanitary Committee on January 2, 1927, PRO CO 733/130/5.

124. "The Malaria Situation in 1927," 29.

125. The Golden Book has now become a historical record. Letter from Granovsky to
Katznelson, July 30, 1942, CZA J1/1731.

126. Letter from Meir to Kligler, July 31, 1942, CZA J1/1731.

ing the mosquito in Palestine and removing the scourge of malaria has been likened to the work of General Gorgas in Panama."[127] The Antimalaria Committee of the Vaad Leumi planned to establish a health center in his honor.[128]

Zionist malariologists believed their work expressed their nationalist commitment. This devotion distinguished them from many other colonial scientists of the time, mostly because the majority of scientists in other settings had imperial interests at heart, not immigrant civilians setting out to achieve political and social renewal.[129] Both Hillel Yofe and Gideon Mer came to Palestine with a firm belief in *kibush ha'avoda* (transforming oneself through physical labor) and integrated that ideal into their malaria work.[130] Before immigrating to Palestine in 1891, Yofe had belonged to the Russian populist revolutionary movement. In Palestine, he was the head of the Hovevei Zion's Office in Jaffa, and in 1908, he argued for the establishment of a land leasing company to settle workers.[131] In 1912, in a lecture to the Hebrew Medical Society, Yofe asserted that Jewish doctors needed to research and study malaria and its local specificity in order to propagate knowledge of malaria protection and endemic areas to the country's Jewish inhabitants. He declared that it was the duty of all physicians to counsel "the nation" by using popular pamphlets as well as to recruit the local intelligentsia, especially physicians' assistants and schoolteachers, to become active in antimalaria activities. Yofe saw malaria activities as a way to mobilize the "the nation" while improving its health.[132] Yaakov Yofe, Hillel Yofe's son, who also worked on Palestine's malaria problem, remarked, "[Mer] hoped that after the area had been cleared of malaria, a unique type of society, based on equality and justice, would be established there. As Jews, he thought we were the bearers of a special ideology and should renew the life of our nation in our own

127. "Dr. Israel Kligler, Malaria Expert," *New York Times* September 25, 1944, biographical file at New York Academy of Medicine Library, New York.

128. Memo to Antimalaria Committee of Vaad Leumi by Dr. Shlifer, October 15, 1944, CZA J1/2284.

129. For the Australian case, see Warwick Anderson, *The Cultivation of Whiteness: Science, Health, and Racial Destiny in Australia* (New York: Basic Books, 2003).

130. Ashbel, *As Much As We Could Do*, 74–75.

131. Shafir, *Land, Labor and the Origins of the Israeli-Palestinian Conflict*, 161.

132. "Tafkid harofe bemilchama 'al haendemiut shebaaretz" [The role of the doctor in the war against endemicity in Palestine], twelfth meeting of the Hebrew Medical Society, June 6, 1912, *Zichronot HaDavarim* 203 (May 1913), 52, CZA Library.

land."[133] This idea—that the eradication of disease and the use of science could create a more just society—was also explored in Herzl's *Altneuland*, as discussed in chapter 1.

Fritz Yekutiel, an active antimalaria activist during the Mandate period and after, wrote, "The importance of the immediate effect of the MRU's work for the survival of the early Jewish settlements can hardly be exaggerated. No less important, however, was the long-term impact of their work. It laid the foundations of malaria research and control in Palestine-Israel for all later generations." Indeed, after the establishment of the state of Israel, the MRU's former chief inspector and prolific writer on malaria affairs, Zvi Saliternik, became the head of the antimalaria department of the Israeli Ministry of Health.[134]

Palestine as a Medicalized Landscape

Knowing Palestine meant controlling Palestine, both cognitively and practically. With possession of scientific knowledge of the land came the power to manipulate it in particular ways with particular results. As the anthropologist Michael Gilsenan notes in his work on the Akkar region of Lebanon, the act of mapping, recording scientific details (in this case, climatological, topographical, anatomical, biological, bacteriological, etc.), and writing reports reordered the landscape and established a "more total domination" over it.[135] If the first part of creating a medicalized landscape involved scientific experimentation, then medical cartography—in this case, the production of maps detailing swamps, malaria incidence, appropriate malaria measures and natural topography—served as a critical tool in the process.

Meade, Florin, and Gesler have noted that modern medical topography/geography began in the late eighteenth century, when medical practitioners, not geographers, described a location's climate and topography in relation to disease incidence.[136] Jewish engineers and malariologists

133. As quoted in Ashbel, *As Much As We Could Do*, 80, 88, 98.

134. Yekutiel, "Masterman, Muhlens, and Malaria," 17–18.

135. Michael Gilsenan, *Lords of the Lebanese Marches* (Berkeley: University of California Press, 1996), 94.

136. Melinda Meade, John Florin, and Wilbert Gesler, *Medical Geography* (New York: Guilford Press, 1988), 19–20; John Eyles and K. J. Woods, *The Social Geography of Medicine and Health* (New York: St. Martin's Press, 1983), 66–114.

during the Mandate period, along with their British and U.S. colleagues, produced an abundance of maps detailing their planned antimalaria work as well as the incidence of malaria in the different governing districts of the country. As Joseph Treidel (1876–1929), a German Jewish land surveyor and hydraulic engineer in Palestine who had previously worked in German African colonies, stated in his address to the Zionist Organization of Germany,

> Just as the physician must diagnose the case of his patient in order that he may find the proper treatment, so it is the business of the engineer to ascertain the causes and origin of the swamps before any effective treatment can be established.[137]

From 1906 to 1909, Treidel surveyed the topography and hydrology of large areas in Palestine, creating a "map of estates" to find out the ownership status of certain estates, whether or not the landowners were willing to sell, and what areas would be desirable for the KKL to purchase. After conducting several experiments in Palestine, Treidel noted that European methods of lowering the level of ground water would not work in Palestine. He could not recommend open ditch drainage because of the growth of swamp grass. Like diseases, he said, swamps have symptoms: a peculiar swamp smell, swamp vegetation, and a high water table. Instead, malariologists and engineers planted eucalyptus trees and aired the ground in Palestine. Installation of stone and clay pipes were other adaptations.

Zvi Saliternik, a well-known Zionist malariologist during the Mandate period, noted that scientists did not map out breeding places of *Anopheles* as part of initial malaria surveys before the Mandate period. In contrast, control measures during the Mandate period systematically included the determination, mapping, and surveillance of vector breeding places. Diagrams of mosquito species, their eggs, wings, and larvae were also produced.[138]

Maps were usually printed in English (although there are some in German and Hebrew) so they could be shared with the Mandatory government and distributed to agencies abroad. This extension of mapping practices reflected an expansion in the role of scientific knowledge during this

---

137. Treidel, "Technical Problems in Soil Drainage"; Arnold, *Colonizing the Body,* 27; Penslar, *Zionism and Technocracy,* 67; and Shilony, *Ideology and Settlement,* 66–69, 96–99.
138. Kligler, *Epidemiology and Control of Malaria,* 53–55.

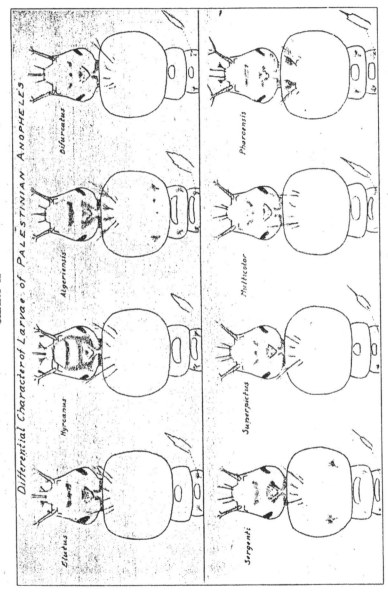

FIGURE 5.3. Diagram of Palestinian *Anopheles*. From Israel Kligler. *Epidemiology of Malaria in Palestine* (Chicago: University of Chicago Press, 1930), 54.

CHART III

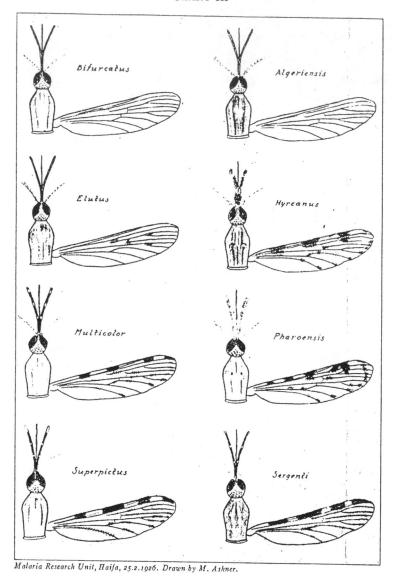

Malaria Research Unit, Haifa, 25.2.1926. Drawn by M. Ashner.

FIGURE 5.4. "Differential Character of Palpi, Thorax, and Wings of Palestinian *Anopheles*. From Israel Kligler, *Epidemiology of Malaria in Palestine* (Chicago: University of Chicago Press, 1930), 55.

time.[139] Although it is impossible to reproduce most of the maps here, several maps that privileged different aspects of malaria control are presented. In addition to the visual effects of the maps, the texts accompanying these maps—often the malaria surveys themselves—descriptively mapped out the swamp areas, water flow in springs and streams, malaria species, soils types, and spleen rates. Scientists gave detailed accounts of drainage projects, including measurements of ditches, canals, and channels. Taken together, the words and visuals paint a vivid picture of Palestine as a land undergoing a "rational" transformation.

Perhaps the most striking example of medical cartography is a map entitled "Jewish Antimalaria Control in Palestine." This map, constructed by the Rosh Pina Malaria Research Station, showed the main Jewish settlement areas with a superimposition of highly malarial infected places for 1920, seven years before the station's establishment. Since the Malaria Research Station did not exist in 1920, the Malaria Research Station constructed this map using older malaria statistics.[140] The Jewish Agency included this map in a document called "Palestine's Health in Figures" submitted to the United Nations Special Committee on Palestine in 1947. The JA used the map to show the reduced malaria incidence in Palestine in 1947 and to prove to the Special Committee how much the Zionists had done to improve the health of the country. Tables accompanying the map showed the decrease in malaria in the Huleh, Beisan, and Emek Hefer areas as a result of Jewish antimalaria work.[141]

The map illustrated the broad scope of antimalaria control work done by Jewish agencies and the epidemiology of malaria among the Jewish population of Palestine. It also reflected Jewish settlement and purchase patterns that came about partially as a result of the low population density of Palestinian Arabs in the valleys and coastal regions. As discussed in chapter 3, such scarcity was due, in part, to the high rate of malaria in those areas. Instead of acknowledging those factors that facilitated certain areas of land purchase, the text stated that these lands were "chosen for Jewish agricultural settlement," highlighting the idea of self-sacrifice and

139. Saliternik, "Reminiscences," 519.
140. "Palestine's Health in Figures," report submitted by the General Council (Vaad Leumi) of the Jewish Community of Palestine to the United Nations Special Committee, 1947, CZA Library. Also reprinted in F. Yekutiel, "Infective Diseases in Israel: Changing Patterns over Thirty Years," *Israel Journal of Medical Science* 15, no. 12 (December 1979): 977.
141. "Palestine's Health in Figures," 13.

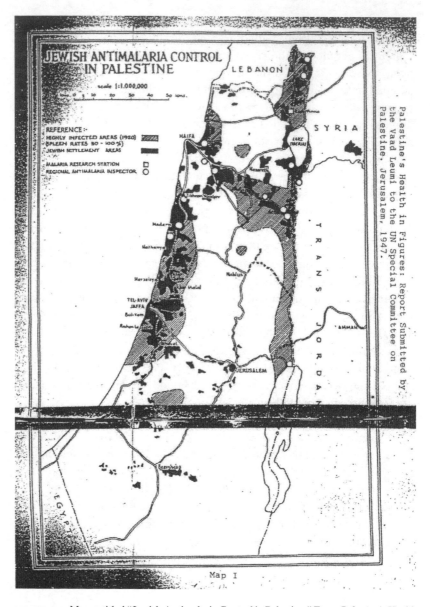

FIGURE 5.5. Map entitled "Jewish Antimalaria Control in Palestine." From *Palestine's Health in Figures: Report Submitted by the Vaad Leumi to the UN Special Committee on Palestine* (Jerusalem, 1947).

devotion in the land's reclamation.[142] Implicit in the map was the agenda of getting rid of the diagonal lines denoting malaria. Palestine was reconstructed here according to the analytical categories of malaria incidence and Jewish settlement, the image of an ideal Palestine brought into reality by drainage measures.[143]

A similar map made by Eli Etkes, an engineer of the MRU, shows swamps, streams, and valleys along with MRU control areas for 1926. Palestine is divided into eleven control areas, each numbered on the map. A table accompanied the map, describing the district, the swamp, the condition of its drainage, and those who drained it, bringing order to what was once thought to be untamable. The MRU submitted a series of drainage maps of Beisan, Balfouria, and Nahalal for this report.

A third map of the Huleh Valley imposes scientific findings about the mosquito onto the landscape. It was drawn up as the result of a mosquito survey completed by Zvi Saliternik and shows the flight patterns of *A. elutus* around the Rosh Pina district. To find out mosquito flight distance, mosquitoes were stained with methylene-blue and then released and traced.[144]

In its attempt to trace malaria transmission, the map showed the position of bedouin tents, the inhabitants of whom were considered by Jewish malariologists, as the main source of malaria infection in Palestine. The spleen rate among the bedouin tribes in the Huleh in 1924 was said to be 80 percent. Movement of bedouin tents was thought to facilitate mosquito distribution, contributing to the idea of the Palestinian Arab population as a significant obstacle in malaria control (see chap. 7). In a report on Herzlia, the placement of bedouin tents as a buffer between the swamp and Jewish settlement was said to diminish mosquito flight to the settlement.[145]

142. "Palestine's Health in Figures," 13.

143. See similar map of Palestine within the League of Nations report that details spleen rates for different years between 1918 and 1925. It is very likely that the MRU produced this map and that it was then submitted to the report, but this is not delineated. League of Nations Health Organization Malaria Commission, *Reports on the Tour*, 20–21. The MRU submitted a series of drainage maps of Beisan, Balfouria, and Nahalal for this report.

144. For similar study, see British Health Department of Palestine, annual report, 1928, 93, M4475/06/1. For MSS work done in Huleh, see letter from Mieldazis to Heiser, August 14, 1924, RF RG5/1.2/207/2656; and letter from Mieldazis to Heiser, June 14, 1924, RF RG5/1.2/207/2656.

145. [Author not given, but likely the British Department of Health of Palestine, MRU Engineering Office], "General Epidemiologic Survey of the Malaria Situation in 1927," 22, ISA M1670/130/33a/6424.

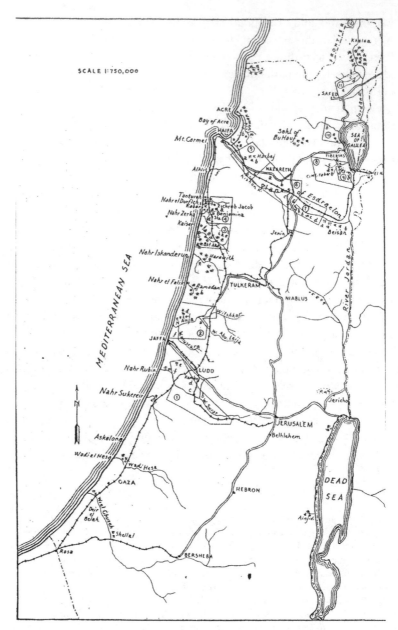

PALESTINE

Showing Swamps, Streams, and Wadis and M.R.U. Control Areas

FIGURE 5.6. MRU map of swamps, streams, and valleys. From Israel Kligler, *Epidemiology of Malaria in Palestine* (Chicago: University of Chicago Press, 1930), 4.

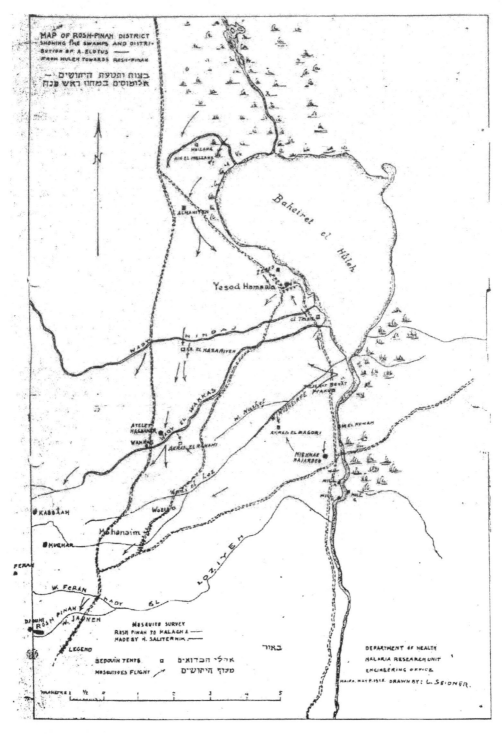

FIGURE 5.7. Map of Rosh Pinah District. Report of 1927 Anti-Malaria Work, p. 17. Israel State Archives M1670/130/33a/6424.

The report that accompanied the map explained that in order to collect data for it, the whole neighborhood from the lower portion of the Huleh to Rosh Pina was inspected each month for adult mosquito counts. The survey found that mosquito flight in spring and summer stayed within the three-kilometer radius of the swamps, but that in fall, some mosquitoes began a longer journey toward Rosh Pina and changed their breeding habitat. Once that was known, Jewish settlements could be placed accordingly and incidence of malaria in Rosh Pina could be explained more precisely.[146]

The MRU prepared another map of the Huleh that charted the canals and channels planned as part of the drainage scheme once the concession was transferred to the Zionist Organization. Scientists used knowledge about the character of the hills (limestone, basalt, etc.), and the flow and level of springs in order to explain the movement of water from the higher altitudes down to Huleh Lake.[147] The straight lines and measurements of proposed intervention contrasted with the winding lines of nature and the wadis that engineers and workers would clear in order to produce flowing waters.

Maps made of the physical landscape were accompanied by surveys and tables that presented the spleen rate and parasites in blood among the people. Countless surveys and analyses of spleens and blood took place throughout the Mandate period. Between 1922 and 1924, scientists examined 30,000 blood films in various laboratories across the country.[148]

Blood examinations almost always compared types of malaria and infection rates between Arabs and Jews, keeping the element of race consistently embedded in data collection and analysis for both British and Zionist agencies. Dr. Shapiro of the MRU specifically reminded Hadassah doctors to keep their monthly case reports on Arab and Jewish malaria incidences separate for reasons of administration, classification, treatment, and drainage implementation.[149] Separating numbers also reflected the

146. Report of 1927 antimalaria work [no official title is given], 12–20, ISA M1670/130/33a/6424.

147. Kligler, *Epidemiology and Control of Malaria*, 44–47.

148. Persis Putnam, Malaria in Palestine: Statistical Review (January 1928), 4, RF, RF1/1.1/825, Persis Putnam folder no. 11. Surgeon Dempster devised the spleen rate as index of malarial infection, which was used until the parassitemia index in late 1940s emerged. Elizabeth Whitcombe, "The Environmental Costs of Irrigation in British India: Waterlogging, Salinity, Malaria," in Arnold and Guha, *Nature, Culture, Imperialism*, 251.

149. Letter from Shapiro to director of Hadassah Medical Organization, October 30, 1929, CZA J113/555; letter from Shapiro to director, Hadassah Medical Organization,

MAP VIII

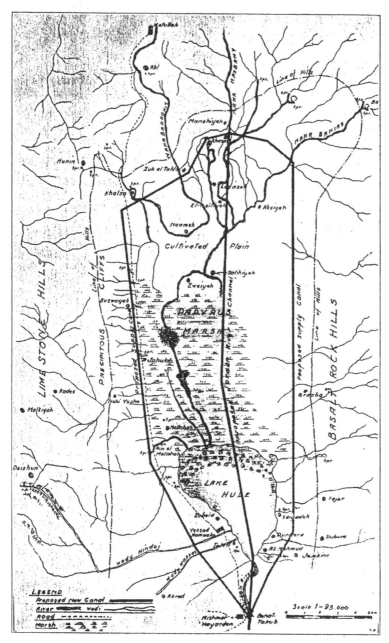

HULEH AREA

FIGURE 5.8. Channeling and planned drainage of Huleh Area. From Israel Kligler, *Epidemiology of Malaria in Palestine* (Chicago: University of Chicago Press, 1930), 46.

fact that different methods were used for determining the prevalence of malaria among the different groups, according to the cultural dictates of each community. Racial division through scientific classification reflected the separatist practices in settlement, the prevailing form of nationalism.[150] Classifying statistics along ethnic lines was also done so Zionist agencies could finance malaria control for Jewish areas in order to improve Jewish health.

Dividing medical statistics according to race was a common colonial practice. Anne Marie Moulin notes a similar practice in Algeria: "The natives were segregated in epidemiological discourse as well as in method of treatment in the field. According to a transparent metaphor, the parasite was the germ, the European was the soil and the natives were the reservoir. Being germ carriers, they came to be identified with the germs themselves."[151] Medical experimentation and categorization strengthened the division between these two communities in Palestine during the Mandate period.[152]

Racial biases created an epidemiological history of the area that was based on scientific data rather than on people's experiences of disease. For instance, in Beisan, except for oral histories gathered by survey writers, hard evidence regarding the extent of malaria incidence in Beisan village before British occupation was hard to find. What was found or told was considered unreliable: "Under Turkish dominion the village had practically no medical supervision and what records there are, if they were available and they are not, would be very untrustworthy."[153] Villagers claimed that malaria was rife every summer prior to 1917 and endemic throughout the whole year, but the British and Zionists often took these narratives as unreliable for two reasons. First, there was no Arabic word for malaria. *Sakhuna*, *hamma*, or *wakhm* were used to describe any

November 4, 1929, CZA J113/555. A complete analysis of the place of race in these surveys is beyond the scope of this chapter. Measurements of incidence surely gave scientific justification to ideas about acquired immunity amongst the Arab population as well as ideas about taking care of one's health. They also revealed levels of treatment of malaria and compliance in each respective community. See "General Epidemiologic Survey of the Malaria Situation in 1927," 29, 33.

150. Shafir, *Land, Labor and the Origins of the Israeli-Palestinian Conflict*, 21.

151. Moulin, "Tropical without the Tropics," in Arnold, *Warm Climates*, 172–173.

152. Shafir, *Land, Labor and the Origins of the Israeli-Palestinian Conflict*, 218; Smith, *Roots of Separatism in Palestine*, 7–8, 59.

153. Buxton, Carley, and Mieldazis, "Malaria Survey of Beisan, 1923," 11, ISA M1503/58/1.

feverish ailment, consistent with the early nineteenth-century medical category of "fevers" before the discovery of malaria transmission and vector
in the late nineteenth century. Consciousness of malaria as a distinct ailment and specific identification by past or present suffers was therefore
considered not possible. During the British occupation of Palestine in
1917, 90 percent of all British soldiers entering Beisan came down with
malaria within their first ten days in the area, thus attesting to its endemicity. But even this experience did not translate into a belief in the "native"
accounts of malaria. In the words of the scientists, "the omnipresent desire of the Arab to please his guest without bringing disrepute upon his
own family" caused him to "tell anything to please—the trait needs no
further elucidation."[154] Such comments indicate the malariologists' racial
views of Beisan's Arab residents and illustrate how these images colored
what was said to be a totally objective, scientific examination.

For medical professionals, the practice of examining spleens also eliminated the need for testimonials from patients, for examination alone was
considered to provide a record of disease through what David Arnold has
called "reading the body."[155] Clinical observation was part of the "discovery" of Palestine and its inhabitants. Deeming personal, medical tes-
timonies valueless also removed the patient as an active agent in his/her
health, denied and invalidated the authority of indigenous narratives of
diseases and place, and further legitimized a strictly medical, intentionally detached approach to gathering information on malaria. In essence,
this approach emptied the meaning of disease because it ignored the patient's appraisal of his/her health as well as his/her social context.[156] This
treatment of local medical histories was compatible with the Zionist vision
of treating Palestine as a dehistoricized and unproblematic space. The use
of science in the medical examination rehistoricized the malaria history of
Palestine by replacing patients' accounts with a new, medicalized history.

Clinical observation was particularly emphasized for the Arab population, whose histories and morbidity data, Kligler thought, were of "little, if any, value. Only the patients with severe attacks of malaria visit a
physician. Personal histories are vague...objective tests yield the most

154. Buxton, Carley, and Mieldazis, "Malaria Survey of Beisan, 1923," 13.

155. Arnold, *Colonizing the Body*, 54.

156. Frantz Fanon, "Medicine and Colonialism," in *Cultural Crisis of Modern Medicine*,
ed. John Ehrenreich (New York: Monthly Review Press), 234; Packard, "Visions of Postwar Health and Development," 101; K. Jaspers, *Allgemeine Psychopathologie*, rev. ed. (1913;
Berlin: Springer, 1923), 5, cited in Canguilhem, *Normal and the Pathological*, 121, 138.

trustworthy results."[157] Considering the prolonged and repeated infections of the *fellahin* and bedouin that went largely untreated, the best measurements were made by spleen exams. These exams were said to measure how enlarged the spleen had become after repeated exposure to the malarial parasite. Jewish narratives, on the other hand, were accepted as a good supplement, but still not a replacement for spleen indexes because Jewish patients were generally new arrivals to the area (so enlargement of the spleen was not always easily detectable) and treatment was more consistent. The MRU in 1924, for instance, "read" the spleens of 5,486 persons in Jewish colonies, finding 1,104 of them with enlarged spleens. Furthermore, Jewish narratives were considered reliable and rational because Jews were considered able to distinguish between malaria and other fevers, something malariologists thought the rural Arab population could not do. Still, Kohn, Weiss and Flatau have noted that in-patient hospital files from the Ein Harod Hospital (a Jewish hospital), doctors did not give patient histories at all in the early 1920s. In the later 1920s, patient histories became a little more detailed. Thus, even in hospital records, clinical results were the most important, rather than patient background.[158]

The practice of regular, extensive clinical observation in malarious areas was therefore another component in the medicalization of the land (and the people). Presentation of spleen and blood data on graphs and maps, found in practically every malaria report, reflected the use of "population" as the unit of analysis in modern medical statistics, particularly in public health. Rather than remaining associated with the person suffering the ailment, data collection was framed either in association with time (seasonal trends or year) or place (regional trends). This association occurred in almost every report.[159] On a map of Palestine, for example, numbers of spleen indices accompanied by the year to which they correspond were placed next to the name of a city or settlement, displacing the biological symptom from person to grid while also privileging the category of medicine as descriptive of the locale.

MRU spleen examinations of both Jewish and Arab rural populations in 1923 followed the N-shaped Jewish settlement schema (running north

157. Kligler, *Epidemiology and Control of Malaria*, 89.

158. "Proceedings of the Twelfth Meeting of the Antimalaria Advisory Commission," May 27, 1926, 7, CZA J1/3116; Kligler, *Epidemiology and Control of Malaria*, 93–98; Kohn, Weiss, and Flatau, "History of Malaria," 179–180.

159. For one example, see "Proceedings of the Twelfth Meeting of the Antimalaria Advisory Commission," 7.

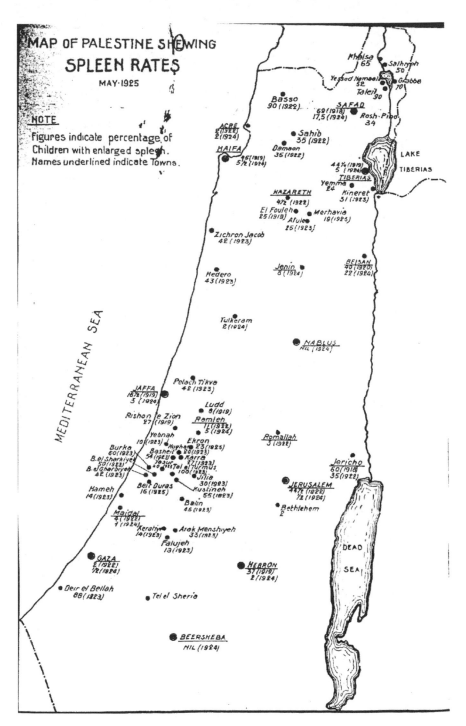

FIGURE 5.9. "Spleen Rates of Palestine, May, 1925." League of Nations, Malaria Commission, *Reports on the Tour of Investigation in Palestine in 1925* (Geneva, 1925).

along the coast, southeast along the Valley of Jezreel axis, and north along the shores of Lake Tiberias): "The villages are grouped in geographical order, beginning in Judea, going north among the coast then east in the Valley of Yezrel and north again along Lower Galilee."[160] Through the process of mapping swamps and spleens, biology was literally imposed onto the physical geography of Palestine. Statistical numbers representing biological processes (enlargement of the spleen) along with swamps, one of the environmental conditions causing malaria and the target of medical and engineering designs, constitute the analytical categories featured in the pictorial representation of Palestine.

"Before and after" maps depicting swamp drainage were used in propaganda pieces by the KKL and the Keren HaYesod to emphasize the Zionist project's ability to produce a healthy and developed land. As opposed to scientific maps drawn for drainage, these maps were constructed in a simple manner to emphasize contrast. In trying to show the results of swamp drainage for Jewish settlement in Nuris, the two maps included in the pamphlet *Jewish Nuris: How the Keren HaYesod Is Populating Historic Rural Districts in Palestine* (1925)[161] construct the landscape of Nuris by what they exclude as much as what they include.[162]

The first map in *Jewish Nuris*, depicts Nuris in 1921 before drainage. It focuses on the presence of swamps to highlight its empty wasteland character and emphasize its contrast to the second map. In keeping with the title, *Jewish Nuris*, the first map erases any sign of *fellah* or bedouin presence or their use of swamps. The first map shows a land untouched and neglected, without logic or rationality. The second map of 1924, after drainage had occurred, displays a vibrant area, divided and organized according to a grid, with controlled springs, fields, forests, and several settlements. Both maps take Nuris in isolation, even though the title of the map notes that Jews were populating "historic rural" districts of Palestine. The only indication of any history or connection to the broader landscape of Palestine was the railroad running from Haifa to Damascus and the arrow pointing to the Shutta Station. Such absence highlights Jewish self-sufficiency but in so doing, purposely disregards the impact upon and interactions with surrounding populations. The juxtaposition of these two

160. "Parasites and Spleen Rates in Rural Jewish Palestine," 1, ISA M1670/130/33a; Khalidi, *Palestinian Identity*, 98.

161. J. Ettinger, *Jewish Nuris: How the Keren haYesod Is Populating Historic Rural Districts in Palestine* (London, 1925), 4–5.

162. Kearns and Gesler, introduction, *Putting Health into Place*, 7.

FIGURE 5.10. Changes to Jewish Nuris. From J. Ettinger, *Jewish Nuris: How the Keren HaYesod Is Populating Historic Rural Districts in Palestine* (London, 1925).

MAP VII

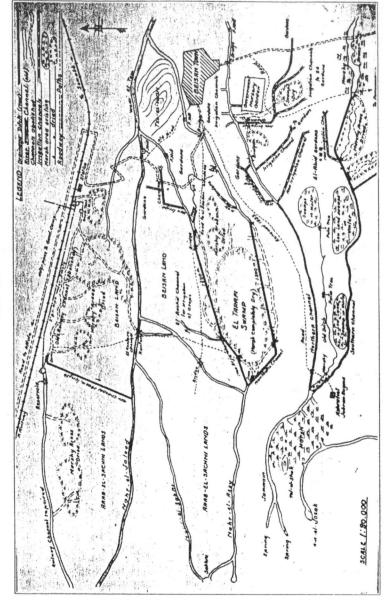

SWAMP AREA, BEISAN, AND PROPOSED DRAINAGE CANALS

FIGURE 5.11. MRU drainage in the Beisan area. From I. Kligler, *Epidemiology of Malaria in Palestine* (Chicago: University of Chicago Press, 1930), 38.

maps illustrates the striking transformation of Palestine's topography and the practical medicalization that came as a result of drainage interventions.

The maps described above are examples of John Kerrigan's description of maps as "site[s] of multiple narratives of production and interpretations: semiotic constructs which share with literary texts an ability to generate the sorts of social consequences which give rise to variations on themselves."[163] Mapping Palestine's topography by pinpointing natural and manmade causes, for instance, marked what Moulin calls a "guilty social space."[164] These spaces can be seen in the description and map of Beisan, where the Arabs' "crude irrigation systems" created shallow stagnant pools. These marshes were caused by "human carelessness."[165] The map of Beisan shows how these guilty spaces were to be repaired with new irrigation channels and dried marshes, with "rational" means.[166] Distinction between rational and irrational spaces and inscription of medical terms onto the landscape also occurred according to drainage. Areas that had been drained and had few new malaria infections were "sanitized districts" whereas other still untamed ones were "unsanitized."[167]

As cultural texts, these maps show the doctors' and engineers' ideal Palestine: a place without malaria, without mosquitoes, without swamps, and in some cases, without Arabs—a place ordered with potential or established flourishing Jewish settlements. Thus, medicalizing the land had practical implications for the fulfillment of the Zionist project, such as increased Jewish settlement, decreased mortality, and, under the umbrella of land purchase, in many cases played a part in the ultimate displacement of bedouin tribes and *fellahin* from their lands.

All types of maps—spleen maps, contoured maps, swamp maps, and drainage maps—helped engineers intercept what they saw as pathological processes and reclaim the land according to modern, scientific principles. Placing scientific knowledge onto a grid was a way to manipulate, transform, and redefine the landscape.

163. John Kerrigan, "The Country of the M: Exploring the Links between Geography and the Writer's Imagination," *Times Literary Supplement*. 4980 (September 11, 1998), 4.

164. Moulin, "Tropical without the Tropics," in Arnold, *Warm Climates*, 170.

165. Kligler, *Epidemiology and Control of Malaria*, 39–40.

166. Kligler, *Epidemiology and Control of Malaria*, 38–39; also inserted in League of Nations Health Organization Malaria Commission, *Reports on the Tour*. In Kligler, this map is accompanied with a table of blood and spleen indexes for Beisan.

167. "The Malaria Situation in 1927," 19.

PART II

Fighting Malaria to Heal the Jewish Nation

CHAPTER SIX

Cultural Landscape

Creating a Culture of Health through Antimalaria Education and Propaganda

Between man and nature stands culture.—Prof. Alfred Grotjahn (1869–1931), professor of social medicine of University of Berlin[1]

Disease Prevention and Education

Zionist antimalaria campaigns transformed the physical topography of Palestine and attempted the same for its Jewish inhabitants and their bodies. Aside from the practical aim of maintaining a healthy settler population, malaria education efforts were part of the creation of a new Jewish nation through the inculcation of hygienic principles and health habits so that what was thought to be an unhealthy, passive people in the Diaspora would emerge renewed and vibrant. As in most hygienic movements of the time, a eugenic message was embedded in a culture of hygiene. Eugenics was favored by many doctors and scientists of the period because

---

1. A. Grotjahn, "Hygiena socialit" [Social hygiene], *Briut* 1, no. 4 (October 15, 1932): 1. According to a photo caption accompanying this article, Grotjahn was the father of modern social hygiene and invented the term "social medicine." For more on Grotjahn, see Paul Weindling, *Health, Race, and German Politics between National Unification and Nazism: 1870–1945* (Cambridge: Cambridge University Press, 1989). The inclusion of his article in *Briut* shows the extent to which the medical community of the *Yishuv* was in constant contact and exchange with European medical developments and discourses, especially those in Germany. A commemorative biography of Grotjahn follows this article.

it advocated the improvement of human society through social interventions that were founded on the philosophy of racial hygiene. During this time (and some would say even today), eugenic measures controlled the reproduction of nonwhite races and the disabled in response to whites' fear that immigrants, the poor, and the unhealthy would either eventually outnumber them or contaminate society with defective or undesirable traits. According to Weindling and other scholars of eugenics, measures such as sterilization constituted "negative eugenic measures," whereas maternity benefits and types of hygiene education constituted "positive eugenic measures." Both types advanced the selective breeding agenda of eugenics.[2] As Gordon Harrison writes, attacking malaria "became in some sense an instrument of social engineering."[3]

Malaria and a Culture of Health in the *Yishuv*

Malaria and general hygiene education cannot be disentangled; general issues of hygiene and health played a part in malaria prevention because other diseases could complicate a malaria attack.[4] Taken together, Zionist educational literature, antimalaria activities, and emphasis on general hygiene produced a Zionist culture of health.[5] The constant propagation, distribution, and circulation of health materials instilled new health habits into people's lives for the "health of the nation." Producing a culture of health in the *Yishuv* was part of the broader nationalist endeavor of creating a new Hebrew culture (*tarbut 'ivri*) in Palestine.[6]

2. Paul Weindling, "Eugenics and the Welfare State during the Weimar Republic," in *State, Social Policy and Social Change in Germany, 1880–1994*, ed. Robert Lee and Eve Rosenhaft (Washington, D.C: Berg, 1997), 135, 157, 160–161, and idem, "Social Medicine at the League of Nations," 136.

3. Harrison, *Mosquitoes, Malaria and Man,* 27.

4. Dobson, *Contours of Death and Disease,* 334; and Harrison, *Mosquitoes, Malaria and Man,* 27.

5. By a "culture of health," I am not referring to the scholarly debate that distinguishes disease from illness, the former being defined biologically and the latter in cultural terms. Bryan Turner, *Medical Power and Social Knowledge* (London: Sage Publications, 1987); Lola Romanucci-Ross, Daniel M. Moerman, and Laurence T. Tancredi, eds., *The Anthropology of Medicine* (New York: Bergin and Garvey, 1991); Martha Loustaunau and Elisa Sobo, *The Cultural Context of Health, Illness and Medicine* (Westport: Bergin and Garvery, 1997); and Goubert, *Conquest of Water,* 120.

6. Letter about Malaria, Vaad HaBriut, 1927–1929, 2, CZA JI/1687. Much of the literature on cultural production focuses on the construction of Hebrew as a daily, national

The numerous professional discussions between Zionist physicians and health bureaucrats about hygiene and malaria issues reveal the importance of public health and medicine for the success of the Zionist endeavor. Responses to the hygiene educational literature are quite hard to assess from available documents, but the sheer number and wide circulation of health publications and activities illustrates the extent to which medical concerns played a vital part in the Zionist cultural project.

Malaria Education, Hygiene, and Zionism

Zionist malaria education campaigns occurred within the context of worldwide public health education efforts of the nineteenth and early twentieth centuries. Western experts on hygiene at the beginning of the nineteenth century understood that if "nation-states wanted to protect their children, it was their duty to make the population healthy and strong and to avoid the procession of avoidable diseases."[7] Health propaganda during the interwar period was typically cast in a nationalist mold.[8] In the United States, for instance, public health campaigns emphasized the reformation of individual and household hygiene.[9] Zionist doctors and scientists joined the effort, attempting to disseminate public health principles, including information on malaria, in order to create "self-improving subjects" through personal hygiene as a way of nation-building.[10] Medical agencies like the Vaad HaBriut and its constituent organizations, Hadassah Medical Organization, Kupat Holim, the Jewish Medical Society, the Nathan and Lina Straus Health Centers, and the MRU, all undertook

---

language and the production of Hebrew literature. See, for example, Itamar Even-Zohar, *'Iyunim besifrut Misrad hahinukh vehatarbut* (Jerusalem, 1965); Itamar Even-Zohar, "The Emergence of a Native Hebrew Culture in Palestine, 1882–1948," *Essential Papers on Zionism*, 727–744; B. Harshav, *Language in Time of Revolution* (Berkeley: University of California Press, 1993); Dan Miron, *Im lo tihyeh Yerushalayim: masot 'al ha-sifrut ha'Ivrit bekhesher tarbuti-politi* (Tel Aviv: Kibbutz HaMeuhad, 1987); Nurith Gertz, *Sifrut ve-ideologyah be-Erets-Yisrael bi-shenot ha-sheloshim* (Tel Aviv: Open University, 1988); C. Rabin, "The National Idea and the Revival of Hebrew," *Essential Papers on Zionism*, 745–762; Zeruhavel, *Recovered Roots.*

7. Goubert, *Conquest of Water*, 109.

8. Weindling, "Introduction: Constructing International Health," 2.

9. Nancy Tomes, *Gospel of Germs: Men, Women and the Microbe in American Life* (Cambridge: Harvard University Press, 1998), 6–7.

10. Comaroff and Comaroff, *Of Revelation and Revolution*, 361; Vaughn, *Curing Their Ills*, 57.

malaria health education efforts.[11] The Vaad HaBriut had a Public Health Education Committee (at times also called the Council for Hygiene Explanation and Education) that organized a variety of public health activities for the *Yishuv*. A social hygiene and medical statistics office also existed under the direction of Dr. Peller, the epidemiologist of the Vaad HaBriut. It was established in February 1926.[12]

U.S. medical discourse and public health practice significantly influenced Jewish Palestine's, mostly because many of the Zionist medical administrators, doctors, and engineers that came to Palestine were originally from the United States or trained there. The influence of the Hadassah Organization, an American Zionist organization very much involved in health affairs in Palestine, also was significant in this regard. Jewish health professionals sent literature on malaria to Palestine from the United States. References to U.S. examples were frequently made in Zionist malaria literature, both scholarly and popular.[13] Interestingly, a project for preserving the health of world Jewry was also in existence during this time. Jewish scientists and health professionals initiated it in Germany, but the project also had branches in Eastern Europe, the United States, and England. The project was called the Union for Preserving the Health of the Jews (OSE). Albert Einstein served as its president. This organization corresponded and exchanged information with professionals active in the Zionist health enterprise.

Although Zionist health bureaucrats drew upon propaganda materials from the United States and Europe, hygienic and malariological principles were adapted to the particular situation in Palestine because of the

11. Although a confusing setup, the Vaad HaBriut was actually under the jurisdiction of both institutions. I thank Shifra Shvarts for clarifying this point; Statutes of the Vaad HaBriut (Health Council) of the Palestine Zionist Executive, adopted by the PZE at the meeting held on January 25, 1926, in accordance with the recommendations of the Sanitary Sub-committee of the Fourteenth Zionist Congress, CZA J1/1047.

12. Decisions of Council for Hygiene Explanations and Education, March 5, 1928, CZA J1/4417. Dr. Brachyahu, Ms. Landsmann, and Dr. Meir were present; report of the Vaad HaBriut of the PZE submitted to the Zionist Executive for the general report to the Fifteenth Zionist Congress, June 1927, 4, CZA J1/1047.

13. Wickliffe Rose, *Field Experiments in Malaria Control* (Chicago: American Medical Association, 1919), 1, CZA J113/1430. Rose was general director of the International Health Board, NY. Reference to Panama Canal is in *Sicha 'al Malaria* [Conversation on malaria], Hadassah National Library, no. 3 (Jerusalem, 1921), 12, CZA Library 4167; American Zionist Medical Unit/Hadassah Medical Organization, *Hamalaria: hitpashtuta, mecholeleya, darkei ha'avarata ve'emtza'ei hahishtamrut mipneiya* [Malaria: Its spreading, its causative agent, means of infection and protective measures], pamphlet no. 2, 1921, Ein Kerem Medical Library.

belief prevalent in colonial medicine and public health education that habits had to be adapted to the more acute tropical and subtropical environments of the colonies. Local conditions and experiences of the Zionist settlers were a central component of Zionist health educational propaganda that was either developed in Palestine or imported from Europe or the United States and then adapted for a Jewish audience.[14] The resulting culture of health campaign was sensitive to the nationalist agenda of the Zionist movement and captured the specific sociopolitical, economic, medical, and ecological conditions of Palestine.[15] Special attention was given to Palestine's mosquito species and swamps.

Teaching malaria prevention was a way to test and gradually concretize the continuing discoveries of malariology. It was also a way to inculcate the science of hygiene into people's hearts and minds, even while the specific formulation and implementation of these two scientific fields were being crystallized in the field of medicine. Historian of medicine Nancy Tomes has remarked that germ theory was not immediately accepted into the popular mind. Tomes believes the end stage of the crystallization process occurred at the end of the 1920s, but the "working out" of hygiene continued into the 1930s in Palestine in part because of the continued waves of immigration.[16] Like doctors in the colonial world who inculcated principles of hygiene and parasitology more easily in the tropics (colonies) because there were no competing institutions as in the metropole, Zionist doctors in Palestine had relative autonomy to initiate and execute hygiene measures on Zionist-owned land for their community.[17]

Malaria Pamphlets, Columns in Newspapers, Popular Health Journals, and Health Weeks

Protect your health for your own good and for the benefit of your sons after you.[18]

14. See Yassky's response to Deutsches Hygiene Museum, Germany, where he thanks them for the typhus posters they sent. He did not use them because the posters did not represent local types (people). Letter from Yassky to Deutsches Hygiene Museum, Germany, December 13, 1931, CZA J113/345.

15. W. Anderson, "Where Is the Postcolonial History of Medicine?" *Bulletin of the History of Medicine* 72, no. 3 (Fall 1998): 526; letter from Yassky and Katznelson to OSE: Agency for the Health of the Jews, Berlin, February 11, 1930, CZA J113/965.

16. Tomes, *Gospel of Germs*, xv, 6, 8. I thank Ann Emmanuelle Birn for originally suggesting the connection between malaria education and consolidation/inculcation of hygiene.

17. Bruno Latour, *Pasteurization of France* (Cambridge: Harvard University Press, 1988), 19–26, 34, 56, 143–144; Tomes, *Gospel of Germs*, 6.

18. Saying at bottom of *Briut* 1, no. 5 (October 1, 1932): 35.

Zionist health propaganda for the general public was available in four different popular media: malaria pamphlets, columns in newspapers, popular health journals, and health weeks. There were two hygiene books that focused on malaria before the Mandate: Dr. Arieh Behem's *Hamilchama, habriut vehamachalot hamidbakot beEretz Israel* (The struggle, health, and infectious disease in *Eretz Israel*, 1915) and Dr. Hillel Yofe's, *Shmirat habriut beEretz Israel* (Protecting health in *Eretz Israel*, 1913).[19]

Malaria pamphlets were mostly produced in the first decade of the Mandate; by the 1930s and 1940s, they were less frequently published or were part of larger collections on general hygiene. The reasons behind this decrease in frequency are because malaria control done in the 1920s led to decreases in incidence in subsequent years and a less intense need to educate the public, and also because Zionist health organizations reprinted, redistributed, and reused pamphlets throughout the period.[20]

Newspapers and journals were another component in teaching the public basic rules of hygiene and sanitation. Each Hebrew newspaper during this period contained a bimonthly or weekly health column. Beginning in 1931 the Hebrew newspaper *Davar*, for instance, published a column called *Shmirat briutecha* (Guarding Your Health) edited by Dr. J. Meyer and then Dr. Shechter.[21] The newspaper *Ha'Aretz* had a weekly health column called *Briutenu* (Our Health) that started in 1936 and was edited by Dr. A. Goldstein. *HaBoker*, another newspaper, printed a health column by Dr. A. Levy. Other newspapers such as *Doar HaYom* and *HaGeh* featured articles on health activities in Palestine but, it seems, did not have regular health columns.

Women's journals and weekend newspaper supplements contained "Ask the Doctor" columns. Articles in these columns and in popular medical journals included discussions on various seasonal diseases like malaria and issues of nutrition and sexuality. They also dealt with questions of physical

19. Dr. Behem was a founding member of the Hebrew-Speaking Physicians' Society of Jerusalem (established 1913). Dr. Hillel Yofe was a central figure on malaria eradication both before and during the Mandate period.

20. Dr. Behem *Hamilchama, habriut vehamachalot hamidbakot beEretz Israel* (Jerusalem, 1915), Ein Kerem Medical Library; Yofe, *Shmirat habriut beEretz Israel*; HaPoel HaTzair, *Ha'am* 62–64 (Jaffa, 1913), Ein Kerem Medical Library.

21. Dr. Meyer expressed to Dr. Brachyahu in 1931 that it was acceptable for contributing authors or doctors to write in another language if they had a hard time expressing themselves in Hebrew. Their contributions were translated into Hebrew or they were rejected; the *Davar* health column did not contain articles in any other language other than Hebrew during the Mandate period. Letter from Dr. Meyer to Brachayahu, May 24, 1931, CZA J113/345.

education, healthy clothing, baby care, and hygiene in social, mental, family, industrial, and school contexts. These articles also covered im- migrant and workers' health issues, historical medicine, medical figures, the medical politics in the *Yishuv*, as well as first-aid and jokes about health. Doctors believed jokes were good for one's health and so some health journals had a regular joke section in their publication.[22] As with malaria pamphlets, once successful malaria control was underway, the topic of malaria figured less prominently in newspaper columns. Topics such as kidney problems, pregnancy, deviant children, eugenics, and ty- phus became more prominent. Still, Zionist health professionals believed that being familiar with a wide range of hygiene topics was important in malaria prevention: creating a healthy body and environment could re- duce the risk of malaria. The topic of malaria returned to the fore when major drainage projects were completed or when malaria epidemics oc- curred.

Popular health journals also carried information and discussions about malaria and hygiene. From 1926 to 1927, the Committee on Public Health of the Jewish Agency published *Briut Ha'Am* (The Health of the Nation). This magazine, edited by Drs. Felix, Peller, and Friedman, was devoted to public health issues and politics. The journal was praised for being written in layman's terms.[23]

Dr. Matmon edited a popular biweekly journal for "health, hygiene, and enlightening" called *Briut* (Health), published between 1932 and 1935. Although the exact profile of readership is unknown, the dispropor- tionate amount of articles on sexuality, the health effects of marriage, and mothering suggest a female audience. *Briut* had a question-and-answer section. Often, instead of publishing the full question and answer, a list would be published where a key word or name of settlement from each letter would be noted and the doctor would write a personal note to that reader without including the original question. Sometimes the doc- tor would write that he sent a letter in the mail the writer would find at their post office. This type of correspondence illustrated the personal attention given in health journals to readers' concerns, made easy be- cause of the relatively small community of Jews in Palestine. From 1932 to 1934, Drs. Friedman and Ben-Raanan edited a monthly called *Shaarei Briut* (Gates of Health). This journal dealt centrally with malaria and

22. See *Briut* and *Shaarei Briut*, Ein Kerem Medical Library.
23. Jewish Agency for Palestine, *Briut Ha'Am*, CZA Library.

the importance of health propaganda for the success of the national endeavor.[24]

In addition to popular health literature, health weeks and health exhibits further consolidated the culture of health. In a spring exhibit during 1932, for instance, the Vaad HaBriut organized a health booth that they called *havra'at ha'aretz* (reclaiming/healing the land) to highlight antimalaria work in Haifa and other places. It also dealt with topics such as the growth of the *Yishuv* and Jews in the Diaspora, preventive and clinical work, medical insurance, hygiene explanations, and convalescent homes.[25]

Hadassah Medical Organization and the British Department of Health and Education sponsored health weeks that were countrywide events. The Straus Health Centers in Jerusalem and Tel Aviv frequently served as the location for health exhibits. Each demonstration or health week covered different themes. Admission was free to the public. Hadassah Medical Organization and the British Health Department of Palestine conceived of them as "intensive propaganda for various phases of public health work throughout the country among all elements of the population."[26] Themes for these weeks and exhibits included, but were not limited to, nutrition, physical culture (1932), or mother and child (1934).[27]

Prior to its organization of health weeks, the Hadassah Medical Organization (HMO) organized a health day in 1921 devoted to malaria education in the Jewish colonies of the Galilee. The HMO handed out thousands of pamphlets, and doctors and sanitary inspectors gave lectures with slides to the settlers. In addition, the HMO organized journeys to the swamps and springs in order to illustrate the development of mosquitoes "on the spot." After the journeys, nurses continued to carry out instructions in individual homes, where they showed settlers how to use mosquito nets, distributed quinine, and gave medical aid.[28]

24. *Shaarei Briut* 1, no. 1 (1931).

25. "The Program for Organizing Beitan HaBriut (Health Booth) in the Spring Exhibit," CZA J1/4355.

26. H. Bowman, *Health Week*, government circular no. 89, October 31, 1924, 1, CZA J113/1446 (also copy in Arabic given). Bowman was the director of education.

27. See critique of this particular exhibit at the Straus Center in Tel Aviv. "'Al hata'arucha letarbut-haguf" [On the exhibit of physical culture], *Briut* 1, no. 6 (October 27, 1932): 49, Ein Kerem Medical Library (the article reads no. 7 but this is a mistake because it is in the edition of no. 6 and is included in its table of contents). See a critique of this exhibit in "Ta'arucha haem ve hayeled," *Briut* 2, no. 9 (July 2, 1934): 78, Ein Kerem Medical Library. This exhibit took place at the Straus Health Center in Tel Aviv.

28. "Jewish Medical and Antimalaria Work in Palestine," memorandum submitted by the Vaad HaBriut of the Palestine Zionist Executive to the Malaria Commission of the League of Nations, Jerusalem, May 1925, 10–11, CZA J1/3116.

SHAAREI BRIUTH

Monthly Popular Medical-Hygienic Journal

EDITORS

Dr. D. Arieh Friedman   **Dr. I. Ben - Raanan**

Vol 1. No. 2   ▮   August 1931

Table of Contents:

Editorial office: „BRIUTH" Tel-Aviv.

Subscriptionnate: 120 mils yearly
Administrative office: Nachlath-Benyamin 5)

FIGURE 6.1. Cover, *Sha'arei Bri'ut: Monthly Popular Medical-Hygienic Journal* 1, no. 2 (August 1931). National Library of Jerusalem.

The first health week took place on November 17–30, 1924. A day was devoted to each of the following subjects: hygiene, infants, food, germs, and recreation. Health professionals gave talks in all schools, both governmental and nongovernmental. They also distributed health leaflets. Two films were shown, one on malaria and one on milk.[29] The director of the Department of Education instructed headmasters of government schools to read the pamphlets so that "the occasion [would be] more impressive."[30] Teachers gave talks in the morning to the students about these topics and then repeated them in the afternoon or evening for the parents and other adults to hear. Doctors thought that by having teachers give lectures in schools, they too would learn principles of hygiene.[31] The HMO and the British Health Department of Palestine set up local committees to implement health week plans. All literature, posters, exhibits, demonstrations, and illustrated talks were performed or published in both Arabic and Hebrew. According to a document written by the deputy director of health, Mr. R. Briercliffe, the government produced 100,000 copies of an Arabic pamphlet for distribution in the schools for the first health week.[32] No material was published in English for the first health week. Cinematographic materials were imported from Europe.[33] When the government found that the distribution of pamphlets to village schools was not efficient, they called upon antimalaria subinspectors and sanitary subinspectors to assist in delivering the pamphlets during their regular village inspections. Scout troops also helped in distribution and the supervision of health weeks.[34]

Several health week committees organized essay contests for elementary and secondary schoolchildren. Students were asked to write about "What I have learned from Health Week" in 1924. In 1936, for a health week whose theme was "Cleanliness: Use It Against Summer Diseases,"

29. Minutes of the meeting of the Jerusalem Health Week Committee, October 29, 1924, 2, CZA S25/590.

30. Bowman, *Health Week*, 2.

31. Letter from Brachyahu, head of Department of School Hygiene to Health Department of Vaad Leumi, January 3, 1940, CZA J17/22.

32. I was unable to find even one copy. R. Briercliffe, circular no. 265, November 4, 1924, CZA J113/1446.

33. Letter from secretary of Health Week Committee to Mrs. Champion, Beersheba, October 14, 1924, CZA J113/1446.

34. H. Bowman to Baden Powell Boy Scouts Association of Palestine, November 8, 1924, CZA J113/1446, Arabic translation given as well, with lists and addresses of Arab scoutmasters attached.

students wrote essays on the question "What is the benefit of cleanliness in the schools or in the house?"[35] Hebrew and Arabic essays were judged separately, and three prizes were given in each district for the best essays.[36]

The 1924's health week encompassed multiple objectives. The Health Week General Committee planned an exhibition in the Russian buildings in Jerusalem. The Department of Education and the Vaad Leumi organized field trips for students in the government schools and in the schools of the Palestine Zionist Executive. The exhibit was open in the evenings for those who worked during the day, and a ladies' day was devoted to Muslim women—a "harem day" where only women were welcome. The exhibit was divided into comprehensive topics: maternity and child welfare, school hygiene, domestic hygiene, housing and town planning, foods and beverages, house sanitation and appliances, municipal hygiene, personal hygiene, a medical section, laboratory and malaria, an epidemic section, institutions, engineering and architecture, and miscellaneous. The laboratory section dealt with milk and water, while blood analyses for the malarial parasite were done on request. The malaria section concentrated on the sources of malaria prevention and clinical treatment. A related section on infectious disease had an exhibit on protecting health in subtropical environments.[37] Dr. Carley of the International Health Board and the Malaria Survey Section served as head of the exhibit on malaria. Drs. Behem and Carley gave talks on malaria, while Mr. Cantor, a sanitary engineer, gave a lecture on general sanitation.[38] Jewish doctors gave lectures

35. Letter from Dr. Y. Levy, head of Department of Education in Public Health, to Dr. Brachyahu, March 24, 1936, CZA J17/667; and letter from Dr. A. Y. Levy, head of Department of Education in Public Health of Vaad Leumi, to Dr. Lurie, Department of Education, Vaad Leumi, April 4, 1936, CZA J17/21.

36. Letter from Director of Education to inspectors of education in each district, November 20, 1924, CZA J113/1446. Hebrew and Arabic translations included the following: letter from Director of Education to inspectors of education (Jerusalem, Jaffa, Samaria, Galilee, Southern), same date, same file; letter from Sharif, inspector of education in Samaria district to Director of Education, December 8, 1924, about essay winners in Nablus, Jenin, Tulkarm, and Beisan, CZA J113/1446; letter from inspector of education, Galilee, to Director of Education, December 10, 1924 for winners in that district, same file.

37. Palestine Health Week: Jerusalem Health Exhibit, November 17–30, 1924, 10–11, National Library. The pamphlet has the Health Week schedule and descriptions of each booth in all three languages: Arabic, English, and Hebrew.

38. The malaria lectures were given on Friday, November 21, 1924, and Saturday, November 29, 1924. Palestine Health Week: Jerusalem Health Exhibit, November 17–30, 1924, 12–13. National Library.

on other topics such as infant welfare, tuberculosis, dental diseases, and eye diseases. A letter to the secretary of the Health Week General Committee noted that lecturers giving their talks in Arabic were harder to find, but a lecture on malaria with cinematograph films for Arab men and women was arranged separately at the YMCA. An additional lecture on malaria given in English was held at the school of archeology with fifty people in attendance. The Health Week General Committee also organized talks on public health, eye diseases, and women's issues.[39]

The first health week was considered a big event, with special decorations and an official opening with dignitaries, notables, and a performance by the police band. A report describing the week noted that two hundred posters in Arabic, Hebrew, and English were posted on the official bulletin boards and at all public centers in Jerusalem. Street banners in three languages were hung across the highway and across the doorway entrance of the exhibition. In addition, a twenty-page program was prepared in all three languages. One thousand copies of the program were distributed free of charge. Thirty-four thousand ninety people of all classes and levels of education attended the first health week. Four thousand eight hundred eighty were school children.[40] The report on health week attributed its success to the exhibit's nonpolitical nature and encouraged the field of health as an arena of cooperation:

> Disease knows no barriers except a united front and hearty cooperation of all agencies to promote preventive measures that are essential for an effective health campaign...The exhibit has shown that individuals are prone to take part, assist and attend health shows, lectures and activities when the disturbing factor who view everything from a political viewpoint are kept in the background.[41]

From time to time, however—at least in Hebron—health lectures were given separately for each religious community.

39. Letter from Clayton to secretary of Health Week General Committee, December 18, 1924, 1–3, CZA J113/1446.

40. Report of the Health Week Exhibition, Jerusalem, November 17–December 1, 1924, 3–4, CZA J113/1446. See also *Hadassah Newsletter* 6, no. 2 (November 1925): 9, CZA Library 1270.

41. "Report of the Health Week Exhibition, Jerusalem," November 17–December 1, 1924, 2–4, CZA J113/1446; letter from Clayton to secretary of Health Week General Committee, December 18, 1924, 2, CZA J113/1446; see also letter to Mrs. Abramson from Ms. Mochenson, November 24, 1924, about successful Gaza lectures; letter from Abramson to

Health weeks were organized by Jewish and Arab health professionals. For instance, the Jerusalem Health Week Executive Committee consisted of prominent Jewish and Arab women and men of Jerusalem: Mrs. Clayton, chair; Mrs. Bentwich, secretary; Mme. Nashashibi, Mrs. Key, Miss Nixon, Dr. Felix, and Mr. Antonius. Yet the potential for a politicized Zionist event did not go unrecognized by some Palestinian Arab doctors. Dr. Canaan, who served on the medical planning committee for the event, warned Ms. Mochenson, secretary of the first health week, that these events should not be used for political or publicity purposes to promote the Zionist endeavor.[42] Dr. Canaan's suspicions were not unwarranted. Zionist doctors and prominent leaders often used health education campaigns to promote the movement's national agendas.

Another health week was organized in 1925, the same year that the League of Nations Malaria Commission visited Palestine. It focused on the prevention of infectious disease, with malaria featured as one of the main scourges. In another health week in 1932, the Advisory Committee for Public Education of the Vaad Leumi (Mr. Ben Ari, Ms. Landsmann, Dr. Meir, Dr. Katznelson, Ms. Berman, Dr. Gruenfelder, Professor Kligler, Mr. Cantor, and Dr. Rokach) used a public trial to circulate information about hygiene. The trial was supposed to explain to the community how summer diseases spread and ways to protect against them. Two trials took place in Tel Aviv and Jerusalem debating different summer diseases.[43] Slogans, placards, and signs on automobiles and in streets supplement the trials. In 1936, for the health week on "Cleanliness," the Arrangement Committee used a new line of propaganda: two houses in

---

Mochenson, Gaza, November 15, 1924, about Muslim doctors' lectures; letter from Abramson to Mochenson, no date, which states that the "population is of the ignorant *fellahin* class—at the same time if their interest in health and hygiene could be awakened it would be excellent." Also letter from Dr. Ayoub, government doctor, Nazareth to Mochenson, November 2, 1924, about willing participation of several Nazareth doctors; letter from Margaret Bailey of Hebron, November 24, 1924; letter from Bailey to Mochenson, November 15, 1924, all from CZA J113/1445.

42. Letter from Mochenson, secretary of Health Week Committee, to Col. Kisch, November 2, 1924, CZA S25/590; letter from Margaret Bailey of Hebron, November 24, 1924, CZA J113/1445.

43. During the planning of the trials, a discussion ensued about whether to have the trials on Shabbat (Jewish Sabbath) due to a possible backlash by the *Haredi* (religious) population. Letter from Y. Levy, head of Department of Public Health Education of Vaad Leumi, to head of Vaad Leumi, June 20, 1932, CZA J1/4355 for quote; minutes on meeting of Advisory Committee of Public Health Education Department of Vaad Leumi, Straus Health Center, May 11, 1932.

each area, one clean and one dirty, were chosen as respective examples of how and how not to live.[44] Educational pamphlets were distributed during health weeks. "How to Watch Your Health" (1933) and "Cleanliness Paves the Way for Health" (1937) are just two examples.[45]

Malaria and Collective Responsibility for the National Endeavor

Zionist health propaganda produced a nationalized version of scientific knowledge for the Jewish public. Promotion of individual and national responsibility through malaria prevention efforts was the first and foremost purpose of these publications. Even in 1947, when malaria was largely under control, the link between individual responsibility and malaria eradication was still emphasized.[46]

Individual subordination to national priorities was a common element in public health planning of this period. Paul Weindling's remarks about eugenic measures in Weimar Germany mirrored the eugenic ideas embedded in Zionist health education propaganda during this time, as he praised "techniques of enforcing orderly behavior with nationalist ideology of a fit and efficient body politic and of devotion to future generations."[47] Zionist doctors believed that personal action toward malaria control had national ramifications for Jewish redemption through the transformation of the land and the nation. Yael Zerubavel has noted that the primacy of the collective over the individual, with a concomitant emphasis on every person's contribution to Zionist nation-building efforts,

44. Report of the Arrangement Committee Meeting for the Health Week, 1936, February 13, 1936, 1–2, ISA M1567/44/12/2373; and draft program for Health Week 5696, April 25–May 2, 1936, ISA M1567/44/12/2373.

45. M. Schechter, "Medical Literature in Palestine," *Medical Leaves* (1939): 168, Ein Kerem Medical Library.

46. Dr. Brachyahu, *Haderekh el briut: hora'ot hahygiena bekitot hayesod shel beit sefer ha 'amami* [The way to health: Hygiene instructions for elementary classes of national schools] (Tel Aviv: Yavneh, 1947), 147, National Library. This pamphlet mentions DDT as a method of control, which shows how malaria propaganda indicated new scientific developments to the public.

47. Weindling, "Eugenics and the Welfare State," 135; Doron Niederland, "Hashpa'a harefuim—ha'olim meGermania 'al hitpatchut harefuah beEretz Israel (1933–1948)" [The influence of the German immigrant doctors: On the development of medicine in Eretz Israel], *Cathedra* 30 (1983): 111–160, Shvarts, *Kupat Holim*, 33–35, Shvarts and Brown, "Kupat Holim," 28–46; Winter and Levy, "Medicine in Palestine following the Flight of Jewish Physicians," 19–23; and Theodore Grushka, *The Health Services of Israel* (Jerusalem, July 1952), 7, for discussions on German influence on Zionist medical thought.

were central educational themes during the prestate period.[48] Health education efforts were no exception.

In fact, the health journal *Briut* opened its first issue with an article describing the importance of health and hygiene for the nation. The article explained that the journal's mission was to teach the "wide avenues of the nation" (i.e., the *Yishuv*) how to live a healthy life based on hygienic principles. The mission was justified by a commitment to individual and national responsibility:

> One of the first certificates [*teudot*] of a cultured nation is the acquisition of health concepts by every person. Every nation needs to be concerned that its sons will be healthy, because only healthy people can bring benefits to their nation and can continue the life and tradition of it. Health—that is to say, the *level of creative power of the land and the nation* [Koach hayotzer shel haaretz ve ha'am] creates the possibility for establishing a greater number of people and for strengthening the quality of individual work and of national production. Not only should the nation be concerned with the health of its sons, but also every individual needs to persevere in the repair of his own health. Because by doing this, he helps not only himself but also his nation...Any defect in the process of [attaining] our health is damage not only to the individual but also to the nation.[49]

The article noted that in Palestine, especially in times of crisis, there was an obligation to strengthen one's body, one's nerves, and one's spirit. Every member in the *Yishuv* was called upon to "recognize his body, the path of his life, the various disturbances that can damage him and the means to protect against them."[50] Malaria and the mosquito were among the causes of disturbances. In another article in *Briut*, Dr. Litvak described the connection between epilepsy and malaria, claiming that in 1926, malaria was the single most important cause of epileptic attacks in the *Yishuv*. This

---

48. Zerubavel, *Recovered Roots*, 81.

49. "Otzar ha'am—briuto" [The treasure of the nation—Its health], *Briut* 1, no. 1 (August 5, 1932): 1, Ein Kerem Medical Library, emphasis added. See also Dr. B. Fink, "Hiyu bri'im" [They were healthy, part 2], *Briut* 1, no. 4 (1932): 4, Ein Kerem Medical Library.

50. "Otzar ha'Am- Briuto," 1. Besides malaria and infectious disease, *Briut* published articles on everything from bad breath, baldness, kidney problems, and nervous children to beauty, pregnancy, headaches, and tonsillitis.

connection caused Litvak to speculate that tropical malaria caused subsequent damage in the central nervous system.[51]

Many malaria pamphlets addressed the issue of collective and individual responsibility. In March 1920, Dr. Arieh Goldberg of the Nathan Straus Health Center in Jerusalem published "Instructions for Fighting Malaria." The four-page, question-and-answer guide stressed that malaria could be caught not by drinking water or by breathing air, but only by mosquitoes that transmit the disease from a sick person to a healthy one. It also raised the importance of guarding against malaria because of malaria's impact upon Jewish workers and the "economy of that nation." The guide reads:

> Malaria weakens the body of the person, deprives him of his vigor and energy to work and brings him to a state of degeneracy.... It causes much damage for the nation's capital by: 1. decreasing its production that comes as a result of the disease that deprives the worker of many days of work; 2. by the large expenditures on medicine.... Malaria delays the *Yishuv*'s work in many places, and therefore it is our responsibility to reckon with our most dangerous enemy.[52]

The guide goes on to explain that it is possible to settle in endemic areas but only on the condition that "everyone understands and knows how to protect his health according to the instructions of the doctor."[53] Furthermore, it is everyone's obligation to be concerned with "their own health, the health of their family, their nation and land." "Everyone who redeems the land (*geulat haaretz*), and acquires things that are dear to him from the nation must participate in the war against malaria and help with all their means to uproot this malignancy from our land."[54]

Another booklet, *Sicha 'al Malaria* (Conversation on Malaria), the third in a series of the Hadassah National Library published by the American Zionist Medical Unit, presented very precise explanations of pathophysiology and parasitology and the role of the mosquito in transmitting malaria. The first two pamphlets of the series were titled *Hishtadel lihiot*

---

51. A. Litvak, "Machalat hanichpah" [Epileptic diseases], *Briut* 1, no. 20 (no date given, but about June 30, 1933): 163, 166, Ein Kerem Medical Library.

52. Dr. Arieh Goldberg on behalf of the management of the Nathan Straus Health Center, *Hora'ot lemilchama beKadachat (Malaria) veleshmira mipneiha* (1920), National Library of Jerusalem, 1, 4.

53. Ibid., 4.

54. Ibid., 4.

naki, bari, vehazak: horá ot beshmirat habriut (Try to Be Clean, Healthy and Strong: Instruction on Protecting Health; 1921) and *Hamalaria: hitpashtuta, mecholeleya, darkei há avarata vé emtzá ei hahishtamrut mipneiya* (Malaria: Its Spreading, Its Causative Agent, Means of Infection and Protective Measures; 1921). The fourth and fifth pamphlets were called *Hazvuvim* (Flies; 1921) and *Sipur 'al Giveret Anopheles* (Story of Miss Anopheles; 1922).[55] Scientists and health professionals thought that giving detailed scientific explanations of the transmission of malaria, the developmental cycle of the parasite, and malaria's pathophysiology, as well as the breeding processes of the mosquito and seasonal trends, would propel the successful evolution of people's habits along with the topographical changes of the land.

Conversation about Malaria addressed different malaria control methods. It compared methods used to eradicate malaria in two U.S. communities, one in a small but prosperous city that focused on clinical treatment only, and the other on a cotton plantation of poor African American residents that treated malaria in a more comprehensive manner. The author believed that it was actually cheaper to invest in antimalaria measures than to treat malaria cases without supplemental environment programs. The pamphlet connected the situation of blacks in the American South and the *Yishuv* by making an analogy between the Zionist emancipation of "enslaved Jews" of Europe (symbolically the enslaved Jews of Egypt in the Biblical story of Exodus) and the emancipation of American slaves.[56] Zionists became "free" in body and spirit by partaking in the national endeavor of hygiene and health.

By the end of the booklet, the emphasis shifted to the topic of mobilization, the responsibility of the individual to participate in the fight against the disease for his own health, the health of others, and the health of future generations.[57] Malaria, it explained, causes feverish attacks that

55. Hadassah Medical Organization, *Hishtadel lihiot naki, bari, vehazak: horá ot beshmirat habriut* [Try to be clean, healthy and strong: Instruction on protecting health], pamphlet no. 1 (1931): 2; Hadassah Medical Organization, *Hamalaria*; Hadassah Medical Organization, *Hazvuvim* [Flies], pamphlet no. 4 (1921); Hadassah Medical Organization, *Sipur 'al HaGiveret Anopheles* [Story of Miss Anopheles], pamphlet no. 5 (Jerusalem: Rafael Chaim Hacohen, 1922).

56. See Jonathan Boyarin, "Reading Exodus into History," *Palestine and Jewish History: Criticism at the Borders of Ethnography* (Minneapolis: University of Minnesota Press, 1996), 40–67.

57. I. Kligler, *Survey of Territory and Plan of Organization of Antimalarial Demonstrations*, no date, 4, CZA J15/7212.

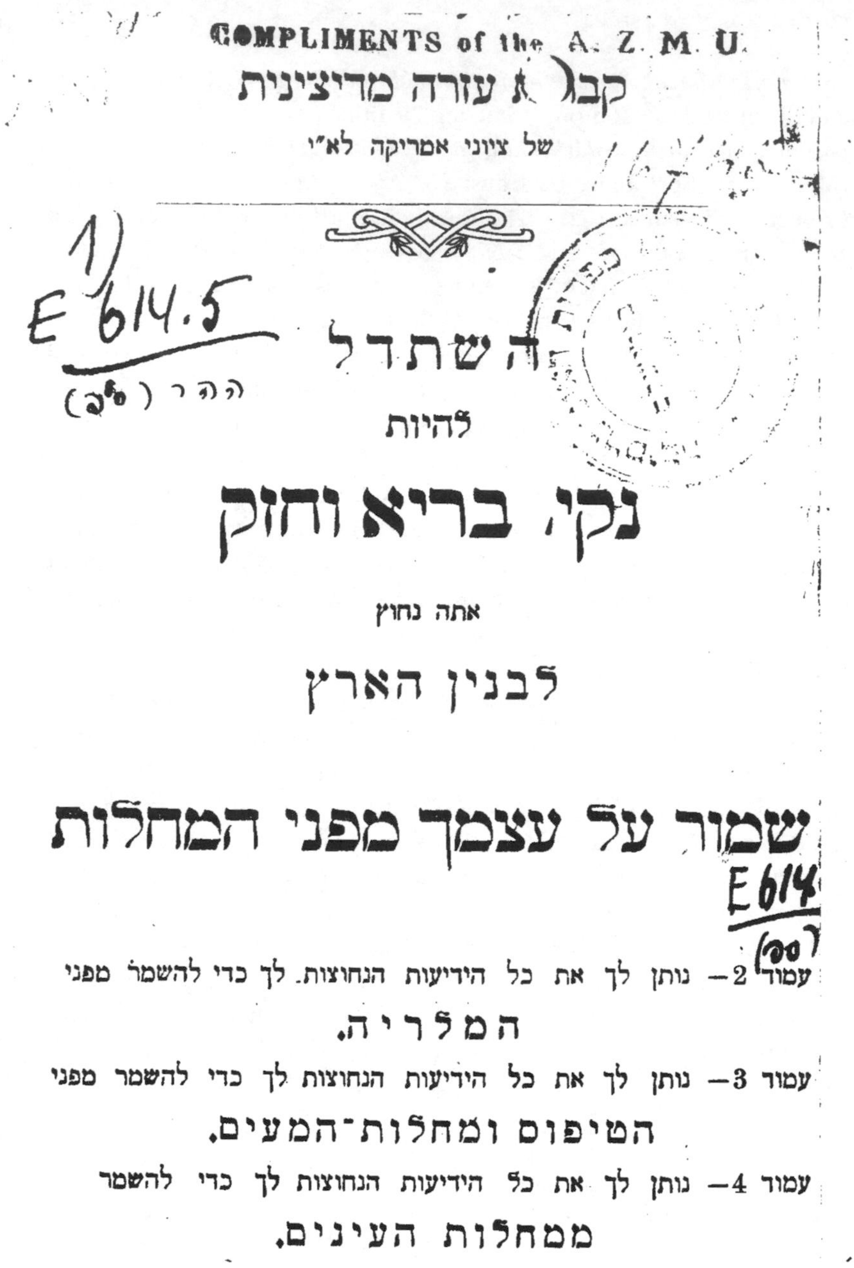
COMPLIMENTS of the A. Z. M. U.

קבוצת עזרה מדיצינית

של ציוני אמריקה לא"י

השתדל

להיות

נקי, בריא וחזק

אתה נחוץ

לבנין הארץ

שמור על עצמך מפני המחלות

עמוד 2 — נותן לך את כל הידיעות הנחוצות. לך כדי להשמר מפני

המלריה.

עמוד 3 — נותן לך את כל הידיעות הנחוצות לך כדי להשמר מפני

הטיפוס ומחלות־המעים.

עמוד 4 — נותן לך את כל הידיעות הנחוצות לך כדי להשמר

ממחלות העינים.

FIGURE 6.2. "Try to Be Clean, Healthy and Strong: You Are Vital for the Building of the Land." From *Hishtadel lihiot naki, bari, vehazak: hora'ot beshmirat habriut* (Try to Be Clean, Healthy, and Strong: Instruction on Protecting Health), pamphlet no. 1 (1931). Ein Kerem Medical Library.

weaken one's body, diminish one's blood, impede the ability to work, and sometimes cause death. The infectious nature of the disease seriously endangered the community at large. The most important warning was about malaria's potential impact upon children, the *Yishuv*'s cherished future. Strong, healthy children were potential healthy soldiers, both literally, for the Zionist nation, and figuratively, in the war against malaria. The pamphlet explained that weak individuals exhibiting blood deficiencies (anemia) caused by malaria could not hope to have strong and healthy children. Believed to have cumulative effects from generation to generation—a belief that reflected the state of knowledge on malariology at the time—the damage from malaria could result in a group of Jews who were "exhausted and broken" and "aged before their time"; in effect, replications of the Diaspora Jew.[58] Inattention to malaria prevention would jeopardize the reproduction of healthy children and therefore frustrate the Zionist project of self-transformation and ultimate redemption. A ten-point program was presented for *Yishuv* members to follow. The last two points read:

> (9) Only with the efforts of all the public will we be able to eradicate malaria; thus, every person from our community (*eda*) should try hard to join the battalion of the war against malaria in order to lengthen the days of our nation; (10) Thus the motto of this battalion will be: Malaria is a national tragedy and a national sin.[59]

Malaria eradication had to be elevated to a national duty in order to ensure complete mobilization, cooperation, and compliance. Even Aaron Aaronsohn, director of the Jewish Agricultural Experiment Station, went so far as to say that every malaria death should weigh on each person's conscience.[60]

Dr. Puchovsky was the author of another pamphlet, *Malaria: mehut hamachala vehalmilchama ba* (Malaria: The Nature of the Disease and

---

58. *Sicha 'al Malaria* [Conversation about malaria], Hadassah National Library, no. 3 (Jerusalem, 1921), 11, CZA Library 4167.

59. *Sicha 'al Malaria*, 11–13.

60. See report of director of the Jewish Agricultural Experiment Station on the Malaria Campaign, appendix no. 1 (no date but likely the early 1920s), 1, 16, CZA J113/1430. See also H. R. Carter, assistant surgeon general, U.S. Public Health Service, 'Resume of Methods for Control of Malaria: Indications, Results, Costs," 1, CZA J113/1430, for an example on how individual versus community factors structured an antimalaria plan.

the War against It). The pamphlet cost 5 grush and was supposed to be published by the malaria station of the Vaad Leumi Health Council in the winter of 1920. The timing was probably set to coincide with the increase in mosquitoes that occurred in the aftermath of the rainy season and an impending wave of immigration, but its publication was delayed until March 1921, due to lack of funds.

Instead of focusing on the *ramifications* of individual action for national survival, Puchovsky made individual action a *prerequisite* for a national endeavor. That is, only through the cumulative effects of individual action against disease could a national endeavor actually take place. Here individual effort in combating disease took on both symbolic and literal significance as the way to carry out one's duty to the nation. Dr. Puchovsky explained that the spread of malaria brings many people to a state of "degeneracy" and "becomes a terrible scourge for the nation in Palestine."[61] He suggested practical ways to educate the public about malaria so they could fulfill their individual obligations—popular lectures, written propaganda, and especially lessons in the schools—because only by transmission of this information into the "brain and blood of the young generation" could the campaign "bring blessing to the land."[62] Elsewhere in the pamphlet, Puchovsky mentioned the socioeconomic issue so common in public health debates by saying that the poor had a greater incidence of malaria than the well-to-do because of their inability to fight the disease and their close living quarters.

Framing malaria control as a moral obligation (or the lack thereof as "national sin") revealed the nature of the Zionist health project as intrinsically reformative, as one not only about physical improvement but also about moral rehabilitation. Malaria propaganda expressed the elevation of moral and physical characteristics in the new, Zionist Jew.

Rationality, Discipline and Changing Habits in the Fight against Malaria

Insistence upon "rational" behavior and "rational" means was frequent in discussions about antimalaria projects and health activities in general. Dr. Puchovsky, for instance, explained that when "participation takes

---

61. Dr. L. Puchovsky, *Malaria: mehut hamachala vehalmilchama ba* [Malaria: The nature of the disease and the war against it], translated into Hebrew by Ezrachi-Krishovsky (Jaffa: Eitan and Shoshani, 1920), 3, Ein Kerem Medical Library.
62. Puchovsky, *Malaria*, 4, 7–8.

place in an active way by all residents of Palestine, from big to small, and the war [against malaria] will be organized according to rational means," then positive results would occur."[63] In his text, Puchovsky stated that there were five ways to control malaria, although none of them provided total defense: two that attack the *Anopheles* and three that target man and promote resistance to the disease. Each method required extreme and uniform, repeated attention, discipline, and supervision by the individual. These mechanisms were based upon scientific knowledge for combating malaria and were therefore considered "rational" approaches.

"Rational" health behavior was also required in the fight against malaria: constantly checking stagnant pools and monitoring other people's health behavior were ways to prove individual responsibility.[64] Discipline within the individual and the community was vital for preventing another outbreak of malaria that would threaten the nation.

In order to decrease the chance of contracting malaria and increase resistance, general hygienic principles, a "rational" medical regimen, had to be followed.[65] This regimen required strict order in all aspects of one's life: economic lifestyle should be sufficient, eating habits should be fixed, a certain number of hours of sleep should be strictly kept. The "rational" approach of hygiene was also a moral prescription: "Damage is done when one comes late from a party or spends time dancing and singing. One needs to go to sleep at an early hour."[66] Tobacco and alcohol were to be used sparingly as any excess was said to weaken the body and make a person more susceptible to infectious diseases like malaria. Finally, the author noted that the cultural state of the *Yishuv* was one of the most important ways of fighting epidemics. By propagating more information about malaria and taking on these new habits, the members of the *Yishuv*

63. Puchovsky reiterates the Zionist claim throughout the Mandate, that the government should actually be the main ones undertaking the reclamation of the land (Havra'at ha'aretz), but he states that all efforts will go to waste without the cooperation of the people. Puchovsky, *Malaria,* 4.

64. Michel Foucault, *Birth of a Clinic: An Archeology of Medical Perception* (New York: Vintage Press, 1975), and *Discipline and Punish: The Birth of the Prison* (Vintage Books: New York, 1979). Foucault discusses how modern hygiene requires disciplining oneself and others.

65. Also Dr. Norman's *Darkei briut (hygiena yom-yom)* [Ways of health (daily hygiene)] (Tel Aviv: Shalom Am Library, 1926), Ein Kerem Medical Library.

66. Puchovsky, *Malaria,* 23.

would bring the flowering of the land that had previously "eaten up the inhabitants thereof."[67]

Exhortations to exhibit discipline were couched in terms of war, even in literature addressed to a young audience.[68] Military imagery was common in depictions of modern medical disease and its eradication. In his classic history of malaria, Gordon Harrison notes that the majority of doctors in the field of malaria control in fact saw themselves at war.[69]

All sanitary work was viewed as a form of war that could be won, in part, through discipline. As the director of the Jewish Agricultural Experiment Station of Palestine recognized, "It [malaria control] requires not only money and efforts, but also stout organization and discipline."[70] The disease and its transmitter, malaria and the mosquito, were considered formidable enemies of the *Yishuv* since they posed obstacles to the central activities of the Zionist project: immigration, settlement, and labor. In this imagining, the *Yishuv* was made up of disciplined soldiers who were to organize into small battalions and, through the militaristic medical methods of treatment and prevention, conquer the enemy and win the war. The millions of parasites in the red blood cells of malaria sufferer were often described as encroaching armies.

Discussions about discipline in malaria control also revealed Zionist images of the Palestinian Arab population and of non-European Jews, circumscribing the intersection of race and health in antimalaria campaigns.[71] Certain racial groups caused great anxiety for Zionist health workers and bureaucrats because they perceived that the groups lacked and even resisted Western hygienic principles. This perspective was not just a Zionist one but also part of the more general colonial mindset of

---

67. Puchovsky, *Malaria*, 23. Puchovsky here takes the phrase straight from the Bible (Numbers 13:32).

68. Fleck, *Genesis*, 59–61; Susan Sontag, *Illness as Metaphor and Aids and Its Metaphors* (New York: Doubleday Books, 1989).

69. Harrison, *Mosquitoes, Malaria and Man*, 2; Deborah Lupton, *Medicine as Culture: Illness, Disease and the Body in Western Societies* (London: Sage Publications, 1994), 61–66.

70. Report of director of the Jewish Agricultural Experiment Station on the malaria campaign, appendix 1 (no date but likely the early 1920s), 17, CZA J113/1430. Also Dr. Matmon, "HaHitui" [Disinfection], and an explanation of germ theory in *Briut* 1, no. 5 (October 1, 1932): 35, Ein Kerem Medical Library.

71. The Sephardim (Jews originating from Spain) and Mizrachim (Jews originating from Middle Eastern and North African countries) were not generally among the leadership of the Zionist movement but, along with the European Jews (Ashkenazim), were the objects of the movement's health and hygiene efforts.

the time. In August 1936, the *HaAretz* column "Our Health" had a two-part article by Dr. Beckman called "The Problem of Malaria and Swamp Drainage in the Northern Sharon." These articles detailed the end of the Mandatory government's Birket Ramadan/Wadi Falik project. According to Beckman, the project was completed because of great pressure exerted by the *Yishuv*. In his description of the transmission of malaria and the way to fight it, Beckman expanded his discussion of malaria strategies to the cultural and economic level of residents living in malaria-infested areas.[72] Supervision of the residents, Beckman explained, was necessary because within malarial areas, bedouin roamed the land with their sheep, providing a constant host for malaria infection. The situation was further complicated by Jewish residents, especially those from Eastern countries, who did not comply with restrictions of sleeping inside and using mosquito nets.[73] Instead, he said, youth danced outside at night in the summer while adults took their symptomatic children to the doctor and then ceased treatment and visits, eventually causing the patient to show signs of anemia. The only way to rid Palestine of malaria, he concluded, was to engage in a comprehensive program that included drainage first and foremost but also addressed resident compliance with preventive measures. Anxiety about compliance, when it surfaced explicitly in this malaria propaganda, usually did so within a discussion of either the Palestinian Arab population or the non-European, Jewish population.

Briut HaTzibur (Health of the Public), written by Dr. A. I. Levy and published in 1935, also revealed Zionist images of the land as desolate and the Palestinian Arab population as unproductive and unsanitary. The author began by noting the *Yishuv*'s strong interest in health matters as evident in their frequency of reading about health in Jewish settlement libraries. According to Dr. Levy, there was a great yearning of the public for a comprehensive guide to health and scientific issues. As a response to this need, *Briut HaTzibur* was written along with four other books published later: *Sanitatzia* (Sanitation), *Briut shel hayechid* (Health of the Individual), *'Ezra rishona* (First Aid), and *Tipul leholim babeit* (Caring

72. Dr. Beckman, "Ba'yat hamalaria veyibush habitzot beSharon hatzfoni," part 1, *HaAretz*, August 18, 1936, 4.

73. Dr. Beckman, "Ba'yat hamalaria veyibush habitzot beSharon hatzfoni" part 2, *HaAretz*, August 25, 1936, 4. Also Sandy Sufian, "Anatomy of the 1936–1939 Revolt: Images of the Body in Political Cartoons of Mandatory Palestine," *Journal of Palestine Studies*, forthcoming.

for the Sick at Home). The Straus Health Center and Hadassah Medical Organization sponsored all of these books. The series was produced as pocket books to carry as constant guides to health and hygiene.

Dr. Arieh Behem's introduction to *Briut HaTzibur* opened with the topic of swamp drainage:

> Our land develops gradually. Draining the swamps, maintaining their destruction, enables the settlement of places either abandoned or desolate, that were neglected for thousands of years without any residence. . . . Then comes the Hebrew settler, pioneer of immigration, the land was in large part not settled, desolate and abandoned. And [in] wide territorial spaces [there] were large swamps, which caused half of the death of those settlers in the area. And thus the residents of Palestine paid, especially the first immigrants, a great tax/duty to infectious diseases in our land, like malaria.[74]

The residents referred to here were not the Palestinian Arab residents but rather the first Jewish immigrants. This description not only ignored the presence of the bedouin who used the swamps as pastureland, but also the *fellahin* in the hills who migrated daily to the swamp for its various uses. When the booklet acknowledged the presence of Arab residents, it did so by identifying their majority status while simultaneously describing them as unhygienic: "our cousins, the Arabs, the majority of the residents in Palestine, are located all over in their villages, in conditions that are unsanitary."[75] Dr. Behem explained that because of poor communications between villages, individual Arab health rather than public health was the primary concern of the Arab community. According to Behem, Arab inversion of Zionist priorities—a supposed emphasis on the individual rather than the collective—explained the underdevelopment of the Palestinian Arab population and highlighted the importance of ensuring the health of the Zionist collective through collective means. Moreover, Zionist settlement, he stated, could change the Arab's unsanitary position by bringing "enlightenment" to both communities. The introduction to the book continued by describing the history of this process as well as the initiative of the *Yishuv* in founding activities that would bring such enlightenment: "The Hebrew *Yishuv* was the first that started a campaign

74. Dr. Arieh Behem, "Hakdama" [foreword], in *Briut HaTzibur: drachim ve'emtza'im shmor neged machalot midbakot* [Health of the public: Ways and means for protecting against infectious diseases], ed. Dr. A. Levy (Tel Aviv: Achiever Press, 1935), 9–10.

75. Behem, "Hakdama," *Briut HaTzibur*, 10.

against infectious disease in order to guard public health.... [where] a war
against malaria was established in the Galilee...there was [also] a need
felt for wide, organized educational work among national authorities in
our land."[76] Zionist health activities would "civilize" the Jewish members
of the *Yishuv* but would also benefit the Palestinian Arab population.

Other Cultural Formations of Malaria and Health

In addition to health booklets and other forms of health propaganda, oral
and written stories as well as films about malaria and its eradication contributed the formation of a culture of health in the *Yishuv*. The Jezreel
Valley and its drainage acquired elite status in the Zionist narrative, for it,
like the blooming of the desert, was the full flower of Zionist pioneering.[77]
The story of swamp drainage in the Jezreel embodied all aspects of the Zionist mythology: a desolate land and uprooted people made anew through
havra'at hakarka vehaYishuv, kibush ha'avoda, and *kibush hakarka.*

Cultural formations about the Jezreel Valley helped heal the people
by expressing their pride, celebrating their self-sufficiency, and connect-
ing the practical healing of the land (through drainage) with the physical and psychological transformation of the Jew's self-image. Zionist love
songs, articles, stories, and pictures of the period all express a love for the
Jezreel; its transformation was considered a "metamorphosis of swamps
into a flowering garden."[78]

Evoking these symbols, Shaul Tzernichovsky's poem "Hazon Nevi-
haAshera" (Vision of a Prophet of the Ashera) begins with the biblical
story of Elijah's struggle against the prophets of Ba'al and invokes the
Jezreel Valley to which Elijah eventually fled.[79] The poet sets up the biblical connection of the Jewish people with the Land of Israel, proceeding

---

76. Behem, "Hakdama," *Briut HaTzibur,* 11.

77. Yoram Bar-Gal and Shmuel Shamai, "Habitzot shel Emek Yizra'el-mitos or metziut?" [The Swamps of the Jezreel Valley—Myth or reality?], *Cathedra* 27 (1983): 163–174,
quotation on 174, for other figures who wrote about the Emek and malaria.

78. Karmi and Talmi, eds.,*The Album of the Valley* (Tel Aviv: Am-'Oved 1965), cited
in Yoram Bar-Gal and Shmuel Shamai, "Swamps of the Jezreel Valley: Struggle over the
Myth," *Sociologia Internationalis* 25, no. 2 (1987): 193–206," quotation on 195. Grove, *Green
Imperialism,* 13–14; Philip, *Civilizing Natures,* 36, 63; "*Shir leEmek,*" no date, CZA J17/7640.
Music notation accompanies this song. Since this song was intended for the twentieth birthday
of the Emek, the date must be around 1942.

79. Kings I:18–19. Ashera was a Phoenician goddess of prosperity and fertility. The word
also means grove. Reuben Alcalay, *The Complete Hebrew-English Dictionary* (Ramat-Gan:
Massada Publishing, n.d.), 175; Shaul Tzernichovsky, "Hazon nevi-haAshera," *Shirim*
(Jerusalem: Schocken Publishing, 1943), 501–504.

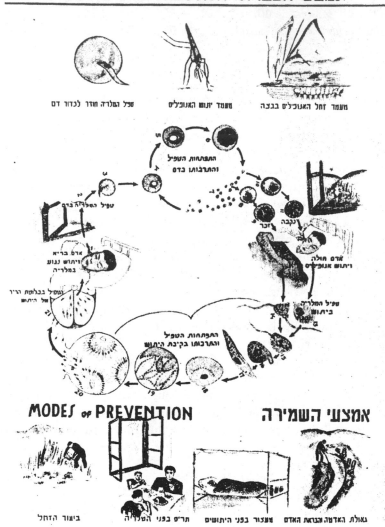

FIGURE 6.3. "Malaria: Modes of Transmission." From Dr. A. I. Levy, *Briut HaTzibur* (Tel Aviv: Achi'ever, 1935), 154. Ein Kerem Medical Library.

to show that the land is in wait of the Jews' return. He personified the past-present Jezreel/Emek/Marj ibn ʿAmir:

> You, you the Emek, large and fertile
> Whose kidneys were blessed with leaves
> Emek Yisrael, ancient Emek!
> As God's reign over you—you will be desolate,
>
> And your servants of the crucified your swords will multiply,
> and in the shadow of the Prophet of the desert, the swamps will cover you,
> A curse to the animals, and a curse to the people who reside in you.
> In the vestiges of your farms swarm (*yishrotz*) a lowly, hated nation,
> In the streams of the mountains that enclose you.
>
> [All of this will continue to exist]
> Until a generation of *maʿapilim* (daring pioneers) will come
> Pioneers of the Diaspora and dreamers of your redemption,
> And I was to them a God, I am the owner of the Emek
> Master of the soil (*karka*), Masters of the Emek.
>
> The smoke from the [sacrificial] incense rose—the smoke of camel droppings
> The fragrance of the plowed soil...
> Zion will be redeemed with the hoe and her fields in cultivation,
> Like the blessing of a new God in the Emek.[80]

While the poem exalted Jewish mastery over the soil and Jewish redemption, it also inferred the erasure of Palestinian Arab cultivation and activity from the area and the likely disappearance of indigenous methods of ridding one's dwelling of mosquitoes through smoke: the "smoke of camel droppings." This custom was replaced with modern work, which produced the "fragrance" of cultivated soil. Of note as well is the word *yishrotz*, which usually refers to the creeping or swarming of insects. Tzernichovsky was probably referring to mosquitoes here in keeping with his reference to swamps two lines before. The choice of this word may have also, however, connoted the Arabs as idle types, not making use of farms and proper cultivation and therefore exhibiting their alleged indifference toward the land, rendering them "a lowly, hated nation." For

---

80. I thank Becky Molloy for helping me translate this poem.

Tzernichovsky, in contrast to the representation of the Arab peasants, the Jezreel lay in wait for the arrival of the daring Zionist pioneers who brought to it a religion of work, a "new God in the Emek."

Another illustration of the Emek's special place in Zionist national mythology and the culture of health was seen in a call to the Zionist youth that made the purchase and reclamation of the Jezreel Valley a lesson and the land (re)productive:

> From this Emek we learned: in the national struggle for revival they overcame the neglect through the power of belief. The cooperative effort along with the power of will of the sons of the nation that answered [the call] fertilized (*yafriach*) the desolation, subjugated the powers of destruction and established a project of life and creation...The Emek is to us the source of bread, the source of vision and belief in a good future for our people. We believe that our future—your future will be of a healthy nation that sucks (*yonek*) the strength of a peaceful life...from the soil of our native land...Brothers, friends from the fields of Emek Yisrael, we call to you: Extend yourselves to a project that redeems the land. Take your hand to a project of building corners of our independent lives in the native land![81]

This call reappropriated the Jezreel as the Jews' native land, a land that was redeemed through activism through hope and mastery over nature. Settlers and Zionist leaders understood this new approach as a way to make the Jewish nation a healthy nation, one that took its sustenance and strength from a land that Jewish settlers had transformed and in so doing they create a deep attachment with it.

A final example of the place of the Jezreel Valley in Zionist culture was shown in the 1935 film *The Land of Promise*. In the middle of this film, we see settlers in the Jezreel Valley singing a song in the dining room about the Emek.[82]

81. The term *yonek* means to breastfeed, thus evoking a common gendered conception of the land as female and as reproductive. "Davar hano'ar beEmek letza'irei Israel," no date, CZA J17/7640.

82. Jewish Agency, film stills from *Land of Promise*, Rossif Collection, RF037, produced in France, Steven Spielberg Film and Video Archive at USHMM, http://resources. ushmm.org/film/search/index.php. The song about the Emek is at minutes 25–28 of the film. See also Hadassah, *Health for Victory* (film), 1942, Hazel Greenwald Collection, VT DA0786, produced in USA, Steven Spielberg films online; Keren Keyemet L'Israel, *This Is Our Valley*, for propaganda film (for Canadians) on Emek Hefer/Wadi Hawarith and swamps, Matis Collection, 1947.

Malaria Education Efforts Targeted to Specific Groups

Malaria education efforts addressed the general Jewish population in Palestine and targeted specific groups that health officials believed were particularly at risk. The immigrant population was a critical focus for malaria propaganda and public health education as it was the most susceptible to malaria infection ("pioneer malaria") and needed to be taught how to acclimate to Palestine's subtropical conditions. According to Henrietta Szold in 1920, the care of the immigrant was the most important Zionist health work.[83] Indeed, at the beginning of the Mandate, malaria and immigrant care were considered two sides of the same coin. A letter to Dr. Bernard Flexner of the Joint Distribution Committee in 1921 makes this point clear: "With the eradication of malaria we will eliminate in a large measure the problems of the immigrant and relief as well."[84]

Malaria was considered the greatest enemy of immigration because of its quick propagation, its frequent manifestations, and its weakening effect on the human body.[85] Upon landing on the coast of Palestine, immigrants were segregated by officials into barracks to "prevent their being infected by the native population" and to "enable the physician to examine their physical fitness for various kinds of labor" as well as to "facilitate education and instruction in regard to precautions to safeguard health." All of this was considered "purely Zionist work." Malaria education efforts began immediately on arrival of the immigrants; quinine, lectures, and instructions on cleanliness and malaria prevention were given.[86] These educational efforts, like those directed to the general public, urged the adoption of new health habits and the use of strict individual discipline in maintaining them as vital to the national project. Instructions were quite ardent and explicit. Because malaria claimed or debilitated so many lives, especially European ones, malaria leaflets for immigrant distribution sought to produce a particular association to one's location and

83. Letter from Szold to Judge Mack, December 26, 1920, 3, CZA J113/554.

84. Letter to Flexner from unknown (likely Kligler or Katznelson), January 4, 1921, 2, CZA J113/554.

85. "Report on Immigration and Sanitation," CZA J113/1431. Also letter from Dr. Noack to the German Department of the Jewish Agency about preparing malaria pamphlets in German for immigrants due in the summer of 1935, April 15, 1935, CZA S7/149/4.

86. *Jewish Morning Journal*, May 2, 1921, as quoted in *Hadassah Newsletter*, 12, ed. Mrs. E. Jacobs (May 1921), 1, CZA Library; letter to Flexner from unknown (likely Kligler or Katznelson), January 4, 1921, 2, CZA J113/554.

community as well as a new relationship to nature and water.[87] The back of a 1922 immigration certificate gave directions to its recipient on how to fight against malaria:

> Try to be clean, healthy and strong, you are vital to the building of the land!!
> The mosquito is your enemy, try to stay away from it.
>
> Malaria: it has the largest incidence in *Eretz Israel*. It causes attacks of malarial [fevers] and in regard to other matters, it requires prolonged treatment and a great loss to the land and to generations.
>
> Guard yourself:
>
> 1. Take quinine every healthy day that the parasite has entered into the blood... Keep quinine at work, in your home and on a journey... quinine is the best way to protect against malaria...
>
> 2. Don't allow the mosquitoes to spread from the surroundings to your home. [Throw away] all necessary tools for water and [wash] metal once every two days.
>
>   Pay attention that pipes and iron don't crack. Once a week, pour a half a liter of oil inside your cistern or well.
>
> 3. Sleep under a mosquito net...[88]

The most common malaria leaflet, published by the Health Department of the Vaad Leumi, described the nature of malaria, its symptoms of fever and shivers, and method of transmission.[89] Suggestions for distributing the leaflet included having leaders of immigrant communities perform the duty at the port cities of Tel Aviv and Haifa and supplying copies of the leaflet to all immigrant houses in those cities. It was important for every immigrant to receive the leaflet, regardless of their location, because it "is impossible to know if the immigrant will go one day to another place... that is the most dangerous in terms of malaria."[90] The leaflet was so popular that in May 1935, and again in October 1935, the Union for

87. Goubert, *Conquest of Water*, 126.

88. The certificate directions gave specific times and medicinal doses for each age group. Z. Saliternik, "Bitzot vekadachat beEmek Yizra'el," 183. A longer, more detailed reprint of this card can be found in Hebrew Aid Society of American Zionists for Eretz Israel. *HaMalaria: Hitpashtuta, meholeleyha, darechai ha'avarata veEmtzai hahishramrut mepnaiha* (Jerusalem, 1921), p. 7, National Library of Jerusalem.

89. Vaad Leumi leKnesset Israel: Machleket habriut [Health Department of the Vaad Leumi], "Hamalaria," CZA J17/648.

90. Letter from Dr. Y. Dodson, medical officer, to Dr. Noack of Health Department of Vaad Leumi, July 1, 1935, CZA J1/3745.

German Immigrants ordered 10,000 copies of it from the Vaad Leumi printed in German and Hebrew. Similarly, the Jewish Agency Department of Immigration ordered a "large load" of these leaflets. The Vaad Leumi leaflet was also sent to the schools.[91]

The short leaflet noted tertian and tropical (malignant) malaria as most common in Palestine. It described their symptomatic differences so that the newcomer could identify which type of malaria s/he had contracted. Immigrants were encouraged to take quick action on medical treatment; otherwise, the condition could turn into a chronic one or even result in the death of the individual or other members of the *Yishuv*. Therefore, the immigrant also was instructed to explain to his/her neighbors about the war against malaria ("because you will not succeed in this war, unless everyone fulfills the doctor's orders"[92]) and to report all noncompliance or dangerous conditions to the sanitarian or doctor. As new members of *Yishuv* society, immigrants were charged with enforcing hygiene and discipline through malaria prevention.

In order to control the environmental conditions that were conducive to mosquito reproduction, immigrants were instructed in this pamphlet, as in the certificate directions, to prevent the formation of water pools near their house or camp. Adoption of the new habit of supervising conditions in order to prevent disease became a national responsibility that immigrants bore from the moment of their arrival:

> Dry all the small swamps, make sure that the faucets/taps of the water pipes do not drip, and if puddles of water form near faucets and pits/holes/cisterns, you must then dry them at least once a week; clean those ditches with standing water and always be careful to cover cisterns. For those large swamps that cannot be drained, pour oil in them in order to kill the larvae. Destroy all the mosquitoes in your house (by using fly swatters or 'Flit') and pay special attention to dark corners in the rooms of the house, which is where mosquitoes usually collect.[93]

The leaflet warns its reader not to go out at night in malaria-infested areas and to take all necessary precautions against mosquitoes indoors (i.e.,

91. Letter from Smilansky to Vaad Leumi, August 15, 1935, and letter from Histadrut Germany 'Olim in Tel Aviv, to Vaad Leumi Medical Supervisory, October 6, 1935, CZA J1/3745; letter from Dr. Noack to heads of schools, June 3, 1936, CZA J1/1542.

92. Vaad Leumi leKnesset Israel-Machleket habriut, "Hamalaria."

93. Vaad HaBriut, "Malaria" (May 1935), 2, CZA J1/1542.

using screens on the doors and windows and covering one's bed with a net). The immigrant was given contact information for organizations that could answer questions regarding malaria, the health stations in Jerusalem and Tel Aviv, all Kupat Holim dispensaries, and the Health Department of the Vaad Leumi. Finally, the leaflet connected malaria with immigration and settlement policies, noting that "if there is no place for your settlement, [then] there must be malaria in the area."[94]

Other educational books appeared that focused on topics relating specifically to malaria. Kupat Holim published another book that was aimed at immigrants, called *Eich tishmore ʿal briutcha beEretz Israel* (How to Protect Your Health in *Eretz Israel*) in 1935. This book was one of a few health education materials in the 1930s that addressed the subject of malaria in a separate chapter. The information in the book was also published in *Yediot* (Information) of the German Immigrant Society in *Eretz Israel*. *Briut haTzibur* (Health of the Public) by Dr. A. Levy was another example of material that treated malaria separately. Because *Eich tishmore* was written for the immigrant, the Hebrew text of this book was fully voweled. Its editor, Dr. Berman, called attention in the introduction of the book to the need for acclimatization of the immigrant to the new subtropical environment of Palestine as well as to the special demands of hard physical labor. Recognizing the incidence of disease among new immigrants within the first years of their settlement, the editor asserted that it is important for them to be aware of conditions in Palestine before immigration and to know how to protect and adapt quickly to their new home.[95] Knowledge about acclimatization and malaria was particularly needed in cases where young people were immigrating without their parents and therefore "lacking the attentive eye and hand of the mother—the caretaker." The consequence of this lack of supervision would be a sick Jewish nation rather than a healthy one: "[O]nly healthy and strong constitutions will be able to survive the many obstacles that stand before the immigrant in our land . . . Our land, in its present state, needs healthy and strong people. The weak and the sick create a burden on others and on themselves."[96]

94. Vaad HaBriut, "Malaria."

95. Dr. Berman, "Hakdama" [foreword], *Eich tishmore ʿal briutcha beEretz Israel* [How to protect your health in Palestine], ed. Dr. Berman (Tel Aviv: Kupat Holim, 1935), 1, 41, National Library, Jerusalem.

96. Dr. Meir, "Al yaʿale holim leEretz Israel!" [Do not immigrate to Eretz Israel sick!], *Eich tishmore al Briutcha beEretz Israel* [How to protect your health in Palestine], 44, National Library.

Dr. Meir, in his chapter titled "Al yaʿale holim leEretz Israel!" (Don't immigrate to *Eretz Israel* sick!), went even further than Dr. Berman by warning people with medical conditions that they would most likely experience a resurfacing of that old condition quickly after coming down with an infectious disease like malaria. Thus, Meir advised, immigrants should not come to Palestine before they were fully cured of any medical condition and should not consider immigrating quickly after having a serious illness. Meir added that people with weak constitutions were usually the most miserable because they were neither healthy nor sick and that they would become lonely and strange to others and ruin the "society and the land." On each immigrant's physical constitution rested the success of the land and its society.[97]

Eich tishmore discussed the inability to be free of the risk of malaria. It added a unique twist to the discussion by specifically referring to Jewish holidays. The chapter on malaria by Dr. Fishel started out by contrasting Diaspora and Palestinian conditions with a quote of the Diaspora adage, "Purim is not our [major] holiday and malaria is not a disease." Purim is the Jewish holiday based on the plot of a government minister named Haman to exterminate the Jews in Persia and overthrow the king. In the story of Purim, Haman's plot is revealed to the king by Esther, a Jewish woman, and the Jews are saved. She becomes a heroine. Esther's Hebrew name is Hadassah, which is also the name of one of the main Jewish medical institutions in Palestine. Perhaps in Fishel's reference to Purim and its importance in Palestine, he implied that malaria and Haman are equal as enemies to the Jews. The invocation of Purim also reminded Jews that only in their homeland could they hope to escape anti-Semitism. After his initial reference to Purim, Dr. Fishel noted that the saying applied to the Diaspora but not to Palestine. On the contrary, he wrote, "In our land, Purim is an important holiday and also malaria—even though it is given to treatment—is a stubborn disease, that causes great suffering to the individual and a great loss in work and in property to the nation as a whole."[98] He considered malaria a serious obstacle to the development of Palestine.

A cyclical pattern of malarial incidence led Dr. Fishel to warn the reader not to assume that once an area was under malaria control, it would be clear of malaria forever. Stringent, repeated maintenance of

97. Meir, "Al yaʿale holim leEretz Israel," 41–44.
98. Dr. Fishel, "Malaria," *Eich tishmore ʿal briutcha beEretz Israel* [How to protect your health in Palestine], 28, National Library.

antimalaria measures had to be kept up. Fishel noted that although the main responsibility for draining swamps and getting rid of mosquito larvae lay with land reclamation and governmental agencies, it was up to each individual to maintain his own health and the antimalaria measures near his settlement by making weekly inspections of closets, dark corners, barns, under beds, and so on. Individual and government cooperation as well as social responsibility were considered key to the complete elimination of malaria.

A pamphlet printed in 1939, edited by Dr. Berman and published by the Council on Immigration of the Jewish Agency and Kupat Holim, also addressed immigrants. It borrowed information and pictures from *Eich tishmore ʿal briutcha beEretz Israel* to produce a smaller *"Shulchan ʿAruch* on the Principles of Health." The *Shulchan ʿAruch*, written by Joseph Karo in the seventeenth century, was considered the most authoritative *Halachic* (Jewish law) guide for daily living accepted by both the Sephardic and Ashkenazi communities. The pamphlet's titular reference to the *Shulchan ʿAruch* proposed a specifically Jewish association to acquiring new habits of health by linking broad health issues and malaria with Jewish religious practices and texts. In effect, it prescribed a new set of secularized *mitzvot* (deeds) that Jews should perform in Palestine. Like the book that preceded it, the 1939 pamphlet discussed the difficulties for Western and Central Europeans in acclimatizing to Palestine to the point that the dream of coming to *Eretz Israel* could "evaporate into despair and disappointment."[99] The pamphlet gave the immigrant directions on how to maximize acclimatization and advised both spiritual and physical training before departing from Europe.

Physical training programs (*hachshara*) started in the Diaspora. In order to arm prospective immigrants with the skills necessary for agricultural labor and prepare them for the harsh, malarial conditions in Palestine, training camps were set up where potential candidates underwent "progressive adaptation." There they would learn how to work the land, study the economic, health, and social conditions of Palestine, and test their individual mental and physical capabilities.[100] The conditions

99. Dr. Davidson, director of the Medical Office of the Jewish Agency Department of Immigration, "Hakdama" [foreword], *Shmor al Briutcha be Eretz Israel* [Protecting your health in Palestine], ed. Dr. T. Berman (Tel Aviv, 1939), 4, National Library.

100. Moulin, "Tropical without the Tropics," in Arnold, *Warm Climates*, 164, for similar preparations by French for settlement in Algeria; "Selection of the Fittest" (1919), reprinted in Ruppin, *Three Decades of Palestine*, 73.

of Palestine were intentionally reproduced in these camps in order to acclimate the prospects to their final destination. Medical examinations checked the physical fitness of applicants before immigration certificates were issued.

Regeneration and rehabilitation of the Jewish soul, mind, and body took on new meaning for the medical community in the 1940s as they prepared for the immigration of Holocaust survivors to Palestine. The absorption of these Jews with their various infectious diseases, especially typhus, alerted health officials to particular medical concerns and needs but also provided an opportunity for the renewed invocation of the Zionist ideology to mobilize funds and people. As Kligler remarked, "There are the basic problems of preventing epidemics and rebuilding health. And last but not least there are the problems of human reconstruction, of rebuilding the soul, of reconstituting these wrecks into self-reliant, productive human beings."[101]

The project of health transformation and health education in Palestine was not one that was all-inclusive but, before the rise of Hitler, one that primarily addressed a preselected group: before an immigrant could come to Palestine, she or he had to meet strict criteria for health. The Hon. J. M. Kenworthy wrote, "The colonists are chosen for their physical and mental fitness and, in the majority of cases, are given some preliminary training and instruction in agriculture."[102]

Screening the health of immigrants was a common practice during this period around the world, one that was based on eugenic principles and which barred people with infectious diseases.[103] The Mandatory government refused immigrants with infectious diseases and the Zionists had to comply, but the Zionists also formulated criteria for their selection of immigrants abroad.[104] Health policy was less documented than other aspects

101. See heightened provisions against typhus (immunization and disinfestation), both in refugee camps in Europe and then at receiving stations in Palestine, in preparation for large numbers of Jewish refugees from Europe after World War II. Notes of Dr. Kligler's Address at Convention on Friday, October 16, 1942, 1, Hadassah Archives, RG1/100/2; memorandum by Kligler regarding typhus, November 11, 1942., Hadassah Archives, RG1/100/ 2; presentation by Dr. Kligler at the Palestine Committee Meeting, PostWar Health Planning, October 27, 1942; Hadassah Archives, RG1/100/2; notes of Dr. Kligler's address at Convention on October 16, 1942, Hadassah Archives RG1/100/2.

102. J. M. Kenworthy, "Revisiting Palestine: Impression of the New Jewish Life," *New Palestine* 12, no. 3 (January 21, 1927): 70, PRO CO 733/135/11.

103. Weindling, "Eugenics and the Welfare State," 135.

104. British Immigration Ordinance, 1925, ISA M1529/13/3/810.

of Zionist immigration policy; immigration criteria are usually analyzed with regard to their economic categorization for distribution of immigration certificates.[105]

Health pamphlets were distributed to emigrants before their departure from Europe to Palestine. Uniform instructions for immigrant medical examinations were demanded at the Fourteenth Zionist Congress.[106] In 1926, the Vaad HaBriut (Health Council of the Vaad Leumi) compiled and the Immigration Department of the Palestine Zionist Executive printed a four-page pamphlet in Hebrew, English, and German detailing instructions for the medical examination of immigrants before embarkation to Palestine and upon arrival. In this pamphlet, moral responsibility was placed on the medical officers of Palestine abroad to closely follow these regulations "because of the eventual damage to Palestine and to the immigrants themselves which may result from the neglect of these instructions."[107] This point was reiterated in the *Shulchan 'Aruch* on the Principles of Health." Regarding selective immigration according to health status, the guide read, "It is not enough to forbid sick people from coming to Palestine, it is our obligation to make sure that the immigrant will be equipped with information on how not to become ill... One whose path is for the benefit of the motherland, must change his direction and adapt to a new way of life."[108]

Changing one's daily habits in clothing, eating, washing, drinking, and general cleanliness were central to disease prevention and thus to transforming the *Yishuv*, especially those new members of the community. Although malaria incidence decreased substantially, this pamphlet recognized the continued existence of many malarial areas in new settlement

105. Selective immigration according health criteria is briefly discussed in Aviva Halamish, "Mediniut ha'aliya vehaklita shel Histadrut hazionit: 1931–1937" (Ph.D., diss., Tel Aviv University, 1995), 142–146; and Davidovitch and Shvarts, "Health and Zionist Ideology."

106. Sanitary Sub-Commission of the Fourteenth Congress, January 19, 1926, CZA J1/2678; letter to chief secretary from A. M. Hyamson, chief immigration officer, September 17, 1927, PRO CO 733/152/4.

107. Vaad Leumi, *Instructions for the Medical Examination of Immigrants*, English version (Jerusalem, 1926), 2, 4, CZA S7/70/4. The immigrants in this pamphlet are considered to be only men, as seen in the text, "women and children of immigrants should be placed in Grade B (of the health criteria categories)," p. 4; "Selection of the Fittest," in Ruppin, *Three Decades of Palestine*, 77–78

108. Davidson, "Hakdama," 5; also "Report on Immigration and Sanitation." The report covers details of the Zionist immigration program, which included selection, reclamation and settlement, and hygiene education. The author (most likely Kligler but not noted in the text) recommends the establishment of a sanitary section within the immigration department. The report is probably written in the early 1920s but no date is specified.

locations. The endemic nature of malaria justified the production of a guide like the *Shulchan 'Aruch* to instruct immigrants in how to avoid this disease. At the end of this booklet, a picture of an antimalaria worker at Nachal Sachni was included in order to impress upon the immigrant the work done already against malaria but also to remind him or her of the omnipresent danger of malaria in Palestine.

Health officials intended for a pamphlet called *Briut Ha'Oleh* (Health of the Immigrant) to be read in the Diaspora in order to prepare the potential immigrant mentally and physically for the change. The introduction to this pamphlet was a reproduction of Dr. Davidson's *Shmor 'al Briutcha be Eretz Israel*. Dr. Davidson was the director of the Medical Office of the Jewish Agency's Department of Immigration. The book's malaria section, written by Dr. Beckman, followed the typical contents of all pamphlets, describing malaria contraction, transmission, symptoms, and antimalaria measures as well as the biting habits of the mosquitoes. Emphasis on supervision of one's self and one's neighbors was made even for those readers in the Diaspora.[109]

Preselection of immigrants also applied to children. Preselection of youth for the Zionist project was explicitly stated as late as 1939:

> The Vaad Leumi exercises supervision not only over the young people's physical and psychic conditions after they are settled in Palestine but also over the selection of candidates abroad. The results of the controlled medical examinations in Germany are submitted to it [Vaad Leumi] before a final choice of the candidates is made. In this way those afflicted with chronic disease, and those lacking capacity for physical work are reduced to that minimum which defies delusion ... Due to screens and the use of nets under careful control even malaria, once so dreaded and so devastating in Palestine appears on our health lists as a vanishing minimum ... About 15% of the Youth *Aliyah* candidates are rejected for reason of health.[110]

Although "great stress was paid to the physical and mental health of prospective candidates" before World War II, at the outbreak of the war,

109. *Briut Ha'Oleh: madrich le'oleh hachadash be'inyanei hygiena vebriut* [Health of the immigration: Guide for the new immigrant on hmatters and health] (Jerusalem: Kupat Holim, 1940), National Library.

110. *Freedom and Work for Jewish Youth: Report for the Third World Youth Aliyah Conference* (Children and Youth Aliyah, 1939), 9, 18, CZA Library. This covers the period 1937–1939.

Zionist leaders and physicians recognized that it was "no longer possible to insist upon full physical fitness" in youth immigration.[111] Still, in a meeting in January 1942, the Antimalaria Council highly recommended health inspection of youth before their work in villages. Members of the council noted the need to turn to the Department of Education of the Vaad Leumi to request their help in the matter of health inspection (prevention of malaria, medical help, accommodation) when recruiting youth for work in villages.[112] Immigrant children and those born in Palestine were supposed to mold to an ideal: the healthy Zionist devoted to the building of Palestine.

Because of the movement of immigrants from settlement to settlement once in Palestine, the MRU set up a system of supervision and surveillance of patients. A doctor was instructed to fill out a malaria card in duplicate for each individual. The card was to be sent along with the settler wherever she went so that she could be kept under observation. Settlement doctors and malaria workers also kept daily diaries. They had to submit weekly reports to Dr. Kligler, the director of the MRU. Just as the health of the land had to be maintained, so too did the health of each immigrant.

Workers were addressed in malaria propaganda but not as vigorously as immigrants. After all, immigrants eventually became laborers. Journals such as *Briut Ha'Oved* (Health of the Worker) contained articles on malaria and how it affected the Zionist sector of labor. *Briut Ha'Oved* (1924–1925), sponsored by Kupat Holim, also dealt with industrial hygiene. Drs. Kligler and Halprin edited another journal by the same name on the physiology and hygiene of work. The Straus Centers sponsored this journal.[113] Kupat Holim published *Briut Ha'Amal* (Health of Labor) for workers in 1938.

The malaria situation was so serious in 1924 among workers in the settlements that the antimalaria committee of Kupat Holim decided to print "malaria cards" like those made by the MRU in order to register all members afflicted with the disease and to assess the progression of their health status. Next to the portion to fill out, the card had a warning that read

111. *Children and Youth Aliyah: An Outline* (Jerusalem: Jewish Agency for Palestine, Department for Child and Youth Immigration, 1947), CZA Library.

112. Minutes of meeting of Antimalarial Council. January 21, 1942, CZA J1/1731.

113. *Briut Ha'Oved: Chapters in the Physiology and Hygiene of Work*, ed. Drs. Kligler and Halprin (Straus Centers, 1938), CZA 6893.

Attention:

1. Malaria is a serious disease—its treatment is harsh and requires time and patience on the part of the individual—and precise information regarding the progression of the disease on the part of the doctor.
2. Save the "malaria card" because your disease history is on it.
3. In every case of fever—take your temperature, 2–3 times a day and go immediately to the doctor.
4. Always bring the "malaria card" with you so that the doctor can record all details of the disease.
5. Completely fill out exactly all the directions from the doctor with regard to treatment and caution.[114]

The same year, Hillel Yofe wrote an article in that journal describing the seriousness of the disease and the essential use of malaria cards. According to Yofe, the cards not only provided a personal malaria history for the patient but also served as a way for scientists to gain information and statistics about types of malaria and their corresponding frequency of attacks, as well as the effects of different methods of treatment.[115] He wrote that malaria was so significant that doctors needed to be concerned with and treat this disease "more than all the other diseases put together" in Palestine.[116] Fighting malaria was one of the first priorities of Zionist work and had to be undertaken with special force, by all residents, and with strong discipline. Yofe focused on the negative effects of tropical (chronic malignant) malaria that had left settlers in the past depressed and sometimes led to emigration from Palestine. Yofe explained that due to the resistance of tropical malaria to available treatments, both patient and doctor had to be exceedingly patient when trying to address it. Consistent treatment was an essential prescription and consistent discipline, once again, was demanded of the patient. Yofe divided the chronic malaria population according to work ability: one group ("weak") who could not work at all and the other who carried malaria and had occasional attacks but could still work. The first group was sent to the hospital

114. "Dapei malaria," *Briut ha'Oved* 2 (Adar Bet/March–April 1924): 42–44, CZA Library.

115. Hillel Yofe, "Leshe'elat haholim bemalaria chronit" [On the question of chronic malaria patients], *Briut Ha'Oved* 1 (1924): 9, National Library. See also Dr. Y. David, "Le'she'elat ripui hakodchim hachronim" [On the question of treating chronic malaria sufferers], *Briut Ha'Oved* 3 (1924): 3–7, National Library.

116. Yofe, "Leshe'elat haholim bemalaria chronit," 6.

and then to a convalescent home in order to hasten the return to work. The convalescent home should be a place, as Yofe explained, in a "therapeutic landscape" like the mountains, with good air and trees, free from malaria.[117] The second group, those who got occasional malaria attacks, were in a more dangerous position, according to Yofe, because they were prone to *cachexia palustra* (loss of weight) or a sudden attack of *malaria perniciosa* (severe malaria attack). The term *malaria cachexia* referred to severe symptoms that developed from untreated chronic malarial infection. These patients became anemic, their spleens enlarged, and their skin became dark yellow.[118] They could pass the parasite into a mosquito if bitten, so they were a risk to their neighbors. This second group was quite common in Jewish settlements in the 1920s. Doing his part for the hygiene campaign, Yofe stressed that showers, kitchens, and general cleanliness were required in all settlements.

Hygiene work in schools was another important part of general health preventive work in the *Yishuv*. Like their colleagues in the United States, by 1928 school educators in Palestine established set hours in the curriculum for hygiene instruction. Until then, hygiene was not set apart as a separate subject of emphasis. A debate about whether or not to integrate hygiene lessons into other daily subjects like geography took place before the decision to separate out hygiene lessons.[119] School health educators instituted a preschool program called Behavior and Habit Clinics that addressed hygiene concerns and the prevention of certain diseases, including malaria. Dr. Brachyahu, a central figure in school hygiene programs in the *Yishuv*, remarked:

> The conquest of hygiene (*kibush hygieni*) will certainly influence in a positive way our war against infectious diseases (that will continue and grow stronger) in Palestine and the hygiene situation of our young generation.[120]

117. Yofe, "Leshe'elat haholim bemalaria chronit," 8.

118. Yofe, "Leshe'elat haholim bemalaria chronit," 8; W. F. Bynum and Roy Porter, *Companion Encyclopedia of the History of Medicine*, vol. 1 (London: Routledge, 1993), 387.

119. Letter from secretariat for health affairs, Dr. Katznelson, to Dr. Berkson, head of Education Department of Vaad Leumi, August 19, 1928, CZA J1/4417. See meeting of Central Committee for Hygiene Affairs, August 5, 1928, CZA J1/4417, Dr. Brachyahu, Dr. Gruenfelder, Dr. Yassky, Dr. Meir, Mr. Mironberg and Dr. Katznelson were present; suggestions of Professional Council for Preventative Work: arranging medical work in schools, decided on June 12, 1935, CZA J1/2101.

120. Brachyahu, Skira shel ha'avoda bemachleket hahygiena bebeit hasefer: October tarp"v-October tarp"z (A Review of the Work of the Hygiene Department in Schools from

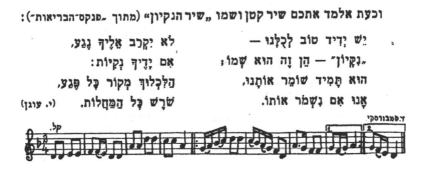

וכעת אלמד אתכם שיר קטן ושמו „שיר הנקיון" (מתוך ״פנקס-הבריאות״):

יֵשׁ יָדִיד טוֹב לְכֻלָּנוּ —   לֹא יִקְרַב אֵלֶיךָ נֶגַע,
„נִקָּיוֹן״ — הֵן זֶה הוּא שְׁמוֹ:   אִם יָדֶיךָ נְקִיּוֹת:
הוּא תָּמִיד שׁוֹמֵר אוֹתָנוּ,   הַלִּכְלוּךְ מְקוֹר כָּל פֶּגַע,
אָנוּ אִם נִשְׁמֹר אוֹתוֹ.   שֹׁרֶשׁ כָּל הַמַּחֲלוֹת.   (י. עוגן)

FIGURE 6.4. "Song of Cleanliness." From Emmanuel Yofe and Sonya Kahana-Charag, *HaDerekh el HaBriut* (The way towards health) (Tel Aviv: Yavneh, 1947), 28. National Library, Jerusalem.

Like *kibush ha'avoda*, *kibush hygieni* (conquest of hygiene) was meant to physically transform the Jew. *Kibush hygieni* also encouraged a particular relationship with the environment (water, mosquitoes, etc.) by encouraging the *Yishuv* to control and subdue it through sanitation. Thus, *kibush hygieni* was a formulation of *kibush ha'avoda* for health; manual labor intended to transform the individual through methods of hygiene. One way to promote *kibush hygieni* was through children's songs such as the following (see fig. 4):

> There is a good friend for all of us—"Clean" is his name
> He always protects us if we will protect him.
> Disease won't get near you if your hands are clean.
> Dirt is the source of all trouble
> The root of all diseases.[121]

The Central Council for Hygiene Explanation also implemented traveling exhibits for the school in the late 1920s, borrowing materials from European institutions such as the Hygiene Museum in Dresden, Germany, thereby maintaining the *Yishuv*'s connection with European medical

1925–6: October 1925–October 1926), *HaRefuah* 2, no. 3 (July 1927): 450, CZA Library; *Hadassah Newsletter* 5, no. 2 (November 1924), 5, CZA J113/1431.

121. Lyrics by Y. 'Ogen and melody by D. Samborsky. Emmanuel Yofe and Sonya Kahana-Charag, *HaDerekh el HaBriut* [The way to health] (Tel Aviv: Yavneh, 1947), 28, National Library, Jerusalem.

trends and institutions.[122] Health educators believed schools were considered "the most fitting institutions to spread all information" about malaria.[123]

The founding of Health Scouts was one way to gain the participation of students in hygiene affairs. The Department of Education of the Vaad Leumi served as the central organizer of these scouts, while local teachers served as club directors.[124] In addition to disseminating basic concepts of hygiene, Health Scouts in 1936 were to promote *totzeret haaretz* (products of [Jewish] Palestine) during the Arab Revolt. This initiative was part of the larger economic politics of the Arab Revolt. Citrus products took on particular significance during this time. Although both Arabs and Jews produced citrus products, oranges had become, by the 1930s, an important crop of the Zionist agricultural economy. The scouts and teachers of hygiene in schools, for instance, were asked to promote "orange juice for health" and explain the hygienic value of citrus fruits, because of their vitamins and their major economic importance to the *Yishuv*.[125] A healthy economy meant a healthy land and healthy Jewish bodies.

Health educators and doctors produced special pamphlets on sanitation and cleanliness for students. Among them were "Malaria," in *Hitbagrut* (Adolescence) by Dr. Brachyahu and sponsored by the Straus Center's popular health library Briut ha'Am, a series of pamphlets entitled *Tarbut haguf* (Physical Culture/Education) by Dr. Simon, and booklets on sex hygiene by Dr. Kahn.[126] One malaria pamphlet written as a story was designed for school-age children. The text began, "Here are two ladies from the family of mosquitoes. The one that stands on the right [accompanied by a picture] is Miss Culex, and the other that stands on the left

122. Protocols of the meeting of Central Council for Hygiene Explanations, May 28, 1928 J1/4417.

123. Letter from Dr. Noack to heads of schools, June 3, 1936, CZA J1/1542.

124. Health scouts were part of a larger program to institute extracurricular activities in the schools in the 1930s. Memo by Department of Education, Vaad Leumi, no date (but file is all for 1934), CZA J17/667.

125. See Letter from Y Cohen, Secretary of Organization for Totzeret Ivrit (Hebrew products) to the Department of Education, Vaad Leumi. August 9, 1936; letter from Dr. Y. Lurie, Head of Department of Education to Executive of Schools, February 2, 1936; letter from Dr. Lurie to Executive of Schools, January 26, 1936, CZA J17/21.

126. Letter from Dr. P. Noack to Department of Education of Vaad Leumi, May 20, 1936 CZA J1/1542.The fact that this pamphlet was still being distributed to students shows that malaria was still a problem. It seems that this pamphlet was not a new one, and, as mentioned before, was probably redistributed to new populations in Palestine when it was necessary; Shechter, "Medical Literature in Palestine," 168.

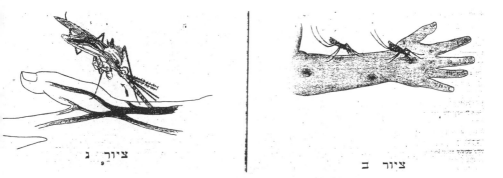

ציור ג

ציור ב

FIGURE 6.5. The bite of Miss Anopheles. From Hadassah Medical Organization, *Sipur 'al HaGiveret Anopheles* (Story of Miss Anopheles), pamphlet no. 5 (Jerusalem: Rafael Chaim Hacohen, 1922), 8–9, CZA 20.682. Courtesy of the Central Zionist Archives.

side is Miss Anopheles."[127] This pamphlet is filled with pictures and the student is instructed, throughout the text's explanation, to examine them closely. The pamphlet discusses, in less sophisticated terms, the life cycle of the female *Anopheles* and the method of its bite. The author tried to relate the case of the mosquito to child's play by asking,

> Surely you were in the *moshava* (settlement) and gathered hen's eggs? Where did you look for them? Under the bushes of the wadi, or inside the cowshed, in the pile of fodder, on the soil or in the water of the wadi? Only in the water can you search for anopheline eggs. I only hope that you will indeed search for these eggs and destroy them without mercy.[128]

In order to inhibit the reproduction of the *Anopheles*, the child was instructed to help his or her fathers and brothers cover cesspits, wells, and buckets, cut reeds and grass near standing water (so that the sun can shine through), petrolize wells and cisterns, and cover their beds with mosquito nets. Everyone, including the youngest of children, was urged to assist in the endeavor: "Miss Anopheles is an enemy so dangerous to man, that we cannot be satisfied with [only] a mosquito net spread out on our beds, we must use all means so as not to give her any access to our houses."

After describing the pathophysiology and symptoms of malaria and adding Miss Parasite to the list of characters, the pamphlet identified

127. Hadassah Medical Organization, *Sipur 'al HaGiveret Anopheles*, 5.
128. Hadassah Medical Organization, *Sipur 'al HaGiveret Anopheles*, 7.

FIGURE 6.6. Child helping his father petrolize well. From Hadassah Medical Organization, *Sipur ʿal HaGiveret Anopheles* (Story of Miss Anopheles), pamphlet no. 5 (Jerusalem: Rafael Chaim Hacohen, 1922), 21, CZA 20.682. Courtesy of the Central Zionist Archives.

national sin as it occurred in the family. *Allowing* a mosquito to bite you increased the chance of transferring those parasites to someone else, perhaps even your "poor" (*miskena*) sister.[129] Familial affiliation personalized the extent and consequences of the sin and increased guilt: "If you had been careful not to get a mosquito bite, then your sister would not have suffered because of you."[130]

Physical and spiritual improvement of youth was a central goal of health propaganda and medical work. As one article stated, "No factor in the building of the Jewish homeland is as fundamental or as important as the rearing of Palestinian youth in good health and healthful surroundings."[131] Hygiene lessons tried to inculcate not only physical health but also attempted a Zionist socialization of children that stressed social and national responsibility. Such "hygienic morality," as scholar Goubert has

129. Hadassah Medical Organization, *Sipur ʿal HaGiveret Anopheles*, 20.

130. Other victims could include, according to the text, the child's friend or uncle who lives a mile away. *Sipur ʿal HaGiveret Anopheles*, 20. There is a play on gender here as the reader is assumed to be a boy, who should look after his sister while the *Anopheles* is always female as well and he needs to look after her.

131. "For the Children of Palestine," editorial, Hadassah Collection 1925, CZA J113/1427.

FIGURE 6.7. Ein Gav: planting trees with children, February 1941." Courtesy of the Keren Keyemet L'Israel Photo Archives

called it, linked physical cleanliness to purity of the mind and the upholding of human dignity.[132]

This principle was also present in the religious sector of the *Yishuv*. In one ultraorthodox girls' school, Agudat Israel, a program of special lessons in *halachot shmirat habriut ve torat nikayon* (principles of protecting health and the laws of cleanliness) was implemented in order to promote independent responsibility and the life of *Torah vé'avoda* (Torah and work).[133] Adoption of this type of educational direction among the ultrareligious community came at the end of the Mandate. In the early years, Dr. Brachyahu noted a refusal by Talmud Torah yeshivas to receive instruction and aid in principles of hygiene.[134] It is not clear if this was a rejection of Western principles of hygiene per se, a rejection of the

132. On school hygiene and "hygienic morality" in France, Goubert, *Conquest of Water*, 150–168.

133. Hanya Gortenberg, "Mosad ledugma" (Institute for Example), Beit Midrash morot velegananot Talpiot beTel Aviv-Agudat Israel, 1943, 18, Rav Kook Institute Archives. There was a section of Agudat Israel that promoted a pioneering ideology of working on the land. This school was most likely part of that movement.

134. Dr. Brachyahu, "Skira shel ha'avoda bemachleket hahygiena bebeit hasefer: October tarp"v-October tarp"z" (A Review of the Work of the Hygiene Department in Schools from 1925–6: October 1925–October 1926), *HaRefuah* 2, no. 3 (July 1927): 450, CZA Library.

greater emphasis in the Zionist project on the body rather than on study of Torah, a refusal to deal with Zionist medical organizations, or all of the above. This refusal is noteworthy because health services and programs of all Jewish communities in Palestine came under the auspices of Zionist health organizations and the Vaad HaBriut. Certainly such a refusal suggests a fissure within the Jewish community with regard to health affairs and reveals that acceptance of Western hygienic principles and practices was not a straightforward, unproblematic venture. It was a process that gradually became consolidated among the *Yishuv*. This gradual process, the anxiety felt by Zionist health professionals to achieve it more quickly, and the practical exigencies behind it help to explain the barrage of hygiene materials addressed to the public and their urgent tone.

Mothers of school-age children were enjoined with enforcing hygienic principles and malaria control. An abundance of material on general hygiene addressed to Jewish women in Mandatory Palestine reflected the domestic science movement popular at the time in the United States, Europe, and in the colonies, where teaching women about germs and household hygiene was a central mission.[135] The *Tipat Halav* program (Drop of Milk, a campaign that distributed milk to mothers and gave them instructions for infant feeding) and nursing education were other ways of enlisting women's participation in the Zionist health endeavor.[136] Malaria's ability to induce miscarriage and its particular threat to children's health drew the concern of health officials, especially given the pronatalist policy of the Zionist project and the general encouragement of an increase in the Jewish birth rate.[137]

As many scholars have pointed out, women were the prime targets for health initiatives during the colonial and early national periods because they were thought to be mothers of the nation and the future generation

135. Tomes, *Gospel of Germs*, 9.

136. Nira Bar Tal, *Hahachshara heteoretit vehama'asit shel ahayot yehudiyot beEretz Israel betkufat hamandat, 1918–1948 bi-re'i hitpathuto shel beit hasefer haahayot 'al shem Henrietta Szold, Hadassah Yerushalayim* [Theoretical and practical training of Jewish nurses in Mandatory Palestine, 1918–1948, as reflected by Hadassah's Henrietta Szold School, Jerusalem]; Rebecca Stockler, "Development of Public Health Nursing Practice as Related to the Health Needs of the Jewish Population in Palestine, 1913–1948" (Ph.D. diss., Teachers College, Columbia University, 1975); Dr. C. Frankenstein, *Child Care in Israel: A Guide to the Social Services for Children and Youth* (Jerusalem: Henrietta Szold Foundation for Child and Youth Welfare), 1950.

137. No author (but most likely Noack), "Health," 1943, 2, CZA J1/7718. See also Noack, Department of Health of Vaad Leumi, "Problems with Health," 1943, 4, CZA J1/7718.

of men.[138] As Weindling said about Weimar Germany, a context from which Zionist medical professionals (especially those who left it in 1933) drew lessons, family welfare measures during this time were inspired by nationalist and biologistic aims, a way to "rebuild the biological fabric of the race and to restore national morale."[139] The author of *Analyse der statistischen Angaben* (Analysis of Particular Statistics), giving merit to the task of infant welfare, wrote,

> [There is a] realization of the physician that he sees that the task of the mother, her character, her education/capacity for education, predominates the future of their children. For a people like the Jewish nation that wants to start under changed conditions of living, on a whole country that has to be conquered once again, the intensive [education] of the women is a significant task that ensures the successful execution of bringing up a group of suitable descendants, the future carriers of this great project.[140]

Yet he also pointed out that acceptance of Western methods of infant care was not unequivocal. Mizrachi, or non-European, Jewish mothers sometimes resisted or did not comply with "proper" (Western) methods of infant care. These women turned to their mothers and neighbors instead for help and advice, so doctors targeted Yemenites, Iraqis, Persians, Turks, and Kurds for mothering education. Their resistance to Western principles of mothering caused great anxiety among infant health care workers and bureaucrats. The author of *Analyse der statistischen Angaben* commented on the frustrations of the nurses and health bureaucrats:

> But this part of our work remains always difficult...The men bring the attitude of the climatic and other living conditions and habits of their different home-lands with them and hold them strongly in huge stubbornness, unconvinced by

---

138. Anna Davin, "Imperialism and Motherhood," *Tensions of Empire: Colonial Cultures in a Bourgeois World*, ed. F. Cooper and A. Stoler (Berkeley: University of California Press, 1997), 87–151; Andrew Parker, Mary Russo, Doris Sommer, and Patricia Yaeger, *Nationalisms and Sexualities* (London: Routledge, 1992); Ann Stoler, *Race and the Education of Desire: Foucault's History of Sexuality and the Colonial Order of Things* (Durham: Duke University Press, 1995); Ann McClintok, *Imperial Leather: Race, Gender and Sexuality in the Colonial Context* (New York: Routledge, 1995); George Mosse, *Nationalism and Sexuality*, and *Image of Man*.

139. Weindling, "Eugenics and the Welfare State," 138.

140. No author noted, *Analyse der statistischen Angaben* (no date noted but seemingly from 1929), 26, CZA F49/1896.

the conditions of Palestine that often virtually press to [hold] the opposite behavior...For the young Oriental woman, the conditions are doubly heavy: she
is torn down from the old tradition and she is not yet connected with the forms,
goals and reasons for modern civilization.[141]

Another health professional noted,

There is no question that the older women think they are much too clever to
visit our clinics. And if one once succeeds in pulling the mothers to us, then all
the instruction is taken ironically. An old mother once said what many actually
think: "I have given birth to seventeen children and twelve have died; I know
how one should raise children."[142]

The author most likely stated this sarcastically, for if twelve out of seventeen children had died, then perhaps the mother, in the eyes of the
author, has not been successful at raising her children. This type of infant
mortality, however, was quite common before and at the beginning of the
Mandate and was due, in large part, to the high incidence of infectious
diseases like malaria. The substantial popular medical literature directed
at women tried to ensure that they raised fit children and adhered to hygienic habits for themselves and for their family. Malaria control was certainly part of that task. Women's committees formed for the purpose of
teaching elementary hygiene and cleanliness.[143]

Pamphlets such as Dr. Sherman's "Health of the Mother," Ben-Ge-
fen's "The Infant in Health and in Sickness" (1929), Gruenfelder's "Care
of the Child in Sickness and in Health," Ahronova's "Feminine Hygiene,"
and Kupat Holim's *Luah haem vehayeled* (The Almanac of Mother and

141. Note the comment about climate and its association with adaptation to health concerns and conditions in Palestine, *Analyse der statistischen Angaben*, 18. The author remarked
about Yemeni women earlier in the report that "the Yemenites push with huge intensity
(probably too much) their first aim: the adjustment to the Ashkenazi culture (and to Eu-
ropeanization). The Yemenite woman, who works in Ashkenazi houses as a servant first of
all, wants to preserve her children in beauty and health like the Ashkenazi women," 5. See
also "WIZO Activities on behalf of immigrant women during period 1941–1945," Written
July 1945, CZA F49/1163; and Sanitary Commission from "Palestine" Organ of the British
Palestine Committee, March 29, 1919, CZA J113/1430.

142. *Analyse der statistischen Angaben*, 6. See also Fanon, "Medicine and Colonialism,"
233.

143. "Report on Immigration and Sanitation," 11; Weindling, "Eugenics and the Welfare
State," 147, 151.

Child) are just a few examples of the popular press targeted at women. In addition, the Committee on Nutrition of the Jewish Agency's Institute for Economic Research published a monthly magazine called *Hamazon* (Food; 1939). Dr. Kligler, the main activist in Palestine's malaria efforts, served on that editorial board.[144]

"A Guidebook for Mothers" was printed in 1933. The Women's International Zionist Organization printed many cookbooks and nutritional guides in order to instruct women in proper dietary and preparation habits. Popular literature directed at mothers about the art of mothering was supplemented by an organized network of infant welfare centers and nurses' visits to mothers' homes. Hadassah, the Women's International Zionist Organization, and Kupat Holim undertook the care of mothers.[145] Infant welfare stations and programs such as *Tipat Halav* (Drop of Milk) made sure that women were instructed properly in Western methods of infant care and that they had enough milk to feed their children. These organizations also utilized the Plunkett system of infant care from New Zealand. In addition to literature and clinical health efforts, organizations such as the Moetzet HaPoelot (Council of Women Workers) addressed the concerns of female immigrant workers with its aim as "the promotion of a new type of healthy and hardy Palestinian womanhood."[146] Another umbrella organization, the Moetzet Irgune HaNashim (Council of Women's Organizations), proposed a joint effort with the Arab women's movement to work toward better conditions for women in Palestine.[147]

Health Education, Malaria, and the Palestinian Arab Community

The Mandatory government sponsored health education activities for the Arab population. Palestinian Arab doctors also emphasized health

144. Letter from Yassky, head of HMO, to secretariat for health affairs of Vaad Leumi (Katznelson), CZA J1/4355; Shechter, "Medical Literature in Palestine," 169.

145. It is interesting to note that Poel HaMizrachi, a Zionist religious group, instituted in August 1942 a Committee for the Purity of the Family in order to promote the "institution of life of family in Israel in the purity and sacredness of the nation." I was unable to find many details about this committee and their actual activities. There was, however, a Council for Family Purity after the founding of the State of Israel. Rav Kook Institute Hativa Pei/Mem.

146. "Agricultural Scheme to Settle Girls on Land," *Pioneers and Helpers* (Women's International Zionist Organization News Bulletin) (June 1927), 6, CZA F49/2464.

147. Irma Pollack's Proposals Regarding the Tasks of the Moatza irgunei hanashim," 3, CZA F49/2104.

education as part of the nationalist endeavor.[148] These types of activities increased in size and nationalist character as the Mandate period unfolded and as the number of Palestinian Arab doctors grew. In 1934, there were forty Arab physicians in Palestine; by 1940, there were 291. In contrast, in 1934, there were 971 Jewish physicians there, and by 1946, there were 2,386.[149]

Health education efforts couched in nationalist terms became fully developed and consolidated among the Palestinian Arab population in the 1940s. The use of medical imagery to describe the nation and evoke nationalist feelings, however, was used early on in the Mandate. Sheikh Farouki, the mufti of Nablus, used medical imagery in his speech at the Arab Congress of the New Arab National Party in Jerusalem on November 9, 1924. He declared, "Gentlemen, we will not use medicines nor a doctor, who will operate on this body as they wish. No, we will not permit them to tear and cut the flesh of our homeland, which is dying. We do not agree to the doctors who will use their own medicines. We will strengthen the body until it is able to stand. We will breathe the spirit of life into its nostrils and return the lost strength to the invalid (applause)."[150]

Palestinian Arab physicians educated in Beirut, Damascus, or Cairo and (less commonly) in Europe adopted Western medicine, hygiene, sanitation, and notions of progress. Courses at American University of Beirut Medical School were given in English and based on U.S. medical practices and philosophies. Dr. Aftim Acra, born in Jerusalem in 1922, was educated there and was a pharmacist from 1946 to 1948. Acra later became professor emeritus of environmental science at the American University, the medical school most attended by Palestinian Arab doctors during the Mandate period. Another doctor, Dr. Elias Srouj, attended medical school there in the 1940s and then returned to Nazareth to work in its infant welfare station.[151]

When asked about public health education in Palestine, Dr. Acra remarked, "During the period under consideration (1920–1947), there were

148. Adnan Abu-Ghazaleh, *Palestinian Arab Cultural Nationalism* (Brattleboro: Amana Books, 1991), for a study of cultural aspects of the Palestinian Arab nationalist project; Nira Reiss, *Health Care of the Arabs in Israel* (Boulder: Westview Press, 1991).

149. Tawfiq Canaan and A. Karam, "Statistical Record of Medical and Hygienic Progress in Palestine" (written in English), *Al-Majala al-Tabiya al-'Arabiya al-Filastiniya* [Journal of the Palestine Arab Medical Association] 2, no. 6 (Sept. 1947): 170.

150. Farouki was president of the Congress. Speech at Arab Congress, November 9, 1924, 2, CZA S25/665. T. Canaan wrote an article called "Al-Saha keasas al-qowmia al-'arabia," *al-Kulliyeh*. I could not locate this article since its original citation did not give a date.

151. Telephone interview with Dr. Elias Srouj, September 26, 1998.

no qualified and properly trained personnel in the field of public health in Palestine, at least among the Arab population. In fact, the entire region lacked such personnel." Historical documents partly support Acra's statements. They show that Arab medical officers and doctors were trained by way of practice in public health measures rather than formal education. In the Jewish sector, Zionist health bureaucrats received formal training in public health either before their arrival in Palestine or as part of postgraduate work abroad.[152] According to Dr. Tawfiq Canaan, president of the Palestine Arab Medical Association, by the 1940s, much of the Arab population (urban and rural) had integrated hygienic concepts and practices into their daily lives:

> Personal hygiene has become a matter of great importance among the greater part of the population. The scientific hygienic care of babies and young children has become more rational and is spreading quickly among the inhabitants. The nursing of the patients at home is nearly free of the old unscientific methods ... A simple visit of the Arab sector of Palestine will show ... the great number of practical lectures given in every city and large village on hygiene and sanitation, the universal desire of the population to help the poor and underprivileged, the interest in the establishment of hospitals and sanatoriums. If one remembers that no subsidies and no material help of any importance are received from abroad, one must but pay tribute to the vision and energy of the Palestinian Arabs.[153]

With regard to malaria practices, Dr. Acra remembered that mosquito nets were used at night by some Arab families to avoid getting bitten by mosquitoes. These nets were provided for use on cribs and beds in highly infested areas. Flit, an insect repellent, was also commonly used (and is still used today).[154] Dr. Acra noted that:

> Many of the traditional customs, activities and practices inherent among the Arabs were abandoned or modified as their lifestyle changed under the Western

---

152. Personal correspondence with Dr. Acra, September 26, 1998, 6, 7.

153. Palestine Arab Medical Association, *The Hygienic and Sanitary Conditions of the Arabs of Palestine*, compiled by Dr. Tawfiq Canaan (Jerusalem: 1946), 8–9; personal correspondence with Dr. Aftim Acra, September 26, 1998, p. 5.

154. Pronounced *fleet*. Personal correspondence with Dr. Aftim Acra, September 26, 1998, 8.

influence brought about by the Britishers and other aliens residing in the country.[155]

As in the Jewish community, Acra explained that full compliance with Western hygienic habits was not always possible because "personal hygiene was not considered by most people to be an absolutely essential practice."[156]

Health education, especially for the youth, was recognized as an important method for achieving Palestinian Arab nationalist goals. Promoting the importance of a healthy nation through the adoption of hygiene was part of a larger process of integrating Western ideals in the pursuit of national "progress."[157]

Palestinian Arab efforts were commonly implemented in tandem with other Arab communities, including Egyptian and Syrian-Lebanese communities in the service of Arab nationalism. The agenda at an educational conference of Muslim students in Beirut called for discussion of, among other things, "(1) the students' duty towards the nation; (2) girls' education; (3) the scout movement; (4) the dangers of alcoholic drinks; (7) the health of the youth; (13) the health of the teeth; and (14) the woman and the art of nursing patients."[158] Materials for health education were frequently borrowed from Egypt and used in Palestine.[159]

Mandatory government health education materials were accompanied by practical instruction given to villagers by appointed medical officers and subinspectors who, depending on the district, were either Palestinian Arab employees or British colonial appointees.[160] The Palestinian Arab community also participated in the health weeks sponsored by Hadassah and the Mandatory government. Drs. Canaan and Dajani served on some of the medical planning committees for these health weeks.[161] Drs.

155. Personal correspondence with Dr. Aftim Acra, 13.

156. Personal correspondence with Dr. Aftim Acra, 13.

157. Abu-Ghazaleh, *Palestinian Arab Cultural Nationalism,* 9.

158. "A Statement on the Students' Educational Conference Held at Beirut," August 20–23, 1925, December 18, 1925, CZA S25/617.

159. It was based on a film of the same name. See "Sunu al-basar: 12 qa'ida liman'a al-'amiya" [Save the Eyes: 12 rules for prevention of blindness; Arabic and English title and text both given], published by the American University of Cairo and sent to the British Health Department of Palestine for use, no date, ISA M1567/44/15/2377.

160. Colonel Heron's statements in "Proceedings of the Eighth Meeting of the Antimalarial Advisory Commission," 4.

161. Dr. Canaan was active as a secretary for the first Health Week planned in 1925. Letter from secretary of Health Week Committee to Mrs. Champion, October 14, 1924, CZA J113/1446; letter from Mochenson to Kisch, July 31, 1924, about Dr. Dajani, CZA S25/596.

George, Khalidi, Malouf, Haddad, and Tannous wrote articles or leaflets for the first health week in 1924. Dr. Kligler wrote an article in both Hebrew and Arabic for Microbe Day.[162]

The Palestine Arab Medical Association organized weekly health and hygiene radio talks in the 1940s. This association began their own medical journal in the last decade of the Mandate and held their first congress in June 1945.[163] The association's journal, *Al-Majala al-Tibbia al-'Arabiya al-Filastiniya* (Journal of the Palestine Arab Medical Association), was mainly a scientific publication that discussed various clinical treatments, but it also featured articles that explicitly linked the association's work to nationalist goals. For instance, articles about medieval Arab physicians, Arab medical traditions, or the creation of an Arabic medical dictionary (perhaps influenced by a similar Zionist project started earlier) illustrated an expression of national consciousness through medical endeavors.

The Mandatory government broadcast talks on health matters before the 1940s in both Arabic and Hebrew, but the Arab program only reached listeners in the villages who were provided with free radio sets. For the initial broadcast, Dr. Dajani spoke on elementary hygiene.[164] Drs. Eid, Bishara, T. Khalidi, Haddad, H. S. Khalidi, Musa, Shuban, and Fares all gave radio lectures on hygiene and infectious disease.[165]

Arabic hygiene literature certainly existed, but it is not readily available today. An article in the Hebrew newspaper *Davar*, on October 1, 1935, mentioned a book called *The Book for Hygiene in Arabic*, published and written by the teacher Mohammed Adib al-'Amari.[166] Arabic newspapers did not have regular health columns like those in the *Yishuv*, but they did announce health exhibits, meetings, and doctor's activities in small articles. Advertisements usually appeared on the last page of the publication.

162. List of articles for Health Week in order of appearance, November 4, 1924, CZA J113/1445.

163. An article in *Filastin* on February 3, 1946, entitled "Al-Jam'aia al-tibbiyya al-'arabiya: majalaha al-jedida wa ba'd anba'ha," advertised the publication of the Palestine Arab Medical Journal and noted the various activities of different doctors. *Filastin* 29 (February 3, 1946): 285.

164. Memo to director of medical services, Colonel Heron, from programme director of Palestine Broadcasting Service, subject: broadcast talks on health matters, March 17, 1936, ISA M1567/44/19/2379; letter dated April 1, 1936, regarding mental hygiene radio talks, ISA M1567/44/19/2379. In Hebrew, Dr. Kligler and Dr. Fishel both gave talks on malaria. Letter dated May 4, 1936, same correspondents, ISA M1567/44/19/2379.

165. Summary of lecturers, station and lectures, folio 18a, June 8, 1936, ISA M1567/44/19/2379.

166. *Davar*, October 1, 1935, CZA J113/348.

As in the *Yishuv*, hygiene was taught in governmental schools attended by Arab students. Cleaning campaigns, rubbish clearing, drainage, oiling of stagnant water pools, and getting rid of mosquitoes were activities of teachers and pupils. The British Health Department of Palestine sponsored and supervised these campaigns, but it is probable that they contributed to nationalist agendas.[167] The campaigns included traveling cinemas with educational films about the fly, the mosquito, and other pests and discussions of more intensive methods of cultivation, irrigation, and drainage.[168] Educated Palestinian Arabs commonly participated in athletic clubs in large villages and towns. Missionary schools and hospitals that commonly served the Palestinian Arab community also attempted to impart Western notions of health and hygiene to their students and patients.[169] The effort to create an "evolution of habits" among the Arab population was primarily overseen by the government, but it did not contain, at least explicitly, the same urgency and gravity as the *Yishuv*'s health propaganda with regard to its role in the success or failure of the nationalist project. From the sources available, it seems that malaria held a minor place compared to general hygiene in Palestinian Arab cultural nationalism.

Professional Medical Literature and Hygiene

In addition to popular medical literature in the service of health, hygiene, and malaria education, there was a vibrant publication of medical literature for the Jewish professional community in Palestine as well as educational programs for health professionals. Before the Mandate, medical literature among the Jewish sector began in 1912–1913, when the proceedings of the Jaffa Medical Association were published. The idea for a medical journal was raised in 1916, but only in 1920 did the publication *HaRefuah* (Medicine) appear under the editorship of Dr. Feigenbaum and then Dr. Doljansky.

The push for the production of a body of local professional medical literature in Palestine was part of a larger project in the medical community to create a Hebrew medicine with a Hebrew medical dictionary as a way

167. Abu-Ghazaleh, *Palestinian Arab Cultural Nationalism*, 79.

168. H. Bowman, *Rural Education in the Near and Middle East, July 1939*, reprinted in *Journal of the Royal Central Asian Society* 26 (1939): 405, St. Antony's College Middle East Center, Bowman Papers, box 2, file 6.

169. Abu-Ghazaleh, *Palestinian Arab Cultural Nationalism*, 81.

to express normalcy as a nation.[170] As editors of *HaRefuah*, the quarterly journal for the Hebrew Medical Society in Palestine, wrote,

> For all of our work in the Diaspora it only had individual value and not national value. Of all the stones that we added to the palace of general medical science [in the Diaspora], no national-Hebrew character was attributed to them. Only when we build our future here in Palestine will there be double and lasting value: national, and individual together.

Likewise, "Our magazine has another important field of work—it must adapt our renascent Hebrew to medical use, providing it with all the theoretical and practical terminology. We must revive old terms and invent many new ones." The indexes of most major medical journals had German and sometimes Russian or English medical terms and their Hebrew equivalents.[171] The drive for Hebrew medical terms was a product of the revival of Hebrew as a language for daily use, integrating medicine and health for the Zionist nationalist project.

HaRefuah provided doctors with articles on tropical morbidity and local diseases in order to keep up with developments in tropical medicine.[172] In 1929, *HaRefuah* became a regular, bimonthly journal under the general sponsorship of the Palestine Arab Medical Association. In 1937, the journal became a monthly publication under the editorship of Prof. Adler, Dr. Friedman, Dr. Rokach, Dr. Halprin, and Professor Rosenbaum.[173]

The journal *Medical Leaves*, funded by Kupat Holim in 1935, contained a section for Hebrew medical terms but also included articles about public health and medical history. The journal was a medium through which Jewish physicians could "aid the rebuilding of Palestine."[174] Its contributors came from the Diaspora and Palestine. Proceeds from sales went to Kupat Holim for their health activities.

In addition to these publications, Hadassah Medical Organization and other institutions offered training courses for medical professionals. These

170. Sandy Sufian, "Defining National Medical Borders: Medical Terminology and the Making of Hebrew Medicine," in *Reapproaching the Border: New Perspectives on the Study of Israel/Palestine*, ed. Mark LeVine and Sandy Sufian (New York: Rowman and Littlefield), in press.

171. "Magamatenu" [Our aim], *HaRefuah* 1, no. 1 (March–May/June 1924): 3; Shechter, "Medical Literature in Palestine," 166.

172. "Magamatenu," 4.

173. Shechter, "Medical Literature in Palestine," 166–167.

174. Foreword, *Medical Leaves* 1 (1937): 7.

courses often embodied the health concerns and ideologies of the *Yishuv*. The program for a theoretical course on health welfare nursing in 1935, for instance, included a lecture to be presented by Dr. Ashbel of the Hebrew University on the climatology of Palestine.[175] This inclusion revealed the continued importance issues such as climate and acclimatization played in the medical life and work of the *Yishuv*. Lectures for teachers, principals, and doctors dealing with school hygiene included discussions of infectious diseases, nutrition, orthopedics, and anthropology of Jews in school hygiene work, to name a few. In 1937, Kupat Holim 'Amamit sponsored a conference in Kfar Saba on hygiene and sanitation for its members.[176] There were also lectures, courses, and conferences for doctors in Palestine to attend that provided them with ways to exchange and discuss their findings.

175. Memo by Yassky, January 2, 1935, CZA J113/1428.
176. "Brachyahu, Skira shel ha'avoda bemachleket hahygiena bebeit hasefer," 449; letter from Dr. Margalit of Kupat Holim Amamit in Moshavot to Dr. Brachyahu, 1937, CZA J113/511.

CHAPTER SEVEN

Contested Landscape

Palestinian Arabs and Zionist Antimalaria Projects

Disease knows no religion or class. Neither does medicine. —David Eder, medical officer of Zionist Commission[1]

In Mandatory Palestine, matters of land and disease brought about the interaction between the Palestinian Arab population and the *Yishuv*.[2] Although the Zionists could establish separate, autonomous health and antimalaria institutions distinct from those caring for the Arab community, they could not prevent the indiscriminate *Anopheles* mosquito from flying from an Arab village or nearby Jewish settlement (or vice versa) and infecting its members, so contact with the Palestinian Arab population was part of the process of remaking the Jewish nation and the land of Palestine. Zionist antimalaria efforts were extended to the Palestinian Arab population primarily when the health of the Jewish community was threatened. In this way, the Arab community was an "inherent aspect of the Zionist [health] project."[3]

1. David Eder, "The Campaign against Malaria and Trachoma in Palestine," May 20, 1918, 4. RF, RF5/2/61/398. Eder was Weizmann's representative in Palestine and became a member of the Zionist Executive in the 1920s.

2. Kimmerling, *Zionism and Territory*, 13.

3. Uri Ram, "The Colonization Perspective in Israeli Sociology: Internal and External Comparisons," *Journal of Historical Sociology* 6, no. 3 (September 1993): 331–333; Penslar, *Zionism and Technocracy*, 3. Zureik also states "Palestinian peasants were not colonized by

The Zionists quickly learned that they would have to deal with the Palestinian Arab community in order to rid their own population of malaria to make the land "healthy."[4] Medical and popular medical literature addressed the complex link between the two communities and the malaria problem.[5] Zionist health officials explicitly recognized the need for antimalaria measures in Jewish settlements and Arab villages and the need for clinical treatment in both communities. They incorporated this imperative into sanitation policy, land buying practices, and medical experimentation. As Dr. Arthur Brunn, an active physician and scientist of malaria in Palestine, remarked,

> From the perspective of protecting health it is impossible to separate the Jewish *Yishuv* from the Arab *Yishuv*. The Arab *Yishuv*, lacking hygiene or rife with infectious diseases, is the source of a constant danger for the Hebrew *Yishuv*. It is not possible to create a simple medical cell that demands that we shut off the source of diseases located, without a doubt, most frequently in the Arab *Yishuv*.[6]

Brunn proceeded to suggest that distribution of medical information to the Palestinian Arab population by the Zionists would equalize the sanitary and hygienic levels of both communities.

Even before the Mandate period, Arthur Ruppin, a leading figure in Zionist settlement affairs in Palestine, commented on the dependence of Jewish health upon Arab health with particular reference to malaria:

> A radical campaign against malarial disease, such as has been successfully conducted in other countries, is only possible when the entire country is unitedly organized for the fight. It is useless to destroy the breeding places of the ano-

---

Zionist settlers in the classical sense of colonialism," Zureik, *Palestinians in Israel*, 66. Also, J. Farley, "Bilharzia: A problem of 'Native Health,' 1900–1950," in *Imperial Medicine and Indigenous Societies*, ed. D. Arnold (Manchester: Manchester University Press, 1988), 189; Packard, "Visions of Postwar Health and Development," 94. This is not to discredit *clinical* medical help given to the Palestinian Arab population by Hadassah and individual Jewish doctors like Dr. Hermann Fogel, Dr. Ticho, and others. Oral History of Said Rabi, national health educator, vice head of Department of Health Education of Israeli Ministry of Health, July 1, 1996.

4. The Campaign against Malaria and Trachoma in Palestine, May 20, 1918, RF, RF5/2/61/398.

5. The interdependence of societies and the extension of hygiene programs to all were understood by hygienists of the period. Latour, *Pasteurization*, 36.

6. Dr. Arthur Brunn, "'Avoda meshutefet berefuah beyn yehudim ve'aravim" [Joint medical work between Jews and Arabs], *Davar*, March 16, 1932.

pheles gnats, the transmitters of the malarial plasma in a Jewish colony, if beside the Jewish colony, on non-Jewish ground, other breeding places are allowed to remain untouched. It is also useless to introduce quinine prophylaxis in a colony and destroy the malarial plasma in its inhabitants, if from the neighboring Arab villages there continue to come people who do not know, and do not want to know, anything about prophylaxis, and who may time and again be the starting point for the transmission of the malarial plasma.[7]

Though Zionist health and political figures recognized that they and the Arab population were fighting "against a common enemy" (the mosquito), they viewed the rural Arabs as complacent and unsanitary,[8] uncivilized, uneducated, backward, and prime carriers of malaria with an immunity acquired by recurrent bouts of it (especially in the Huleh region). Jews of Middle Eastern origin were cast in the same light as the Arabs. A letter by the Kinneret Yemeni community noted in June 1925 that Dr. Meir of Kupat Holim often passed their settlement, refused to treat their sick, and thought of them not as Jews "but acted as though we were Arabs." The community lived next to a swamp and was in dire need of visiting doctors who did not require a payment. The matter was resolved in August 1926.[9]

The bedouin, the shepherd, and the "primitive irrigation system" were targets of blame. Combating malaria and conquering the swamps, as Ruppin opined in 1927, meant struggling against the bedouin and "hostile neighbors."[10] From the beginning of malaria work by the American Zionist Medical Unit (1919), health officials feared that the Arab population

7. Quinine prophylaxis in the Mandate period was generally not used in Palestine except in emergency cases in Jewish settlements. Ruppin, it must be remembered, is referring to clinical practices in the period before the Mandate. A. Ruppin, "Sanitation of Palestine," reprinted in *Palestine* 4, no. 20 (n.d. [c. 1914–1918?]): 1–3, CZA J113/1430.

8. Letter from Dr. Bluestone to MRU, Haifa, January 27, 1927, CZA J113/555.

9. Letter from Kinneret Yemeni community to Kupat Holim, June 20, 1925, CZA J113/754; letter from head of branch, Dr. Yashan, to Hadassah management, CZA J113/754. It is likely that these Zionist doctors thought of the Yemeni population as having acquired immunity to malaria like they thought of their Palestinian Arab counterparts. Meir's passing over the settlement, however, most likely has its roots in the history of the Kinneret farm. For that history, see Shafir, *Land, Labor and the Origins of the Israeli-Palestinian Conflict,* 114–121.

10. Ruppin, *Three Decades of Palestine,* 153–154. Falah mentions Ruppin's comments on the bedouin and malaria but he misdates the quote, citing the publication date of the book rather than the date on which this speech was delivered in September 1927 at the Fifteenth Zionist Congress. Ruppin's comments refer particularly to the drainage of the Jezreel Valley and are situated within a larger discussion of the places of *halutziut* (pioneering principle) and *baalebatiut* (principle of middle-class capitalism) in Zionist colonization. Falah, "Pre-State Jewish Colonization," 293.

would infect the Jewish population until Jewish health organizations could alleviate the situation.[11] Zionist health professionals did recognize that Jewish immigrants were particularly susceptible in the settlements in malarious areas. Settlers, on the other hand, believed that the Palestinian Arab population was the *main cause* of malaria transmission; in a survey in 1925, the primary reason given for the spread of infectious disease was their Arab neighbors.

> In most of the settlements, they [the Arabs] live near the colonies. In a fourth of the colonies [surveyed], they live in the settlement itself. The wandering bedouin are not considered in the survey. Close neighbors, such as the Arabs, who stand in a much lower level of cultural and sanitary development, can be a source for every kind of infectious disease, especially smallpox, malaria, intestinal diseases, typhus and dysentery. The second reason for infectious diseases is bodies of standing water.[12]

Results from the survey covered 44 colonies with a total of 12,200 people. The survey was distributed to 11,172 Jews and 1,028 Arabs. More than half of the settlements responded to the survey (it was sent to 81). An article on the survey by Dr. Ratner, a doctor in Haifa, noted that 82 percent of the Jewish settlements who responded to the survey reported that they lived near Arabs.[13] He concluded from the responses that the settlers did not have the most elementary knowledge of hygiene, the ways in which diseases spread, or how to protect themselves. Ratner called for the medical establishment, especially Jewish colony doctors, to complete sanitary work in order to decrease the incidence of disease and institute necessary clinical treatment for *Havra'at chaiyei hayishuv ha'ivri*, the healing of life in the Hebrew *Yishuv*.[14]

The idea that Jews contracted malaria primarily from Arabs rather than from each other echoed other colonial settings[15] in which colonists assumed

11. Synopsis of conversation with Dr. H. R. Carter, U.S. Marine Hospital, to Szold, October 22, 1919, CZA J113/1430.

12. Dr. A. Ratner, "Letotzaot hamishal hasanitari bemoshavot beshnat 1925," *Briut ha'Am* 1, no. 1 (1926): 42, CZA Library.

13. Ratner, draft of "Letotzaot hamishal hasanitari bemoshavot beshnat 1925," CZA J1/1637.

14. Ratner, "Letotzaot hamishal hasanitari," 44–45.

15. The idea that mosquitoes infecting Europeans came not from other Europeans first but from "natives" is echoed earlier in the African context and justified residential segregation. Watts, *Epidemics and History*, 262.

"natives" had an acquired immunity to malaria. However, acquired immunity to malaria is stage-specific, species-specific, dependent on continuous infection over twelve years or longer, and is gradual, usually occurring in children. It does not last long, and it does not negate infection.[16] (Actual tests of malaria immunity were done later, in Nazi medical experiments.[17]) In West Africa, Watts reported that British colons believed that black Africans were immune to certain infectious diseases but could transmit them to white Britons, so British settlers built far away from black African communities at a distance that exceeded the range of the *Anopheles* mosquito.[18] In the Philippines, it was commonly held that "the greatest source of danger to the white man in a malarial locality lies in the native population, especially in the native children"; consequently, it was "futile" to attempt to "rid any locality of malaria so long as the native element in the question is neglected."[19]

According to Gordon Harrison, Europeans in the colonial age viewed acquired immunity to tropical diseases (malaria in particular) by blacks and "natives" as evidence of the hierarchy of the races. "Natives'" lessened susceptibility to malaria was compared to that of animals' and believed to prove the "natives'" lower station on the racial ladder. According to Harrison, this supposed racial difference in terms of disease helped "define the mission of empire and the role of the doctor in it." Dr. Victor Heiser of the International Health Bureau stated (about Filipinos), "As long as the Oriental was allowed to remain disease-ridden, he was a constant threat to the Occidental who clung to the idea that he could keep himself healthy in a small, disease-ringed circle." Heiser visited Palestine in 1925 with the International Health Bureau.[20] This attitude was repeated in Freetown,

16. For acquired immunity, see G. Corbellini, "Acquired Immunity Against Malaria as a Tool for the Control of the Disease: The Strategy Proposed by the Malaria Commission of the League of Nations 1933," *Parrasitologia* 40, nos. 1–2 (June 1998): 110.

17. Tests were undertaken on humans in concentration camps during World War II for knowledge about acquired immunity. Inmates of Dachau were infected with malaria and then treated with different various drugs to test their negative efficacy. See Corbellini, "Acquired Immunity," 112–113.

18. Watts, *Epidemics and History*, 262.

19. Shapira reiterates this attitude with regard to eye diseases in the *Yishuv*. See Shapira, *Land and Power*, 59; Warwick Anderson, "Immunities of Empire: Race, Disease and the New Tropical Medicine, 1900–1920," *Bulletin of the History of Medicine* 70, no. 1 (1996): 109–110; Charles Craig, "Observations upon Malaria: Latent Infection in Natives of the Philippine Islands" (1906), 525, as quoted in Anderson, "Immunities of Empire," 116–117.

20. Harrison, *Mosquitoes, Malaria and Man*, 4–5, 128; Worboys, "Germs, Malaria and the Invention of Mansonian Tropical Medicine," in Arnold, *Warm Climates*, 193; Ken DeBevoise,

Sierra Leone, where the removal of Europeans to the hills was an attempt to isolate Europeans from the supposed parasites in African blood.[21] Zionists were not immune to the racism of the colonial era.

Zionist health and settlement officials believed that the practices and beliefs of the indigenous inhabitants were the most serious challenges to malaria control measures.[22] They were critical of what they saw as the Arabs' improper and backward cultivation techniques. An alleged failure to care for the land caused areas in Palestine to become marshes that were breeding grounds for mosquitoes. In effect, the Palestinian Arabs' supposed neglect of the land created favorable conditions for the presence and spread of malaria in Palestine.[23] In drainage surveys of Migdal (just south of the Arab village of Mejdel and also of Tiberias), for instance, Dr. Kligler noted that the *fellahin* would dam the wadi and flood the land in the dry season. The excess water would stand on the clay soil and create mosquito-infested swamps.

Rice growing by the *fellah* posed another problem for Zionist malariologists. They maintained that the routine method of watering in the rice fields of Arab lands caused increased malaria infection of Jewish settlers nearby.[24] Dr. Kligler opposed the "primitive" flooding techniques of the Arabs (which was less labor-intensive). As chair of the Antimalaria Committee of the Vaad Leumi in the early 1940s, he alerted the government to the increased danger of malaria infection due to Arab methods of flooding and offered suggestions for "rational" (scientific) methods of watering. According to Kligler, the government did not respond adequately. This was most likely because the British Health Department of Palestine had ordered an end to antimalaria work in the Huleh area in 1942 in order to initiate a rice-growing experiment there.[25] In this case, the economic

Agents of Apocalypse: Epidemic Disease in the Colonial Philippines (Princeton: Princeton University Press, 1995), 145–147; V. Heiser, *An American Doctor's Odyssey: Adventures in Forty-Five Countries* (New York: W. W. Norton, 1936), 37, as quoted in Anderson, "Immunities of Empire," 112.

21. For Jews, Blacks, and Arabs, see Gilman, *Disease and Pathology*, 30–35, 135. For debates on Jewish immunity to disease in late nineteenth-century America, see Sander Gilman, *Picturing Health and Illness: Images of Identity and Difference* (Baltimore: Johns Hopkins University Press, 1995), 108–112.

22. Anderson, "Immunities of Empire," 118. See also *Survey of Palestine*, 2: 609–634.

23. Kligler, *Epidemiology and Control of Malaria*, 11, 114–116.

24. Dr. P. Noack, Department of Health of Vaad Leumi, "Health Report," 1942, 4, and "Health Problems in 1943," both of CZA J1/7718.

25. Noack, "Health Report." See also Kitron, "Malaria, Agriculture and Development," 301.

interests of the government prevailed over health needs. Interestingly, however, in a survey of the swamps in Nuris in 1937, Zvi Saliternik noted that the Arab practice of watering lands allowed the alternation of quantities of water in a channel in the day and at night, thus *lowering* the chance for mosquito breeding. Nevertheless, Saliternik recommended hiring an Arab inspector whose salary the Jews would help finance to supervise bedouin practices and regulate water flow in the channels.[26]

Grazing buffalos in swamps, utilizing water holes as drinking locations for livestock, growing papyrus, reed cutting, and fetching water in malaria-infected waters caused concern among Zionist [and British] health officials.[27] Cattle and water buffalo that grazed in the swamps frustrated ongoing antimalaria works. Particularly in the Huleh, sanitary engineers noted that

> [t]here is nothing more destructive to artificial channels than uncontrolled cattle; they trample down the banks to get to drinking water, the buffalo likes to wallow and creates nasty muddy pools. These pools and hoof prints are all potential breeding places for mosquitoes.[28]

Mosquito larvae on the hooves of the animals were transported from one area to another, which spread malaria. Engineers built concrete dams for grazing in order to accommodate the shepherds and their livestock. Similar accommodations were made at Ein Semonia for shepherds. They recognized that once drainage took place in the Huleh, however, the number of cattle and buffaloes would decrease. And it did.[29]

Migratory practices of the bedouin also interfered with the work of Zionist malariologists. Nomadic bedouin tribes with high rates of malaria

---

26. Zvi Saliternik, "Hamatzav beEmek Nuris" [The situation in Nuris], 2, CZA J1/1726; Zvi Saliternik to Kupat Holim, "Hamatzav besviva Tel Amal" [The situation around Tel Amal], April 21, 1937, 3, CZA J1/1726.

27. See the example of Bussa village and the Bussa dam as a source of *Anopheles*. "Proceedings of Seventh Meeting of Antimalarial Advisory Commission," 11; Cantor, "Wadi esh-Shahm," September 2, 1921, 1, CZA KKL3/99aleph; Drainage Investigation—Khulde District, CZA. KKL3/99aleph; American Zionist Medical Unit, "Ma darush letakanat hamatzav beKhulde," September 15, 1921, 1, CZA KKL3/99aleph and KKL3/99bet; "Technical Description of the Scheme for the Drainage of the Swamp near Khulde," 1–2, CZA KKL3/99aleph.

28. Rendel, Palmer, and Tritton, "Huleh Basin," 30, 31.

29. The dams would usually have sluice gates to avoid mosquito breeding. "Drainage Operations of the Malaria Research Unit," 3, part of British Health Department of Palestine MRU annual report, Haifa 1923, ISA 1670/130/33a/mem/6420; "Drainage Operations of Various Jewish Agencies," 9.

may have become infected outside Palestine, making locally acquired malarial indexes hard to construct and further frustrating malariologists who tried to keep statistics on malaria.[30]

Palestinian Arab methods of cultivation challenged Western capitalist concepts of productivity. Zionists and the Mandatory government imagined productivity as agricultural production for the primary purpose of revenue and expansion of markets.[31] They saw natural resources as commodities. Furthermore, during this period, productivity was seen as a primary measure of health; the greater the yield of workers, the "healthier" they were.[32] The emphasis on labor and agricultural productivity was as much a response to the anti-Semitic image of Jews as parasitic as it was an expression of Zionist capitalism.[33] Palestinian Arab agricultural practices clashed with the government and Zionist ideas of intensive cultivation. Arabic farmers worked within the natural limitations of the land in order to provide a livelihood for the *fellahin* and bedouin of the area. Zionists strove to conquer and improve it.

The rift over agricultural methods revealed the conflict between capitalist and noncapitalist modes of production. As Talal Asad has explained, the Mandatory government ensured the growth of the capitalist mode at the expense of the noncapitalist mode but prevented the total elimination of the noncapitalist mode until the Jewish capitalist sector gained prominence, leading eventually to the destruction of the noncapitalist (Arab) mode of production.[34]

Despite Zionist claims to the contrary, *fellah* sharecropping had its own logic, privileging communal economic security over immediate monetary rewards.[35] As opposed to Jewish mixed farming that depended

30. MRU annual report, 1923, 16, ISA M1670/130/33a/6420; League of Nations Health Organization Malaria Commission, *Reports on the Tour*, 36; Bunton, "Land Law and the 'Development' of Palestine," 6.

31. Cronon, *Changes in the Land*, 162, 166, 167.

32. Packard, "Visions of Postwar Health and Development," 94.

33. Shafir argues that the concept of productivization was a mirror image of the anti-Semitic portrayal. Shafir, *Land, Labor and the Origins of the Israeli-Palestinian Conflict*, 81.

34. Talal Asad, "Anthropological Texts and Ideological Problem: An Analysis of Cohen on Arab Villages," *Review of Middle East Studies* 1 (1975): 14, 21.

35. Yaakov Firestone, "Production and Trade in an Islamic Context: Sharika Contracts in the Transitional Economy of Northern Samaria, 1983–1943," *International Journal of Middle East Studies* 6, no. 2 (April 1975): 189–190; Levine, "Discourse of Development in Mandate Palestine," 114–115. On adjustment of the *musha'a* system, see also Metzer, *Divided Economy of Mandatory Palestine*, 95–96.

on capital from abroad, Arab agriculture more often relied upon self-produced inputs.[36] The agricultural product per worker for the Arab sector grew annually throughout the Mandate period, and by World War II, accounted for seventy-two percent of the total increase in noncitrus production of the entire country.[37]

Though colonists saw it as barren and stagnant, swampland did, in fact, provide a place for Arab production of papyrus, rice, and other products.[38] Papyrus production in Palestine was the largest in the world at that time and was used to make huts, mats, rafts, and baskets. Peasants used the rhizomes of the papyrus for fuel. Swamp areas also supported the growth of certain flora and fauna in Palestine that contributed to a stable ecosystem and were used by the *fellahin* for various medicinal and nutritional purposes. Zionist bureaucrats did recognize the value of papyrus in private deliberations on the transfer of the Huleh concession. Mr. Singer of the Zionist Organization wrote Colonel Kisch of the Palestine Zionist Executive in early discussions about the Huleh Valley, "Kindly make enquiries to ascertain . . . what would be the possibilities of a commercial exploitation of the papyrus in the Huleh marsh as and when the Keren HaYesod obtains the concession."[39] An exchange of letters between Singer, Kisch, and Henriques, an engineer involved with British colonial and Zionist swamp draining efforts, dealt with the profitability of such an endeavor, issues of taxation, and whether any Arab would have to be compensated. Productivity was determined by *who* was overseeing the production of papyrus and whether it could be used to capitalist ends.

Zionist health officials cited another significant obstacle to the fulfillment of antimalaria plans; that is, the demonstration and resistance of Arab villagers against swamp draining when a project encroached upon their land. Palestinian Arabs had, in scholar Zachary Lockman's words, a "capacity for agency which often intruded upon and altered Zionism's

36. Metzer, *Divided Economy of Mandatory Palestine,* 96, 152–153.

37. Metzer suggests that this growth was due to rising labor-land ratios. *Divided Economy of Mandatory Palestine,* 145, 149, 154.

38. Rod Giblett, *Postmodern Wetlands: Culture, History, Ecology* (Edinburgh: Edinburgh University Press, 1996), 222.

39. Letter from Paul Singer, Financial and Economic Committee of Zionist Organization, to Colonel Kisch, September 24, 1925, CZA S25/595; letter from Storrs, director of lands, to the secretary of the Palestine Zionist Executive, October 20, 1925, CZA S5/595; letter from Singer to Kisch, November 3, 1925, CZA S5/595; letter from Henriques to Kisch, no date, CZA S5/595.

conceptions of itself and its mission and always registered itself on and helped shape the Zionist project."[40] Indigenous resistance to land management activities was common in the colonial world. As Richard Grove notes, colonial forest and land management programs

> brought about frequent clashes and contests over land use. These typically involved the colonial state, private companies and local people as separate and competing actors in contests of governance, protest and manipulation. In some instances colonized peoples succeeded in blocking the advent of state forest or soil conservation and then turning the colonial programmes to their own ends... the enforcement of colonial attempts at environmental control resulted in a whole typology of resistance and reaction by indigenous peoples and colonial inhabitants.[41]

The imagined detachment the Arab population felt about the land allowed Zionist drainage engineers to feel justified in sanitizing Arab land. Sanitization disregarded prior use or value placed on the swamp. A compulsion to make the wasteland productive created a sense of deep frustration when Zionist land reclamation agencies faced resistance on Arab-owned or Arab-leased land. Instead of viewing this resistance as proof of Arab attachment to the land and an alternate vision of the swamp as useful, Zionist health professionals decided that the *fellahin* and bedouin were complacent about their health. Arab resistance to attempts to transform the physiography of their property and which threatened grazing or water rights reinforced the view of "local populations as inherently unhealthy and incapable of caring for their own health needs."[42] Indigenous resistance also confirmed the Zionist belief that the Arab community had no real attachment to the land because they blocked "improvements" on it.

Fellahin and bedouin resistance occurred when drainage schemes extended into Arab land. Engineers wanted to prevent mosquitoes from flying from Arab village swamps to Jewish settlements where swamps had just been drained.[43] When such measures would commence, issues of

40. Lockman, *Comrades and Enemies*, 371.
41. Grove, *Ecology, Climate, and Empire*, 3.
42. Packard, "Visions of Postwar Health and Development," 95.
43. Outstanding malarial problems in neighboring Arab lands were constantly discussed. See problem of ditches on Arab land near Kfar Yehezkel and discussion about areas near Nahalal in "Proceedings of Seventh Meeting of Antimalarial Advisory Commission," 12.

borders would sometimes be raised, Palestinian Arab villagers would voice opposition, and the antimalaria project would have to be delayed until border disputes were resolved.[44] This frustrated Zionist land and malaria agencies considerably because it prolonged the attainment of malaria control. It also posed potential contractual problems between the engineering company and land agency, sometimes resulting in financial loss. Agitation was evident in the minutes of the KKL:

> The third difficulty is that [our] Arab neighbors don't want at first to agree to the drainage of swamps that lie on the borders of our land with theirs. In cases where the swamp is dangerous ... where the swamp was created by a spring that is located on Arab land bordering our land—we were compelled to drain that part of the swamp—if the neighbors themselves will not drain it. On the other hand, in places where neighboring villages give us their help—they profit from our "betterment" (hashbacha) work which is done on their land that borders ours.[45]

The KKL asked the government to intercede, but disruptions continued. Resistance to Jewish drainage projects arose where there were border disputes, threats to the economic livelihood of the fellahin, and questions of water rights. One such incident transpired in 1922 when drainage work was stopped in the area between Ein Taboun and Nachal Jalud until a border dispute between Kumia and Nuris/Hodia by Arab residents was resolved.[46] Demonstrations by Arabs neighboring the settlements of Givat Yczekial and Nahalal occurred the same year, causing interruptions in the work of KKL.[47] The 1923 MRU annual report to the British Health Department of Palestine described the Ein Taboun swamp as the northern

---

44. Confusion for the British about which land category swamps fit into reflected the conflicting conceptualization of land ownership with rural practices. Bunton, "Land Law and the 'Development' of Palestine," 2–3; Smith, Roots of Separatism in Palestine, 90–91; Swearingen, Moroccan Mirages, 48–50.

45. "Minutes of the Central Executive of the KKL," Jerusalem, 1921–1923, 2, CZA A246/441.

46. Telegram to Pevsner Jochevedson of Nazareth from Vriesland, Ettinger, and Thon, September 6, 1922, CZA KKL3/56; "Drainage Operations of Various Jewish Agencies," 7; British Health Department of Palestine MRU annual report, Haifa 1923; Giblett, Postmodern Wetlands, 150, 170–171, 205–227.

47. Letter from Granovsky of Joint Committee for Public Works to head of KKL, September 18, 1922, CZA KKL3/57aleph. About providing a local person in AbuShosha so that negotiations with the Arab community could be eased for the process of evaluating the malaria situation, see letter from Yosef Shapira, head of Department of Health, MRU, Haifa, to Zionist Executive Settlement Department, April 23, 1928, CZA KKL5/3331.

boundary of the Nuris area in the Jezreel Valley. "An open concrete-lined canal to carry off the water was constructed from the Jalud River up to the swamp, but its drainage was held up due to a disagreement with the neighbors regarding boundary demarcations and water rights. Through the intercession of the MRU, an open ditch has been made...[and] is at best only a temporary arrangement until a more satisfactory agreement is reached."[48] Despite the claim by Colonel Heron, the director of the British Health Department of Palestine, that there was considerable co-operation in the Jezreel and Beisan areas, other sources show that in the case of Nahalal, the Arabs of Malul would not allow KKL workers to dig in their fields because of the damage they felt it would cause to their livelihood.[49] They protested,

> What will happen to the water? The water gushes from the springs and makes swamps; swamps from which our herds drink! If you concentrate all the water in deep channels in pipes—from where will they be able to drink water? [It is possible] to leave one pipe exposed to give them water but we don't want that. These springs are in our fields and if the herds start to graze on our fields, it will be a tragedy for generations.[50]

Arguments between the governor of Nazareth and land settlement leader Yehoshua Hankin ensued. In this case, extensive negotiations took place and a special area was finally made for pumps and pipes [on Jewish land] where the Arab neighbors could bring their herds to drink.

In Beisan, Palestinian Arab residents opposed an antimalaria scheme, claiming that it deprived them of the swamps that they used as a bathing ground for the buffalo.[51] In other villages of the Valley of Jezreel, work ceased when neighboring Arabs broke drainage pipes. Occurrences such as these occasionally prompted the KKL to seek government assistance:

48. Letter from David Stern, head of Colonization Department of KKL, to Cantor, sanitary engineer of government, "Havra'at Khulde," October 5, 1921, CZA KKL3/99 bet; response by government, to director, Zionist Commission, "Antimalarial measures-Khulde," November 4, 1921, CZA KKL3/99 aleph.

49. "Proceedings of the Eighth Meeting of the Antimalarial Advisory Commission," 2.

50. "Nahalal" (Tel Aviv, 1929), as quoted in Shmuel Din, "Yibush habitzot beNahalal," 93–94, CZA Yellow Folder Swamp Drainage.

51. For Arab families drawing their livelihood from the Kabbara marshes, their legal dispute, and the government's invocation of sanitary reasons as way to justify expropriation, see Bunton, "Land Law and the 'Development' of Palestine," 2, 7–18; League of Nations Health Organization Malaria Commission, *Reports on the Tour*, 61.

We would in the first place respectfully draw your attention to the great stress that is laid by the Commission [of inspection of KKL drainage works[52]] throughout its report on the adverse effect upon our work caused by the non-drainage of swamps on land not belonging to us which lie adjacent to our colonies. The reason for this non-drainage is to be found in (1) the indifference of the Arab villages to such work and (2) administrative orders prohibiting the execution of such work.

The Jewish National Fund [KKL] is unable without the help of the Government to overcome the indifference of the neighbors of our colonies to the danger of malaria arising out of undrained swamps and affecting the whole environment. It can show that where the Arab neighbors have consented to our drainage work being extended over their land, they themselves have reaped much benefit for which they have been very grateful. This occurred at Ein Sheikh, Nahalal, where by carrying out canal[ization] beyond the border of our land, a considerable stretch of neighboring Arab territory was reclaimed.

On the other hand, at Ein Beda, Nahalal, the non-cooperation of the Arabs caused delay in our work and until today only the Jewish side of the swamp has been tapped. It is obvious that such a state of affairs is bad for the whole District.[53]

Ein Sheikh was the center of collection of water from all the swamp areas in Nahalal. Ein Beda spread along both Arab and Jewish property. As a drainage report stated, "Like most border swamps, its drainage was long delayed by disputes over water rights, swamp grass, etc. and it was not until late this year that permission was obtained to proceed with the work." Ein Beda was considered one of the worst swamps in Nahalal.

The KKL saw Arab resistance as confirmation of their general image of the Palestinian Arab community. They even argued that where Arabs agreed to drainage, they were not complacent about their health or their land, but rather "grateful."[54]

52. The commission was composed of Mr. Louis Cantor, sanitary engineer, Dr. Kligler, Sanitarian of MRU, and members of government personnel. The report was to be kept confidential.

53. Letter to director of Department of Public Health of Government, from KKL, official translation, November 5, 1923, CZA A246/440; "Drainage Operations of Various Jewish Agencies," 10.

54. Giblett, *Postmodern Wetlands*, 141, 147; also, Granovsky and Ettinger's confidential report to Heron detailing several points of frustration by the KKL toward their Arab neighbors with regard to drainage. One particular instance noted the governor of Nablus calling by

Acknowledging the interdependence of Jewish health and healthy Arab lands, yet fearing that indiscriminate application of the Antimalaria Ordinance would produce discord between the two, the government suggested that

> the best means of obtaining permanently improved conditions on Arab land near new [Jewish] colonies is for the colonists to make efforts to cooperate with Arab owners in improving the malaria conditions. Such action, with the aid of the district medical officer of health and the Malaria Research Unit, has already proved its value in several instances without resort to the law.[55]

These confrontations forced Zionist malaria officials to come to terms with the Palestinian Arab community in the field of health and recognize their dependence upon them.[56] However, in the end, *fellahin* and bedouin opposition to Zionist drainage interventions was used to justify the Zionist view that the Palestinian Arab population posed obstacles to development.[57] Charles Kamen notes,

> The Jewish analysis of Arab agriculture necessarily included an evaluation of the peasants' capacity to adapt themselves to the expected transformation of the countryside, to recognize the opportunities it provided them, and by implication, to accept the inevitability of Jewish colonization.[58]

The KKL expected the Arab inhabitants in the area to welcome Zionist extension of drainage works because it was supposed to improve their health. Yet for those Arabs whose livelihood and herds were threatened by swamp drainage efforts, the benefits of health did not outweigh the

---

administrative order the cessation of JNF work in Ain Tabun in the Jezreel. May 11, 1923, 5 CZA A246/440.

55. Letter from Colonel Heron, director of British Health Department of Palestine, to JNF, June 13, 1923, CZA A246/440.

56. It also notes the government's hesitation to take firm action on behalf of one group, perhaps reflecting larger political policies of the Mandate period. Kamen has argued that the government did not encourage Jews and Arabs to jointly develop the land. I would agree with him in a general sense but the quote noted above points to a case where this is not so. Charles Kamen, *Little Common Ground: Arab Agriculture and Jewish Settlement in Palestine, 1920–1948* (Pittsburgh: University of Pittsburgh Press, 1991), 4.

57. Additional complaints to the government by Zionist officials included the distribution of free quinine to the Arab population but not to the Jewish colonies, despite the fact that Zionist officials felt they were the primary ones taking active steps to combat the disease.

58. Kamen, *Little Common Ground*, 82.

losses. As the Kherkes and Zeita Arabs claimed, "It's better to die from malaria than from hunger, because our existence depends on the herds which need the swamps."[59] The *fellahin* and bedouin increasingly saw that development schemes such as swamp drainage eventually led to the massive transformation of the the land on which they lived.[60]

Chronicles of health efforts during various uprisings of the Mandate period enrich the social history of Palestine because they are insights into mundane concerns of a troubled time.[61] Health reports and primary documents on health issues are very rarely given as bibliographical sources with which to reconstruct the daily occurrences, fatalities, or political consequences of science. The events of 1921 and 1929 and the 1936–1939 revolt have never been fully understood in light of their influence on the everyday health conditions in Palestine.[62]

The Arab Revolt of 1936–1939 was a hindrance to antimalarial works.[63] The annual report of the British Health Department of Palestine of 1938 stated that malaria control was the most difficult to carry out compared to all other spheres of departmental work because of the unsettled state of the country.[64] Owing to the violence and problematic financial situation of the country, no new major schemes of reclamation

59. Yoram Nimrod, "'Ma'asecha yekarvum' yachasei yehudim-'aravim bepeulo hatziburi shel Zvi Botkovsky" ['Your actions will bring them close': A new approach to Jewish/Arab relations: the economic initiative of Zvi Botkovsky], *Tzion* 57, no. 4 (1992): 433, Historical Society of Israel, Jerusalem.

60. Levine, "Discourse of Development in Mandate Palestine," 114.

61. For one, rather uncritical, article on Jewish doctors and the 1929 Riots, see Nissim Levy, "Harofim ha'ivriim beEretz Israel beme'oraot tarpa"t" [Hebrew doctors in Palestine during the 1929 riots], *HaRefuah* 117 nos. 3–4 (August 1989): 92–95; T. Bowden, "The Politics of Arab Rebellion in Palestine, 1936–1939," *Middle East Studies* 11, no. 2 (May 1975): 147–170; J. Jankowski, "The Palestinian Revolt of 1936–1939," *Muslim World* 63 (1973): 220–233; B. Kalkas, "The Revolt of 1936: Chronicle of Events," in *The Transformation of Palestine: Essays on the Origins and Development of the Arab-Israeli Conflict*, ed. Ibrahim Abu-Lughod (Evanston: Northwestern University Press, 1971), 237–274; Y. Porath, *The Palestinian Arab National Movement: From Riots to Rebellion; Volume Two*; Ted Swedenburg, "The Role of the Palestinian Peasantry in the Great Revolt (1936–1939)," in *Islam, Politics and Social Movements* ed. E. Burke, and I. Lapidus (Berkeley: University of California Press, 1988), 169–203.

62. On details of health care, wounded, and refugees during the 1929 Riots, see the archival material "Hadassah in Palestine," 7–10, Hadassah Archives, file 57/5.

63. Penslar notes that Zionist settlement engineers conceived of Arab political animosity as one of many obstacles that would disappear with careful planning. I did not find this attitude among sanitary engineers. Penslar, *Zionism and Technocracy*, 4.

64. British Health Department of Palestine, annual report 1938, chapter 20, PRO CO 733/399/24. Exceptions to limiting completing projects were the Birket Ramadan and Lake

were undertaken during the revolt. Delays and discontinuation of some malaria work resulted in a fluctuation in malaria cases in the *Yishuv* in the mid-1930s and an epidemic situation in 1939. Statistics of the Kupat Holim Klali (General Kupat Holim) of the Histadrut and the Kupat Holim Amami (Hadassah) confirmed the epidemic state of the *Yishuv*. Dr. Mer of the Rosh Pina Malaria Research Station approved the analysis.[65] As Zvi Saliternik, a Zionist malariologist, wrote, "In the present political situation, there is no way to carry out the necessary steps to prevent the malaria dangers from the said settlements with swamp drainage, daily work, etc."[66] In response, Dr. Mer set up a centralized service for supervision of antimalarial activities.

In areas where tension was particularly high, both Palestinian Arab and Jewish workers faced repeated difficulties getting to work at both Zionist and government reclamation schemes. In the area of Herzliya, for instance, no Jewish drainage work could be done, and malaria infection increased.[67] To prevent a malaria epidemic in the Wadi Rubin and Wadi Musrara, arrangements were made for an Arab constable to accompany an Arab engineer so that he could carry out his work there in August 1936. When the constable was not available, laborers on the project had to be discharged, causing frustrating delays on the project.[68] Mixed Arab-Jewish areas were also problematic, but Arab and Jewish volunteer laborers were accompanied by guards or by governmental labor.[69] In addition to problems with workers' access to project areas, the depressed economic condition of the country prevented many Jewish settlements from spending large sums of money for drainage. Chemical control and other less expensive, more temporary methods were used instead. These measures did not fully prevent *Anopheles* breeding unless they were carried out continuously. Political tensions, Palestinian Arab agricultural

Tiberias, which were completed in 1939 by the government. British Health Department of Palestine, Annual report 1939, 12, Ein Kerem Library.

65. Health Project in Palestine 1937–1938, CZA J1/7718; letter from Dr. Yassky to PICA Executive, November 7, 1939, CZA J1/1825.

66. Letter from Saliternik to Kupat Holim, April 21, 1937, 9, CZA J1/1726.

67. Letter from head of Sanitary Committee of Herzliya Colony to Vaad Leumi Department of Health, September 21, 1936, CZA J1/1542.

68. Letter from N. M. Maclennan, senior medical officer of endemic diseases, to director of district superintendent of police, southern district, Jaffa, August 17, 1936, and August 30, 1936, ISA M1552/25/18. Maclennan was transferred in 1938 to the British Guiana Colony.

69. British Health Department of Palestine, annual report 1938, 12.

methods, and resistance to Zionist drainage projects on Arab lands converged with Zionist images of the Palestinian Arab population as unproductive, primitive, and unattached to the land to deepen the Zionist view
that the Palestinian Arab population was an obstacle to malaria control
and to the emergence of *their* Palestine.[70]

Practical Implications of Representation

Concepts about the Arab population influenced plans for malaria control
and the placement of settlements. Swamps on Arab land, for instance, affected the type of drainage recommended for the Jews and determined the
precise placement of Jewish settlements.[71] In a preliminary report about
the possibilities of draining the area of Malul (Nahalal) in August 1921, I.
Gutmann noted that a swamp causing malaria infection was situated on
Arab land. Recognizing the improbability of draining Arab-owned land
in this case, he recommended palliative measures such as quinine prophylaxis and planting of eucalyptus trees to decrease the potential incidence
of malaria. He also instructed the KKL to build the Jewish colony near
the "northwestern corner of the property.... This is a fairly convenient
geographical location in reference to the most valuable lands of the property. Most important of all, it is protected from the swamps by the ridge
running along the line Ein Mudawarra-Ein al-Sheikh."[72]

Race-based medical notions also promoted either a separation from or
the exclusion of mixed Arab-Jewish areas for clinical treatment. For example, an American named Dr. Carter suggested to the American Zionist
Medical Unit in their initial campaigns against malaria during the period
of civil administration that it was best to apply the "sterilization method"
(giving quinine to those with a history of malaria infection or positive blood
findings) to Jewish communities that were not mixed with the Arab population because "this method is an individualistic one and the cooperation

70. Kimmerling, *Zionism and Territory*, 1, 7; Gershon Shafir, "Changing Nationalism and
Israel's 'Open Frontier' on the West Bank," *Theory and Society* 13, no. 6 (1984): 805, as
quoted in Ram 337.

71. See examples of Ein Harod, Gevah, and Kfar Yehezkel as well as Nahalal in the
Jezreel Valley in the early 1920s. "Proceedings of the Seventh Meeting of the Antimalarial
Advisory Commission," 12.

72. I. Gutmann, "Preliminary Report on Water Supply, Irrigation Possibilities and
Drainage of JNF Lands at Dilb, Yajur and Malul," August 1921, 13–15, CZA KKL3/56.

of Jews may be easier obtained."[73] The assumption was that Arab communities would not readily cooperate.

The idea that swampland was unproductive, along with incidence rates of malaria, justified drainage work, but the positive effects of antimalaria campaigns were sometimes outweighed by disruptive results. Antimalaria measures altered the material conditions of the Arab village and bedouin populations, and thus, raised their resistance. Kligler, in his report on the malaria situation in Palestine, observed that the annual migration of bedouin tribes from winter to summer quarters caused a distribution of the malaria parasite and a constant extension of infection to other areas. During this migration, the "heavily infected bedouin roams over the country," trying to find new places for grazing and watering their animals.[74] As more and more swamps were drained, the area for animal grazing (particularly for water buffalo and sheep) became increasingly restricted. In most cases, bedouin tribes had to find new grazing spots for their buffalo and other animals in still-existing swamps. Professor Yekutiel, a Jewish doctor and public health activist who worked on malaria research at the Rosh Pina Malaria Research Station during the Mandate, noted that this migration actually spread malaria, albeit not drastically, to farther areas because bedouin tribes had to travel longer distances to find grazing spots. In addition, he remarked that the constant influx of immigrants from malaria-free countries into a situation of malarial Palestine created a greater risk factor for spread of the disease.[75] According to Falah, reduction of grazing land, including that in the areas of Marj ibn ʿAmir/Emek and the Huleh Valley, accelerated the processes of sedentarization and detribalization for the bedouin population and contributed to the depopulation of indigenous Arab inhabitants in northern Palestine. He argues that "[t]hese experiences [of sedentarization and detribalization] brought about significant changes in bedouin perception of land and their social and political organization... the sedentarization process in this case

73. Suggestions obtained from Dr. H. R. Carter during interview at Baltimore, CZA J113/1430.

74. Kligler, *Epidemiology and Control of Malaria*, 159–160; British Health Department of Palestine, annual report for 1934, 34–35, ISA M4475/06/1. See El-Asmar for a discussion of the bedouin as the typical Zionist representation of the Arab before 1948, Rouzi El-Asmar, "Portrayal of the Arab in Hebrew Children's Literature," *Journal of Palestine Studies* 16, no. 1 (1987): 85.

75. Interview with Fritz Yekutiel, Jerusalem, Israel. April 2, 1997; Kligler, *Epidemiology and Control of Malaria*, 158–160.

was essentially shaped by the conflict arising between the indigenous pastoral bedouin and the new immigrant community of Jewish cultivators."[76]

Land purchase that often included swamp areas in many cases led to Arab displacement. In the 1930s, provisions by the government in land sales reserved some land for the Arabs, but bedouins had no legal right to the land, and no provisions were made for them. The KKL and other land-purchase agencies sometimes faced strong resistance by the bedouin to leaving their land for Zionist reclamation projects; in 1926, members of the Ghawarna in Jidro attacked colonists and police after the legal transfer of their lands had occurred.[77] The actions of this tribe were part of a continual bedouin resistance to the loss of grazing land through land sales.[78] Resistance also occurred in Wadi Hawarith,[79] where Zionists complained that the government was "prepared to let an area fit for orange growing remain grazing land merely because its present occupants claim to need it for their flocks, know nothing of irrigation, and because, in case of transfer, it seems likely that the tribe will lose its identity as a tribe and become a scattered community."[80]

Removal of the *fellahin* and bedouin from the lands on which they lived to areas away from Jewish settlement increasingly separated the physical, economic, and social world of the Arab from the Jew. In fact, Zionist settlement officials explicitly used the terms "farther away" or "farther away from our land" in discussions about arranging land compensation for the

76. Falah, "Pre-State Jewish Colonization," 289–290, 293. Falah includes a discussion of afforestation and its impact upon depopulation and sedentarization.

77. Telegram from Zionicom to Eder, February 23, 1926, CZA S25/517.

78. Falah, "Pre-State Jewish Colonization," 305; Stein, *Land Question in Palestine*, 36, on the antagonism caused to the bedouin by Zionists unfamiliar with grazing rights and patterns. Swedenburg adds that until today, Israeli nature lovers still chase bedouin off their lands. Swedenburg, *Memories of Revolt*, 59; Smith, *Roots of Separatism in Palestine*, 96–97, for complaint by bedouin to Hope-Simpson about the loss of their livelihood to Jewish purchases in Valley of Esdraelon; Smith, *Roots of Separatism in Palestine*, 102–103, for Ghawarna resistance in Zawr al-Zarka/Kabbara lands.

79. Extract from *Daily News Bulletin* 61 (August 3, 1932), letter to chief secretary from Arlosoroff, executive of Jewish Agency, December 30, 1931. Ref. 453/31, and letter to Cunliffe-Lister, principal sec. of state for colonies, from H. C. Wauchope, December 24, 1931, PRO CO 733/218/1.

80. From the Shaw Commission report. Quoted by the author of the pamphlet, L. Stein, memorandum on the "Report of the Commission on the Palestine Disturbances of August, 1929," 102, cited in Kamen, *Little Common Ground*, 82.

fellah in Nuris.[81] The gradual separation of these communities was conjoined by a broader exclusivist policy of labor and settlement.

Agents of Development: Zionist Claims as Mutual Benefactor

Diagnosing the Palestinian Arabs as obstacles to malaria control buttressed the Zionist assertion that the Jews were the most effective agents of development in Palestine, that they were the ones transforming the land from an empty wasteland to productive, fertile, lucrative land.[82] The British and Zionist settlement and medical officials felt that the *fellahin* and bedouin hindered the development of Palestine, the attainment of Jewish transformation and, in turn, the realization of the Zionist project.[83] The British saw malaria and malarial symptoms as one of the main causes for apathy in the Arab population and for the decades of neglect of Palestine,[84] but the British were not driven by the nationalist fervor of the Zionists.

Lethargy due to anemia, a symptom of malaria, may have contributed to the image of the Arab as apathetic, but it was not the basis for the entire stereotype. An article called "Palestine Hygiene and Disease," published in 1919, quoted Dr. E. W. G. Masterman, a British physician who spent time in Palestine:

> The bedouin are popularly credited, on account of their entirely open-air life, with great soundness of constitution, but it cannot be said that this is the case with the nomads of Palestine. They are exceedingly scantly clad...and their skins are exposed to all the extremes of heat and cold...while during the latter [winter] season, the atmosphere of their dwellings is commonly saturated with

81. Letter from Ettinger to the head council of KKL, "Hakarka'ot beEmek-Yizrael," December 13, 1921, 1–2, CZA KKL3/59. This document also states that the KKL would try to decrease "as much as possible" the amount of compensation provided to the *fellah*.

82. Lehn and Davis, *Jewish National Fund*, 55–56; Packard, "Visions of Postwar Health and Development," 101; Giblett, *Postmodern Wetlands*, 226.

83. Malaria Control Demonstration: Breeding Places of Anopheles, British Health Department of Palestine, MRU, Haifa, 1923, 16, for details of how bedouin and shepherd contribute to the spread of malaria, ISA M1670/130/33a/6420; Dr. Hammer, *Skira 'al ha'avoda haanti-malarit beshnat* 1935 [Survey of antimalaria work in 1935], Emek Zevulun, which cites the bedouin as a particular concern. October 15, 1935, 1, CZA J1/3749; Kligler, *Epidemiology and Control of Malaria*, vii.

84. "Proceedings of the Eighth Meeting of the Antimalarial Advisory Commission," 4; Harrison, "'Hot Beds of Disease,'" 11.

the irritating smoke from wood or dung. It might be supposed that the smoke would at least afford some protection from insect pests, but the truth is that, under such conditions ... mosquitoes and other insect pests are found in abundance. Doubtless in the days when the Bedouin ... were [able] to keep themselves in good physique by martial exercises, they enjoyed greater robustness; but now a large proportion of the bedouin of Palestine are *sallow in complexion and constantly suffer from malarial fever* ... from which it might be thought their outdoor life would save them. Even their nomad habits do not relieve them from epidemics.[85]

Masterman continued by describing the "mongrel bedouin in the neighborhood of the towns of Palestine" as having very low morals, being fatalists, and leaving their ill uncared for.

The Zionists' description of the Arab stood in stark contrast to their view of the Jew. The latter was taking the initiative to revive the land from its slumber and was increasingly enlightened, developed, productive, and constructive. The "primitive" agricultural methods of the *fellahin* coupled with bedouin nomadism served as justification for Zionist health professionals to request that some lands be turned over to the Jews:

> We think the Government should see to it that the work of the commission is instituted with the object of liquidating its lands and handing them to bedouins who keep their cattle in a very primitive manner. Lands fit for the most intensive rearings should be suspended early because of this tendency. The government must substitute [this with a] policy of handing over state lands to Jewish settlements.[86]

Zionist propaganda featuring Jewish initiative was frequently presented in film series and books. A 1922 series of films called *Eretz Israel Hamitchadeshet* (Land of Israel in the Process of Renewal) intended for commercial distribution to the "civilized West" and to the United States, for instance, depicted Jewish activities in Palestine. Hospitals, swamp draining, and eucalyptus trees were included as central images. The political and

85. Statement by President, Colonel Heron, "Proceedings of the Eighth Meeting of the Antimalarial Advisory Commission," 2, emphasis added; "Palestine Hygiene and Disease," in *Palestine: The Organ of the British Palestine Committee* 6, no. 4 (September 6, 1919): 1–3, CZA J113/1430; Aref el-Aref, *Bedouin Love, Law and Legend*, 154, about bedouin illness stemming from mosquitoes.

86. "The State Lands Question," CZA S25/526.

moral consequences of these images were explicitly recognized: "From the outside, the films should not carry a Jewish imprint. They should look, externally at least, quite neutral." However, every "thoughtful" person who viewed the films would be sure to "draw the necessary conclusions." In addition, as "important moral instrument[s] in the realization of our aims," it was expected that the contrast between the "half-civilized Arab and the highly civilized and energetic Jew will not fail to produce a deep impression upon the cinema frequenters."[87]

In *The Palestine Issue: A Factual Analysis*, the Jewish Agency claimed,

> The Jews have in reality augmented the territory of Palestine. They have not extended its area in square miles but they have so increased its productivity that they may be said to have enlarged the land itself...The Jews have increased that capacity by their own zeal and enterprise as surely as if they had added to the land new provinces.[88]

Zionist financial aid to Arab agriculture and health also strove for positive political results. For instance, when Nathan Straus, a prominent American Jew who was active in health affairs in Palestine, gave agricultural implements to the Arab Agrarian Party in 1925 and then later supplied relief to the Arab community after the 1927 earthquake, he commented to Gershon Agronsky of the Palestine Zionist Executive, "I am particularly pleased that this modest gift has had such a favorable effect on the sentiments of the Arab population towards the Zionists. This and the help extended by other Jews will certainly go far towards showing the Arabs that their attitude heretofore has been quite unjustifiable." The Arab Agrarian Party was most likely among the organizations in the Nazareth, Nablus-Jenin, and Hebron regions that called themselves *Hizb al-Zurra'* (Party of the Farmers) that the Palestine Zionist Executive attempted to create or support in the 1920s as a counter measure to the Arab Executive. These farmers' parties relied on Zionist funds and were led by influential village sheikhs who had contacts with urban organizations. The parties varied in their political stance on Zionists, but many

---

87. Letter from secretary for trade and industry of PZE to Dr. Berthold Feiwel, managing director, Keren HaYesod, January 13, 1922; also letter from BenDov to Keren HaYesod, January 11, 1922, CZA KKL3/29/1–2 and CZA KKL3/31, which is a file on pictures and films about the KKL.

88. Jewish Agency, *The Palestine Issue: A Factual Analysis*, submitted to members of UN by Jewish Agency for Palestine, 1947, 19–20, CZA Library.

of them accepted the terms of the Mandate and the Balfour Declaration. Scholar Yehoshua Porath has noted that the Zionists took these organizations seriously because they had gained more support than the *al-Hizb al-Watani* (the urban opposition party), but by 1927, their dependence on the Zionists caused their downfall.[89]

The Zionist claims on Palestinian Arab progress was part of a wider effort to encourage both international and British support for further Jewish settlement.[90] Zionist medical professionals and land agencies claimed that by undertaking antimalaria work, they directly "benefited all Arabs in the neighborhood."[91] There is no doubt that drainage work decreased malaria incidence in all populations in Palestine, but that benefit was inherently linked to the elevation of the *fellahin*'s economic condition. Early in the Mandate, the idea that the low standard of living of the rural Arab community was raised by Jewish presence was shared by both Zionist and British leaders. Following the "malaria blocks development model" widely held in malariological circles of the time, Zionist leaders, like their colleagues in the League of Nations Malaria Commission, saw drainage works as "permanent measures of social hygiene." Both thought that "agricultural reclamation will improve the economic conditions of the people, will increase their prosperity and general well-being, will enhance their resistance to disease, and will tend to make them favorably disposed towards other, possible more direct, antimalaria methods."[92] The Joint Survey Commission Report of 1927 stated, "Rural contact is raising the *fellahin* in efficiency and manner of living rather than lowering

89. Neil Caplan, *Futile Diplomacy: Early Arab-Zionist Negotiation Attempts, 1913–1931*, vol. 1 (London: Frank Cass, 1983), 64; Yehoshua Porath, *The Emergence of the Palestinian-Arab National Movement, 1918–1929*, 229–230; letter from Nathan Strauss to Gershon Agronsky of Palestine Zionist Executive, September 2, 1927, CZA S25/704; see letters from R. H. to Straus (September 25, 1925), Kisch to Straus (September 15, 1925, and September 23, 1925), and from Jalal and Mohammed AbuGhosh to Kisch (July 6, 1925), CZA S25/704. Concerning earthquake relief, the following newspapers noted their thanks to Nathan Straus and the Yishuv: *al-Jamia* (July 18, 1927), *Mir'at al-Sharq* (July 20, 1927), *Filastin* (July 19, 1927), *Al-Muqattam* (July 16, 1927), and *Aliph-Ba* (July 20, 1927), Arab Papers on Straus's Donation for the Earthquake Victims, CZA S25/704; "The Arabs Thank the Jews," translation of article from *HaAretz*, July 24, 1927, *Hadassah Newsletter* 7, no. 11 (October 1927): 4, CZA Library.

90. Kamen, *Little Common Ground*, 4–5, 69, 83–84.

91. Jewish Agency for Palestine, "Trend of Economic Development," March 1946, 27–28, CZA Library.

92. League of Nations Health Organization Malaria Commission, *Reports on the Tour*, 26, about bonification works in Palestine and the role of land drainage within them. Packard and

the Jews, especially in the hill country."[93] Elwood Meade, commissioner
of the Bureau of Reclamation in the United States during the 1930s,
chaired this commission, which addressed malaria as an obstacle to colonization and evaluated drainage projects.

Zionist leaders claimed that benefits imparted by the Jews were said
to reflect the uniqueness of the Zionist project in general and its method
of colonization. Such claims continued to be expressed by Zionist officials
until the end of the Mandate. The *Palestine Issue* stated:

> What has been the effect of Jewish settlement on the Arab population in Pales-
> tine? The Jewish return to Palestine is unique in the history of colonization—it
> is one of the few instances on record where European colonization raised the
> standard of life of the native population. It has not been conducted through the
> exploitation of native labor. Instead of rich and fertile land being acquired for
> a few strings of beads, marsh and uninhabited desert were purchased at exorbi-
> tant prices. These are striking departures from the usual pattern of colonization
> in a backward area...But most revealing of all this process as already stated,
> has been accompanied by a great increase in the native population...chiefly,
> however, the increase is due to the improved health conditions introduced by
> Jewish nursing services and sanitation...So much for the myth of the "dispos-
> sessed" Arab.[94]

It is important to note that the author placed the Zionist project explicitly within the scope of European colonization here. The idea that there
was a direct correlation between health and economic prosperity was part
of a trend in interwar public health work that moved away from chari-
table work to finding a scientific solution for poverty and disease. Ruth
Fromenson of Hadassah reiterated the sentiment in the *Palestine Issue* in
a discussion of the benefits of Hadassah health care provision when she
proclaimed, "Very soon the Arab must recognize that his Jewish neighbor

Brown, *Medical Anthropology*, special issue on malaria and development, 17, no. 3 (1997); Peter Brown, "Failure as Success: Multiple Meanings of Eradication in the Rockefeller Foundation Sardinia Project, 1946–1951," *Parassitologia* 40, nos. 1–2 (June 1998): 118; and Packard,
"Visions of Postwar Health and Development," 93–115.

93. Joint Survey Commission Agricultural Colonization Advisory Commission, December 1927, 46, Elwood Meade, chair, PRO CO 733/156/3; R. Hendricks, *A Model for National
Health Care* (New Brunswick: Rutgers University Press, 1993), 21.

94. *The Palestine Issue: A Factual Analysis*, submitted to the members of the United Nations by the Jewish Agency for Palestine, 1947, 21–22, CZA Library; Weindling, "Introduction: Constructing International Health," 4.

has come not to drive him out but to help him toward a happier day."[95]
Setting up an equation in which the Jews were primarily the providers
and the Arabs the receivers reflected the Zionist construction of Arabs
as passive and backwards and the Jews as the harbingers of civilization.[96]
While privileging the beneficial effects of Jewish settlement, this framework obscured its disruptive effects.

In 1937, the British Health Department of Palestine began to question
the claim of Jewish benefits in general and the uniqueness of the Jewish
contribution in particular:

> The antimalarial measures executed or maintained by Jewish bodies or own-
> ers or occupiers of land have in general not affected any very large part of the
> total Arab population. The lands in question were for the most part cultivated
> by Arabs whose villages were some considerable distance from the mosquito
> breeding areas. In fact the principle throughout has been that both Arabs and
> Jews have as circumstances necessitated been called upon to do their proper
> share of antimalarial work throughout the country irrespective of the commu-
> nity which might as a result be indirectly benefitted.[97]

Like the British Health Department of Palestine, Dr. Tawfiq Canaan,
president of the Palestine Arab Medical Association, challenged the
Zionist assertion of improved health conditions in the Arab population in
a report, *Hygienic and Sanitary Conditions of the Arab of Palestine* (1946).
The report was submitted to the Anglo-American Commission of Inquiry.
Interestingly, the report was organized in a very similar manner to Jewish Agency reports submitted during the Mandate period to various commissions before the promulgation of various White Papers. The government published White Papers in an attempt to reformulate British policies
regarding immigration and land purchase issues. These White Papers
were necessitated by increased political tensions between the Zionists and
Arabs during the Mandate and were an attempt by the government to
"uphold" the Mandate promise of being "in favour of the establishment
in Palestine of a national home for the Jewish people" while also taking
care not to "prejudice the civil and religious rights of existing non-Jewish

---

95. Letter from Ruth Fromenson, *Hadassah Newsletter* 5, no. 2 (November 1924), CZA
Library.

96. Kamen, *Little Common Ground*, 5–6; Packard, "Visions of Postwar Health and Development," 101; Berkowitz. *Western Jewry and the Zionist Project*, 103–105.

97. British Health Department of Palestine, comments on JA memorandum, 11–12.

communities in Palestine." Perhaps the Palestine Arab Medical Association, after reading so many Jewish Agency reports in previous years in response to White Papers and seeing their effect, thought it beneficial to use the same framework to present their views.[98]

In addition to being the president of the Palestine Arab Medical Association, Dr. Tawfiq Canaan was an acclaimed medical doctor in Palestine and a scholar on Palestinian folklore. After providing a description of health problems and institutions for the Arab community, he emphasized in the report the significant participation of the Palestinian Arab population in antimalaria measures, stressing that Arabs were indeed involved in and promoted the development of the country:

> The Arabs have greatly contributed to this [government antimalarial] campaign, either by financial help (as in the case of Birket Ramadan, Wadi Rubin, al-Zreqat, the swamps of Tell al-Rish, Wadi al-Kabbani and Wadi al-Sheikh) or by offering unpaid labor (as in the case of al-Ladjun, Nasmin, Tanturah, Domarah, Der al-Balah, Na'meh and the scores of wadis scattered all over the country).

Likewise,

> The Jewish organizations have also drained some swamps, especially those in their colonies. But it must be said on the basis of the above-mentioned data and figures that the Arabs have done, under the leadership of the P.H.D. [Public Health Department], their share in this sanitary work.[99]

Canaan's accent on Palestinian participation in malaria control compares with what Silva describes in the case of colonial Sri Lanka as a nationalist stance toward antimalaria measures; that is, for nationalist leaders, malaria control was an issue of "self-realization, particularly in a context of their ambiguous relationship with the West."[100]

98. W. Laquer and B. Rubin, eds., *Israel-Arab Reader: A Documentary History of the Middle East Conflict* (New York: Penguin Books, 1984), 34–35, 45–58, 64–77; "The Arab Case for Palestine-Evidence Submitted by the Arab office, Jerusalem (1946)," *Israel-Arab Reader*, 98–99; Palestine Arab Medical Association, *The Hygienic and Sanitary Conditions of the Arabs of Palestine* (Jerusalem, March 1946), 14–15.

99. Canaan presented statistics on the number of labor days offered by the Arab community from 1931–1939 for antimalaria efforts. Palestine Arab Medical Association, *The Hygiene and Sanitary Conditions of the Arabs of Palestine*, 6. Confirmation of Canaan's claim was made in comments on memorandum, 3–4, 8.

Canaan's claims about the Palestinian community's active participation in development schemes were not unfounded. Even though the Zionists saw them as an obstacle to antimalaria measures, in many cases Palestinian Arabs carried out swamp drainage projects and took an active part in land transformation. Arab local residents were often laborers in government malaria control measures, as called for by the Antimalaria Ordinance (1922), which made participation in control measures compulsory; some Arab sanitary engineers and doctors were employed as foremen and medical officers.[101] Leaders of the village were given the responsibility for supervising the work and reporting malaria cases. Prisoners and railway workers were also employed for governmental swamp drainage.[102] In addition, the government sometimes employed workers from Egypt in drainage projects before and at the beginning of the Mandate period. It seems, however, that this importation gradually decreased after the Antimalarial Ordinance in 1922. In many cases, the *effendi*s who owned the land supervised the quinine distribution to the men as they returned from doing drainage work or work in the field. Under the Antimalaria Ordinance, Arab owners of the land had to refund the costs of antimalaria work originally financed by the government. In areas such as Wadi Tom and Mashra'or Wadi Rubadyeh (Migdal area), local inhabitants contributed to the cost and labor of the schemes.[103]

In Beisan, a canal, the only water source for the village, formed the Shatta swamp. The canal took water from the Jalud River to the village of Shatta, but after passing the station, it spread, forming a pool of stagnant water. Egyptians dug a new canal at Shutta Station. This was done at the expense of the Shutta villagers, under the supervision of Dr. Adib Haddad, the medical officer of Beisan, and Rafik Bey Beydoun, district officer

100. K. T. Silva, "'Public Health' for Whose Benefit? Multiple Discourses on Malaria in Sri Lanka," *Medical Anthropology*, special issue on malaria and development, 17, no. 3 (1997): 196.

101. I. Kligler, "Report of the Antimalarial work for May and June 1921," 1, RF, RF5/2/61/398; British Health Department of Palestine, comments on JA memorandum, 10.

102. See the case of Meithalun village near Sanour swamp. "Proceedings of the Seventh Meeting of the Antimalarial Advisory Commission," 14; eighth meeting of same commission, November 22, 1923, 4, JDC Archives, file 280. See Kishon scheme in "Proceedings of the Eighth Meeting of the Antimalarial Advisory Commission," 5; Katriel, "Remaking Place," 160.

103. Schedule of proposed antimalarial drainage scheme, no date (but most likely 1935), ISA M1503/1/86(59); confidential letter from Heron to chief secretary, May 23, 1935, ISA M1503/1/86(59); letter from J. H. Hall, chief secretary, to Ben Tzvi of June 1, 1935, entitled, "Memorandum on certain questions regarding Government antimalaria measures raised in the Vaad Leumi's memorandum," May 16, 1935, ISA M1503/1/86(60).

of Beisan.[104] The Rai al-Baladiya and Arab al-Sakr tribes worked on the Mushra'and Madna streams. In Damaria Marsh, local bedouin provided the labor for the antimalaria scheme (in their area) while in Musherifa, Zeeb, Tel al-Shok, and Wadi Bireh, villagers carried out the work.[105]

The Supreme Muslim Council, a Palestinian Arab political organization that administered *waqf* (religious endowment) properties during the Mandate, was involved in drainage measures on land that it owned, sometimes in cooperation with the Malaria Survey Section of the Rockefeller Foundation. In the Birket Ramadan scheme, the Supreme Muslim Council, owner of 5,000 of the 8,000 dunams of this area, shared its financial obligation with the Hanotaiah Company, a Jewish land reclamation company, and the government. The Supreme Muslim Council wanted to reclaim the area to settle nomadic tribes there. The Hanotaiah Company took an interest in reclamation in order to protect its large holdings in the surrounding area.[106]

Arab laborers also participated in swamp drainage projects initiated by Zionist agencies. Arab labor for Zionist drainage endeavors was often a way to avoid further loss of Jewish life,[107] a pragmatic act supported by the belief that Arabs had acquired immunity to malaria. Mr. Yehoshua Hankin, a well-known land-purchasing agent, recommended that the KKL employ non-Jewish labor to safeguard the health of Jewish workers who were "less acclimatized and less used to living under such conditions."[108] Arab labor also reduced drainage costs. A time sheet for antimalaria work at Kishon Jeida in July 1931, showed the discrepancy between Arab and Jewish pay; Zvi Kavsviner was paid 250 mils per day, whereas Said Saleh was paid 100 mils per day by the KKL.[109] Since the July 1920, Zionist con-

104. "Proceedings of the Eighth Meeting of the Antimalarial Advisory Commission," 9. See also "Drainage Operations of Various Jewish Agencies," 8.

105. Letter from district commissioner, Haifa District, to chief secretary, on Antimalarial measures, May 27, 1935, 3, ISA M1503/1/86(59).

106. Confidential letter from Heron to assistant commissioner on special duty, Information required by Royal Commission, March 29, 1937, 2, PRO CO733/345/10. I was unable to find out more about this company.

107. For case of Barbados, see Watts, *Epidemics and History*, 228.

108. Yehoshua Hankin was active in acquiring lands for the Zionist movement through such agencies as the Jewish Colonization Association (ICA), KKL, and PLDC (Palestine Land Development Company). He is most known for the acquisition and settlement of large tracts of land for the Zionist endeavor, including the Jezreel Valley.

109. Letter from Paul Singer to KKL, January 21, 1926, CZA S25/595; time sheet for Kishon Jeida, 1931, CZA KKL5/4688.

FIGURE 7.1. Arabs and Jews canalizing Kabbara, 1926. Courtesy of the Keren Keyemet L'Israel Photo Archives

ference in London, the KKL had had an official policy of employing only Jewish labor. Zionist separatism, as Barbara Smith has pointed out, began to transform itself into a socioeconomic reality in the 1920s. According to Lehn and Davis, this policy was tightened after 1929. Many infractions, though, did occur.[110]

In fact, as Baruch Kimmerling has remarked, although the Zionists' relationship to the local population was basically exclusivist, it could not be entirely so because of problems with land acquisition.[111] Zionist officials employed Arab laborers in May 1935, in drainage works for the area of Wadi Jeida and in June 1936, in Gesher-Nahalim (Wadi Bireh, Wadi Arab, the Jordan River from Kibbutz Gesher to Wadi Bireh).[112] The Palestine Electric Corporation, run by Zionist entrepreneur Pinchas

110. Lehn and Davis, *Jewish National Fund*, 49–50; Smith, *Roots of Separatism in Palestine*, 4. For a Zionist perspective, see Granovsky, *Land Policy in Palestine*, 107–109.

111. Kimmerling, *Zionism and Territory*, 13; Metzer, *Divided Economy of Mandatory Palestine*, 131.

112. Note on inspections made by district commissioner, Trip 1, May 7, 1935, ISA M1503/ 1/86(59).

Rutenberg, sponsored this work, and Kligler and Eshner of the Hebrew University supervised it.[113] On the Khulde project, the inhabitants of Khuldeh el-Islam supplied ten free laborers for drainage work done by the KKL, most of which took place on Arab land. A government foreman was requested to act as an intermediary and supervise the work of the Arab laborers for two to three months. "Otherwise," wrote the director, "it will be impossible for our workmen to cooperate with these laborers."[114]

Jews and Arabs did sometimes work together. In Beisan, Jewish colonists and the Arabs of Shutta and Arab al-Sokhneh constructed a new channel for the Jalud stream to dry a large swamp there.[115] Similar cooperative labor took place in 1923 for the MRU schemes at Merchavia and Solim, Jalud swamp (Bet Alpha and Sachni villages), and Khulde. At first, the drainage was delayed because Ein Sachni and Wadi Sachni were located outside of the Nuris property and because of resistance by villagers who used the water for irrigation, but in the end, a drainage scheme that addressed their irrigation concerns was implemented.[116]

In the EMICA scheme of the Huleh drainage, work done by both Arabs and Jews was distributed according to "skill"—the assumed ability and scientific understanding of each respective community.[117] This division of labor reflected a split in the labor market, resulting from what Shafir has described as an adaptation in the campaign of Conquest of Labor made in the early years of the twentieth century as a result of its Zionists' inability to completely exclude Arab labor. Although it still failed to conquer the labor market, this split involved a monopoly on skilled labor

113. Letter from Palestine Electric Corporation to PICA, June 25, 1936, CZA J15/5589.

114. Letter from D. L. M., director of KKL Colonization Department, to governor, Ramleh, subject: drainage, Khulde village, November 4, 1921, CZA KKL3/99 aleph. Also letter from DLM, director of KKL Colonization Department, to district governor, Jaffa, subject: drainage, Khulde village, November 21, 1921, CZA KKL3/99aleph.

115. "Proceedings of the Eighth Meeting of the Antimalarial Advisory Commission," 7–8.

116. "Drainage Operations of the Malaria Research Unit," 4; and "Drainage Operations of Various Jewish Agencies," 9; letter from PMO Jaffa District to President Khulda Jewish Colony, October 19, 1921, CZA KKL3/99aleph. See also work done side by side (i.e., Jews in north, Arabs in south of swamp area or visa versa), cooperating or working separately in different degrees in Ein el-Maity, Wadi Zerganiyeh, Ein el-Asawir and Wadi Ara, Wadi Sindiani, Wadis Malul, and Mujeidel. Note on inspections made by district commissioner, trips 1 and 2, May 7 and 23, 1935, and confidential letter from Heron to chief secretary, May 23, 1935, 2, both of ISA M1503/1/86(59).

117. EMICA was a company established in February 1933 between the ICA and the Palestine Emergency Fund for economic development work in Palestine. Tyler, *State Lands*, 99, 108, 110.

by Jews that created a hierarchy of labor for Arabs and Jews before the Mandate period. The division was ironic, as Gershon Shafir has pointed out, because most Jews had no experience in agricultural work at all before coming to Palestine.[118]

In the projected EMICA scheme, Jews were to be the majority of employees but Arab laborers would also be employed. In addition to an entire staff of technical advisers, contractors, designers, and administrators, approximately 300 Jewish manual laborers would be hired to work at a mechanical plant doing excavation of the Jordan and the lake and constructing a barrage. Workers would live in mosquito-proof quarters away from the marsh area but where they would not be at risk of malaria infection. Secondary work considered suitable for the Arab sector was construction of small drainage and irrigation channels, reclamation work in the marsh area, subsoil drainage for springs, and road making.[119] Provisions for their lodging and health were not considered. It was estimated that it would take about five years to complete the proposed scheme. At the Kabbara swamp project, Egyptian and Palestinian Arab laborers slept near the swamps, increasing their exposure to malaria, whereas Jews returned to Zichron every night after work, protecting themselves from infection.[120]

Canaan's assertions about Arab participation in swamp drainage in the memorandum to the Anglo-American Commission of Inquiry were presented as a counter to Zionist claims of being chief actors in developing Palestine. It made a strong case against the Zionist claim of raising the Arab's standard of health. Mortality figures proved his point. He noted that although the death rate, both general and infant, gradually decreased throughout the Mandate, this was not entirely attributable to Zionist influence. Canaan explained that mortality rates fluctuate naturally and that Arab villages located far from Jewish colonies also saw a marked decrease in infant mortality. He also contested Zionist assertions about birth rates:

> The Zionists claim also the increased birth rate to be due to the great sums of money they have brought into the country. This is not the case, for the height of the births had reached its peak before the Jewish influence could have affected the same, namely in the years 1926–1930. The highest birth rate of the Arabs and at the same time of the whole of the Palestine population was not in Arab cities or

118. Shafir, *Land, Labor and the Origins of the Israeli-Palestinian Conflict,* 66–72.
119. Rendel, Palmer, and Tritton, "Huleh Basin," 35, 36.
120. League of Nations Health Organization Malaria Commission, *Reports on the Tour,* 65.

villages near Jewish colonies, but in distant localities. This proves conclusively that the Jewish influence had no effect on the birth rate.[121]

In an earlier article, separate from this report, Canaan refuted the Zionist claim in more general and dramatic terms, bringing in a critique of Zionist morality as well:

> Another supposed blessing of Jewish colonization is the "higher standard" of life which the peasants are said to have attained. But has this so-called blessing not also been a curse to the Arabs? A great many of our young men have lost their pure patriarchal customs and their simple manners. They neglected their work, ruined their health and lost their money. Has this not always been the natural result of imposing forcibly Western civilization on a primitive people? ... And what about the blessings of Bolshevism whose doctrines have affected some of our young men? What a distressing example do some of the communistic colonies with their loose moral principles and lax family ties present to the peasant![122]

Canaan's words reflect a view commonly held by the Arab nationalist elite that peasants were simple, primitive people, but like other Arab medical figures, he actively challenged the Zionists. The doctor disputed the value-neutral notion of development by pointing out the unequal consequences of those efforts. Canaan questioned, especially in his discussion on the natural fluctuation of infant mortality and birth rates, the basic assumptions of the Zionist belief recently critiqued by contemporary scholars Randall Packard and Peter Brown "that improvements in health will lead to improvements in social and economic conditions [which] assumes that health can be conceptually and programmatically separated from processes of social and ecological change, and that health is a bounded entity, subject to manipulation and management." These scholars have noted that the period of 1920–1950 was a crucial one for the development of a bounded concept of health and that during this period, malaria was the focus of this debate.[123]

---

121. Palestine Arab Medical Association, *Hygiene and Sanitary Conditions*, 10–12. A table entitled "Infantile mortality in places far from Jewish influence" follows the discussion. According the table, these places included Khan Yunis, Hebron villages, Hebron, Bethlehem, Bet Djala, and Jericho village.

122. T. Canaan, "Zionist Ambitions and the Palestine Crisis," lecture given at the Newman School of Missions, *Palestine and Transjordan* 1, no. 13 (August 29, 1936), 6–7.

123. R. Packard and P. Brown, "Rethinking Health, Development and Malaria: Historicizing a Cultural Model in International Health," *Medical Anthropology*, special issue dedicated

Canaan also recognized the lack of capital resources among the Arabs for health compared to the Zionists. Finally, Canaan noted the strong emphasis in the Zionist public health project on improving Jewish health, with Palestinian Arab health conditions addressed only when they might affect Jewish health status.[124] He denied the claim of extensive medical care assigned to the Arab sector. Although he acknowledged the help Jewish institutions gave to the Palestinian Arab population, especially in private hospitals and by Jewish medical specialists, he believed that the Zionist claim was overstated. As evidence, he presented data showing that the number of Arabs treated in Jewish hospitals was minimal, that Arabs commonly paid for their treatment in Jewish hospitals, and that even sanitary institutions such as the Straus Health Centers, which explicitly expressed their mission as extending medical service "irrespective of race or creed," had actually done very little for the Arab population. A report by Dr. R. Bachi of the Central Committee for Medical Statistics of the Hadassah Medical Organization in 1943 confirmed Canaan's claims by stating that the ethnic separation of health institutions intensified right after the Arab Revolt but weakened again soon after. Bachi noted that even in the early 1940s, this separation still existed among most of the population and that only the wealthy Arabs visited Jewish health institutions, while poor Jews visited non-Jewish hospitals. This justified Bachi's call for more Jewish doctors in Arab health care.[125]

Malaria as an Arena for Jewish-Arab Understanding

Despite the ultimate demise of exchange between the Jewish and Arab medical communities, antimalaria and development schemes throughout the Mandate period served as an opportunity to promote medical and political contacts to further Jewish-Arab understanding. Although there was some effort to establish closer relations between Arabs and Jews in the field of

---

to malaria and development 7, no. 3 (1997): 183–185; Packard and Brown, "Rethinking Health," 186; Packard, "Visions of Postwar Health and Development," 103–104.

124. Falah shows as well that Jewish colonization had "little positive impact" upon bedouin economic stability and improvement of living standards. If anything it had the opposite effect. "Pre-State Jewish Colonization," 290.

125. Palestine Arab Medical Association, *Hygiene and Sanitary*, 12–14; R. Bachi, "Statistical Information on Doctors in Eretz Israel" (Hebrew), pamphlet no. 2, Central Committee for Medical Statistics-Hadassah Medical Organization. 1943, 13, CZA J1/8427; Zureik's inclusion of G. Mansur, *The Arab Worker in the Palestine Mandate* (Jerusalem: Commercial Press, 1936), 25 as quoted in Zureik, *Palestinians in Israel*, 51.

health, especially after the Arab Revolt and during World War II, it was
motivated primarily by the recognition that an improvement of Arab
health would secure Jewish health.[126] Proponents of each proposal described it as a practical way to solve Palestine's health problems.[127] In
1939, Dr. Brunn raised an idea he had proposed years earlier to deter
infectious diseases, including malaria, between Arabs, Jews, and the British.
Recognizing the tenuous political situation, Brunn wrote, "All the nationalist feelings, conflict of nation against nation, cannot stand up to this clear
work. Preventing medical cooperation could bring catastrophe." Brunn
felt that medical cooperation was especially important because of war
conditions in Palestine. Brunn sent the same proposal to Chaim Kalvar-
isky, head of the League for Arab-Jewish Rapprochement, for his perusal.
Abraham Katznelson, head of the Department of Health for the Vaad
HaBriut, ultimately disagreed with Brunn on grounds that there was no
autonomous Arab medical agency with which to coordinate activities at
that time.

Dr. Mandelberg of the League for Fighting Tuberculosis in Tel Aviv
also wrote to the League for Arab-Jewish Rapprochement. He posited
Arab-Jewish cooperation in the campaign against infectious disease, particularly tuberculosis, in 1941. Similar to the sentiments made by Brunn
and others about the need for Arab health care in order to protect Jewish
health, Mandelberg wrote,

> There are strong connections between them [Jews and Arabs]. First of all,
> among the borders of the Jewish *Yishuv* and the Arab one there is always
> a close linkage, there are neighborhoods where there are mixed populations,
> there are groups of people as well as individuals of one nation that live among
> the other, Arabs in great numbers visit the Jewish communities, and also visit
> the same administrative offices, food and other places as well as work together.
> In short, there are many possibilities for infecting one another.[128]

126. This is not to say that they were not sincere attempts but only that they were firstly
motivated by Jewish health interests.

127. These cooperative efforts were consistent with the resumption of economic relations
between Arabs and Jews after the Revolt and during the World War II, where Owen argues,
"new patterns of association [were] called into being." Roger Owen, introduction, *Studies in
the Economic and Social History of Palestine in the Nineteenth and Twentieth Centuries*, ed.
Roger Owen (Carbondale: Southern Illinois University Press, 1982), 5.

128. Letter from Katznelson to Brunn, November 27, 1939, CZA J1/3490. See Brunn's
letter and official proposal in the same file; letter from Mandelberg to the League of Arab-
Jewish Rapprochement, March 4, 1941, CZA J113/1661.

Drs. Mer and Kligler also prepared a program for the establishment of mixed health welfare stations in the Huleh Valley. They planned that the health centers were to maintain close contact with the Rosh Pina Malaria Research Station. They intended the mixed health centers to be combined with infant welfare centers, dispensaries, social service offices, and a traveling nurse service. It is unclear whether or not this program was ever implemented.[129]

Negotiations for the transfer of the Huleh Concession from Mr. Salam (Beirut) to the Zionist Organization were briefly framed as an experiment in Arab-Jewish cooperation in developing Palestine. Even Chaim Weizmann, an important Zionist leader, recognized the potential and offered to discuss the matter more fully with Mr. Salam.[130] Sir Hope Simpson called for all development schemes to be supported by Arabs and Jews to advance the "neglected but historic country in the path of modern efficiency, by the joint endeavor of the two great sections of its population, with the assistance of the Mandatory Power."[131] However, many of the projects cast the Jews in the role of benevolent developers; cooperation usually meant Zionist initiation and Arab participation.

By the end of the Mandate, the idea of cooperation between Zionists and Arabs had spread to include all Arab countries of the Middle East. A proposal for a Middle East Malaria Institute by Lt. Col. Yoeli, a Zionist who worked on antimalaria control for the *Yishuv* and then served in the British army during World War II, was submitted to the government. This plan called for regional cooperation in antimalaria activity centered in Palestine. The proposal received mixed reactions from the Mandatory government and from London. Mr. Edelman, a member of Parliament and a strong proponent of the project, wrote a series of letters to the colonial secretary, Mr. Creech Jones, expressing his support for it:

129. Letter from Hebrew Women's Zionist Organization of America) to Dr. Magnes, March 14, 1940, CZA J113/1661. See also minutes of twentieth stated meeting, Hadassah Emergency Committee, February 25, 1941, 3, CZA J113/1661. See random notes on Upper Galilee Trip, "Health Centres for the Use of Jews and Arabs in Some Central Place in the Hule," February 16–18, 1941, 1–2, CZA J113/1661.

130. Note of interview at Great Russell Street between Mr. Shoucair and Mr. Stein, March 12, 1925, CZA S25/595; memo on the Huleh Concession, note on meeting at Hotel Cecil, January 22, 1925, CZA S25/595. Present at the January 26, 1925, meeting were Dr. Weizmann, Mr. Stein, Mr. Salam, and Mr. Salam, Jr.

131. John Hope-Simpson, "Palestine Report on Immigration, Land, Settlement and Development," October 1930, 162, ISA M4381/1/3/73.

> I feel very strongly that it is men like Yoeli and the sort of work which they are trying to do, which can act as a cement to unify the Jews and Arab peoples of Palestine[132] [and] I believe that you are of the same view as myself in the matter of Arab/Jewish cooperation: namely that it can be hastened and encourage by joint institutions in Palestine. Can there be a more desirable joint institute than a research center to abolish malaria throughout the Middle East?[133]

Creech Jones, Colonel Heron, director of the British Health Department of Palestine, and others did not agree with Edelman for practical reasons.[134] Heron believed that most countries could act on their own and that other projects in the colonies of the Middle East were more worthy. Although Creech Jones recognized the value of using constructive works as a method to promote mutual understanding, he noted that in addition to potential technical problems, political circumstances of the time would impede or even prohibit the center's success. In addition, he noted that it might be taken as an attempt to sidetrack the already established Regional Health Bureau of the Arab League.[135] From the extant documentation, it seems that the plan was never realized.

Conclusion

Frazier and Scarpaci have argued that material landscapes are where power relations are practiced, challenged, and reconstituted.[136] Malaria and the

---

132. Letter from Edelman to Right Honorable A. Creech Jones, colonial secretary, Colonial Office, December 6, 1946, PRO CO 733/493/4.

133. Letter from Edelman to Creech Jones, October 7, 1946, PRO CO 733/493/4.

134. Letter from John Higham to Mr. Campbell, January 23, 1946, PRO CO 733/493/4. Also comments made by Heron and Subcommission of Colonial Medical Research Committee, July 2, 1946, PRO CO 733/493/4, in comment section of file. Mention is also made about the Rockefeller Foundation's malaria work in the Middle East and thus the redundancy of the Yoeli plan. See letter from high commissioner of Palestine to secretary of state, December 21, 1946, that considers the financial problems of the plan and need for prevention rather than research, PRO CO 733/493/4.

135. A note on the memo entitled "Malaria Institute of Palestine in Middle East, a Proposal" by Heron, April 4, 1946, 45 PRO CO 733/493/4; letter to Edelman from Creech Jones, March 4, 1947, PRO CO 733/493/4; see letter to Mr. Campbell from John Higham, January 23, 1946, PRO CO 733/493/4.

136. Lessie Jo Frazier and Joseph L. Scarpaci, "Landscapes of State Violence and the Struggle to Reclaim Community: Mental Health and Human Rights in Iquique, Chile," *Putting Health into Place: Landscape, Identity and Well-Being*, ed. Robin A. Kearns and Wilbert M. Gesler (Syracuse: Syracuse University Press, 1998), 56.

work to eliminate it provides an instructive example of how Zionist representations about the land and the indigenous Arab inhabitants informed measures to control this parasitic disease in Mandatory Palestine. Zionist images of the Palestinian Arab population in antimalaria measures reveal the place of the Arab community within the wider ideological (and practical) project of transforming the health of the *Yishuv* and of Palestine's landscape. In order to transform the land into one that was "healthy, productive and renewed," the Zionists had to face and respond to their Arab neighbors, particularly with regard to their health status.

The nature and scope of drainage interventions were also affected by Zionist beliefs about the Palestinian Arab community. These projects had several implications, some detrimental and some positive for each respective community. Besides improving the health of the *Yishuv*, drainage projects substantiated the Zionist claim that they were prominent actors in the development of Palestine and that their work benefited the health and economy of the Arab sector. In fact, a more complex process of implementation, contestation, and negotiation took place against the backdrop of tension between what the Zionists believed and how their plans were realized.

Ecological Landscape

Old Paradigms, New Meanings

After the Mandate: 1948 to Today

In 1948, health services collapsed for the Palestinian Arab community. That community relied heavily on Mandatory governmental health services and missionary agencies during the Mandate period, neither of which was available after the departure of the colonial rulers. Further, the Mandatory government closed missionary medical agencies that were associated with the Axis powers of World War II, leaving the Palestinian Arab community in 1948 without major sources of health care. Private medical services and initiatives with limited money and medical resources, assistance from surrounding Arab neighbors, and international health organizations were the primary source of health care for Arabs during and after the war. The 1948 War and resulting refugee problem exacerbated the lack of care.

For the *Yishuv*, health services during this period were adapted to war conditions but were severely strained. The autonomous position of Zionist medical services during the Mandate proved advantageous because it established a firm medical infrastructure that became the basis for today's Israeli health care system.

During the War of 1948, local outbreaks of malaria occurred, but none reached epidemic proportions, even though very few antimalaria projects were carried out.[1] When the state of Israel was established in 1948, malaria was still a problem; 1,172 new cases were reported that year. The Ministry of Health established an Antimalaria Division in 1949 under the direction of Zvi Saliternik, one of the Zionist malariologists of the Malaria Research Station during the Mandate period. As a national institution, this division consulted with the Israeli government on the supervision of agricultural settlements, the use of pesticides, water development projects, and other agricultural endeavors.[2] The legacy of Kligler's insistence on such coordination during the Mandate was evident in this arrangement. By 1951, when the draining of the Huleh began, malaria was largely under control. By 1959, Israeli scientists only reported thirty-six new malaria cases.

Between 1960 and 1966, Israel participated in the World Health Organization's (WHO) Global Malaria Eradication Program.[3] As part of the WHO endeavor, Herzl's prediction in *Altneuland* of exporting Steineck's scientific knowledge to Africa eventually became a reality. In the early 1960s, a team of Israeli experts went to Africa to share their knowledge of malaria and ways to combat it.[4] By 1967, malaria had become virtually eradicated in Israel. Given the continued presence of *Anopheles* vectors in Israel, antilarval activities, and monitoring continue to this day. For the time being, however, Israel remains relatively free from malaria, except for a small amount of imported cases.

Current World Malaria Problem

Aside from its limited profile in Israel, malaria today presently poses a significant health challenge, killing one million people every year. There are approximately 300 million cases of malaria globally and about half of the world's population lives in malarial areas.[5] Malaria has reached epidemic proportions in Africa, Latin America, parts of Asia, and the

1. Kitron, "Malaria, Agriculture and Development," 301.

2. Kitron, "Malaria, Agriculture and Development," 301–302.

3. This endeavor recapitulated the stress put on clinical examination in the pre–World War II era, while ignoring the social, cultural, and economic contexts in which DDT spraying occurred. Saliternik, "Reminiscences," 520; Birn, "Eradication, Control or Neither?" 146; Packard, "Visions of Postwar Health and Development," 110.

4. Greenblatt, "Historical Trends," 516.

5. Edwin Nye and Mary Gibson, *Ronald Ross: Malariologist and Polymath: A Biography* (New York: St. Martin's Press, 1997), 30; "Malaria in Africa" at Roll Back Malaria Web site, available at http://www.rbm.who.int/cmc_upload/0/000/015/370/RBMInfosheet_3.htm.

southern part of the former Soviet Union.[6] The disease has reemerged in places like India and, because of increased travel, in the United States. All in all, 300 to 600 million new cases arise every year. According to the Roll Back Malaria report of the WHO, 90 percent of all malaria deaths occur in Africa south of the Sahara. Malaria is one of the top killers of children in the tropics at a rate of one death approximately every thirty seconds.[7] This mortality rate is higher than the current statistics for AIDS.

Malaria has increased in part because of road building, dams, mining, logging, and irrigation projects that move humans into uninhabited areas infested with malaria-infected mosquitoes.[8] Vector resistance to insecticides has also frustrated efforts to eradicate malaria, and attempts to find a vaccination have so far failed. Describing the seriousness of the present situation, Dr. David Nabarro of the World Bank remarked, "We're dealing here with just about the cleverest pathogen that's known to human medical science."[9]

Announced in 1998, the Roll Back Malaria initiative of the WHO, in cooperation with the World Bank, the United Nations Development Program, and UNICEF, was budgeted at 19 million dollars for the first year and targets Africa, the world's most endemic area for malaria.[10] Retaining the "malaria blocks development" discourse of the colonial period, James Wolfensohn, president of the World Bank, noted that financiers today are recruited on the basis that the proposed antimalaria program "is central to the economic development of Africa."[11] The result has been the Global Fund to Fight AIDS, TB, and Malaria commitment of 942 million dollars for the first five years for malaria control. The program funds research for new drugs, as the malaria parasite has quickly become resistant to drugs used in the past. As the Roll Back Malaria information sheet explains,

6. Barbara Crossette, "UN and World Bank Unite to Wage War on Malaria," *New York Times*, October 31, 1998, A4; Gary Taubes, "A Mosquito Bites Back," *New York Times Magazine*, October 28, 1997, 40–46.

7. Kitron, "Malaria, Agriculture and Development," 295; Watts, *Epidemics and History*, 279; Crossette, "UN and World Bank Unite to Wage War on Malaria," A4; Roll Back Malaria, "Facts of ACTs," http://www.rbm.who.int/cmc_upload/0/000/015/364/RBMInfo sheet_9.htm.

8. Craig Turner, "UN Launches Joint Campaign to Fight Spread of Malaria," *Los Angeles Times*, October 31, 1998, A3.

9. Crossette, "UN and World Bank Unite to Wage War on Malaria," A4.

10. Ibid.

11. Ibid.

inappropriate use of antimalarial drugs during the past century has contributed to the current situation: antimalarial drugs were deployed on a large scale, always as monotherapies, introduced in sequence, and were generally poorly managed in that their use was continued despite unacceptably high levels of resistance. In addition, there has been over-reliance on both quinoline compounds (i.e., quinine, chloroquine, amodiaquine, mefloquine and primaquine) and antifolate drugs (i.e., sulfonamides, pyrimethamine, proguanil and chlorproguanil), with consequent encouragement of cross-resistance among these compounds.[12]

Scientists are deploying a new group of antimalarials—the artemisinin compounds, derived from the plant *artemesia annua* (the medicines are called artesunate, dihydroartemisinin, and artemether). So far, these drugs are producing positive outcomes. As a result, the Roll Back Malaria (RBM) Partnership is encouraging combination therapies of drugs containing artemisinin compounds as well as the use of insecticide-treated nets. Thirty-two countries have adopted the organization's recommendations since 2001. According to RBM current predictions, about 132 million courses of artemisinin-based combination therapy (ACT) drugs will be required worldwide by the end of 2005.

Pharmaceutical companies have contributed to the problem of malaria in the world. Nye and Gibson, biographers of Ronald Ross, the discoverer of the malaria vector, explain that the "pharmaceutical industry is not encouraged to develop products for the treatment of epidemic diseases because the countries in most need are the least able to pay for their products."[13] Pharmaceutical companies did not invest in malaria drug research in the late 1990s, but the RBM partnership is now working to encourage production of antimalarials through internationally recognized good manufacturing practice standards.

Reformulating Healing the Land and the Nation, Revisiting the Huleh

Just as the malarial parasite still poses a serious threat for the worldwide malaria problem, malaria and its control have continued to be a focus of concern in Israel throughout the twentieth century and into the twenty-first.

12. Roll Back Malaria Partnership Consensus Statement: Ensuring Access to Effective Malaria Case Management, March 2004, 1, available at http://www.rollbackmalaria.org/partnership/wg/wg_management/docs/RBMConsensusStatement_ACT.pdf.
13. Nye and Gibson, *Ronald Ross*, 29.

Indeed, nationalism, health, and the environment remain key components of Israeli society even now. When examining this phenomenon today, we see once again that disease and its treatment provide a window through which to view nations' (and now states') ideologies about development and disease and their corresponding practical methodologies. Again, we observe the tension between ideology and materiality; again we witness the synchronous consequences of disease upon demography, topography, politics, and ecology. In the case of Israel, these themes continue that society's dialogue with the past, especially the conversation with the powerful mythologies that Zionist history has produced, while at the same time they forge possibilities for ideological reformulation according to new circumstances. The paradigm of "healing the land and the nation" still exists—perhaps not in those exact words—but it has taken on new meanings.

As we discovered in the introduction to this book, drainage interventions of the Mandate period and early state period have had significant ecological consequences for the Jewish state, particularly in the Huleh region. Urit Navo, representative of the Society for the Protection of Nature in Israel (SPNI), explains draining the Huleh as the embodiment of a modern dream to rule nature, to create something new, and as emblematic of newly independent nations who wanted to "prove that they were developed and advanced in the tools of great modern projects."[14]

Recognizing the negative consequences of its drainage project in the 1950s, the Nature Reserves Authority of Israel introduced a long-term plan to restore the Huleh Reserve in 1971. The Israeli government originally made this land available as a reserve because of conservation efforts initiated by the Society for the Protection of Nature in Israel, who insisted that 300 hectares be set aside from the drainage project. The site was officially preserved in 1964 and became Israel's first nature reserve and largest aquatic reserve. In 1971, the Nature Reserves Authority began to reconstruct the swamps and lake and recreate some of its habitats. Among other things, it cleared a meadow to reintroduce water buffalo and built an observation tower for public use. The work was completed in 1978.

The continued damaging, unanticipated ecological effects of the Huleh drainage caused the government to decide in 1992 that areas outside of the reserve should be "restored." Reflooding this part of the Huleh area extended over an area of 800 hectares and includes a 110 hectare lake called Lake Agmon. The venture was a combination of a wetlands

14. Rachel Yona, "Martivim et haHuleh" [Flooding the Huleh], *Davar*, May 20, 1994, 12.

conservation project and a tourist site. The Huleh Nature Reserve became one of Israel's two Ramsar sites under the Ramsar Convention on Wetlands of International Importance (particularly waterfowl habitats) in May 1996.[15] The other Ramsar site is En Afeq, the largest remaining swampland on Israel's coastal plain. En Afeq is the remnant of the Kabbara swamp (drained during the Mandate period), the Tanninim River, and Enot Gibbeton. En Afeq is designated for tourism and recreation as well. According to Israel's National Report on the Implementation of the Ramsar Convention, the En Afeq Nature Reserve (about 12 kilometers northeast from Haifa) is meant to "conserve the plant and animal life which was associated with the vanished coastal swamp and to protect the scenic beauty of the last remnants of coastal wetlands of the region."[16] In other words, it is a powerful reminder of Israel's topographical, environmental, and historical past. Other wetlands (some of them constructed) have more practical purposes: they are expected in the future to help solve the water scarcity problem in Israel by becoming part of the water reuse and purification process.[17]

As swamp draining and malaria symbolized the history of Zionist settlement, its dire consequences and responses symbolize a new context of changed ideas, disillusionment, and negative assessment of the Zionist project. Today, less than 3 percent of the original wetland of Palestine survives. Recognition of this drastic topographical change in the land is evident among many Israeli experts. Giora Shaham, the Hula Project director for the KKL, saw the original drainage and contemporary reflooding efforts in the Huleh in this way: "Draining the swamps was the correct thing to do to meet the needs of the pioneering days. Today the world has changed and the needs are different so we are making what we call a 'surgical correction.'"[18] Shaham's words remind us of the "corrections" made to the

15. Ramsar Convention on Wetlands, National Report of Israel for COP7: National Report prepared for the Seventh Meeting of the Conference of the Contracting Parties to the Convention on Wetlands (Ramsar, Iran, 1971). Implementation of the Ramsar Convention in general, and of the Ramsar Strategic Plan 1997–2002 in particular, during the period since the national report was prepared in 1995 for Ramsar, http://www.ramsar.org.

16. Reuven Ortal and Shoshana Gabbay, "Conservation of Wetlands in Israel: Israel National Report on the Implementation of the Ramsar Convention, February 1999," 3, available at http://www.sviva.gov.il/Enviroment/Static/Binaries/index_pirsumim/p0379_1.pdf, 3.

17. National Report of Israel for COP7, 8. For more on the riparian politics of the Middle East, see Water in the Middle East: Conflict or Cooperation? ed. Thomas Naff and Ruth Matson, MERI Special Studies, no. 2 (Boulder: Westview Press, 1984).

18. Collins, "After the Flood," 14.

FIGURE C.1. Bridge at the Huleh Nature Reserve, December 2005. Photograph by the author.

swamps during the Mandate period by engineers and malariologists in the name of healing the land and the nation. The general vocabulary used is the same in recent years, although the aim and meaning of these "surgical corrections"—of healing Israel and Israeli Jews—are now reversed.

Reflooding the Huleh Valley continues the ethos of "healing the land and the nation," but the principles underlying the phrase's meaning are now vastly different. As the Israel National Report on Wetland Conservation states, "The entire restoration program is meant to implement the principle of sustainable use which sees the management of wetlands as part of a complex system which transforms wetlands into assets rather than obstacles to sustainable development."[19] Israelis now try to "heal the land" by utilizing a conservation and biodiversity discourse; "redeeming" the land now takes place by attempting restoration to a predrained status; scientists have gathered flora and fauna samples to reconstruct the lake.[20] Just like the Zionists returning to the Holy Land in Zionist nationalist dis-

19. Ortal and Gabbay, "Conservation of Wetlands in Israel," 2, 36.
20. They gathered names of over seven hundred species of aquatic invertebrates, fish, and higher vertebrates that lived in the lake. Dimentman, Bromley, and Por, *Lake Hula*, 6–7; Collins, "After the Flood," 14.

course, the swamps and Israel's ecosystems at large are returning to their original state. The words "rehabilitate," "restore," "conserve," and "preserve" are central to discussions of Israel's remaining wetland areas and nature reserves. Instead of transforming swamps, Israeli policymakers speak of the "wise" or "rational" use of wetlands: "The wise use of wetlands is the sustainable utilization for the benefit of mankind, in a way compatible with the maintenance of the natural properties of the ecosystem."[21] "Wise" means prioritizing the ecosystem, "restoring and conserving the remnants" of swamps, classifying and delineating wetlands in Israel, and completing the national inventory of wetland flora and invertebrates in the country, among other things.[22] The word "rational" appears again but this time with a different connotation. Humans, the land, water, and animals are all interconnected in this vision.

Although Israel does not have a specific national policy on wetlands, it does have a variety of laws on nature conservation and protection of natural habitats, wildlife, and biodiversity. Wetland policies are integrated into these other policies. The country possesses an overall strategy for sustainable development whose approach can be summarized as aiming to "protect, conserve and rehabilitate the diversity, quality, quantity and function of natural systems and to ensure integrated management of coastal resources."[23] As part of that strategy, environmental impact statements must be prepared when an area is developed. The assumption now, in contrast to the early years of the Zionist movement, is that swamps (rather than their destruction) are an essential part of development. Reformation of the land and the people has been drastically reformulated.

But as Francis Dov Por of Hebrew University's Life Sciences Institute stated, "It's now clear beyond any doubt that we're not talking about 'rehabilitating' the Hula.... but about an entirely new body of water with new hydrological and limnological characteristics... With the ecological changes that new body of water will undergo, we can expect to see certain species of the old lake and swamps returning, but mainly this will be a new, artificial body of water."[24] Indigenous plants and birds have returned, but the new ecosystem is still a manmade one. Just as "healing

21. Ortal and Gabbay, "Conservation of Wetlands in Israel," 3.

22. Ramsar Convention on Wetlands, National Report of Israel for COP7. Implementation of the Ramsar Convention in general, and of the Ramsar Strategic Plan 1997–2002 in particular, during the period since the national report was prepared in 1995 for Ramsar. http://www.ramsar.org/cop7/cop7_nr_israel/html, 4, 5.

23. Ramsar Convention on Wetlands, National Report of Israel for COP7, 20.

24. Collins, "After the Flood," 14.

FIGURE C.2. Turtles on reeds, Huleh Nature Reserve, December 2005. Photograph by the author.

the land and the nation" conveyed the social engineering of a new Jew and the scientific engineering of a new Land of Israel in the Mandate period, the "restoration" of the Huleh in the state of Israel actually creates similar, manmade ecological relationships.

A belief in science and technology continues to hold a central place in healing Israel's land and the Israeli nation. Part of the Huleh wetland project started in 1994, for instance, included a three-year research program. This program included the fields of agriculture, hydrology, soil conservation, ecotourism, and recreational development. The agricultural research component included discovery of the best organic agricultural methods. Hydrological research focused on the monitoring of former fish, vegetation, and invertebrates as well as geochemical surveys. Soil research included peat-soil fertility. Ecotourism research surveyed shade trees and grass and introduced water buffalo and waterfowl into designated areas. Recreational development studies investigated the needs of wetland areas. Landscape surveys are part of these development schemes. But instead of surveys for rechanneling the land and streams, as done in the Mandate period, planned work should "not destroy the ecosystem, wildlife and landscape features of the river" or the area being surveyed.[25] Policymakers claim that ecological balance is now essential for the health of the Israeli land and the sustainability of the Israeli nation-state.

Student education is another method that has continued in the Huleh reflooding project. Students, for instance, track eagles marked with transmitters as part of school projects. Endangered bird species such as eagles live in the Huleh now along with the black stork, the pygmy cormorant, the common crane, and the imperial eagle. The students aid scientists in correlating data with surveys on fauna in the area. This project is integrated into an internet program that helps students from Israel, the Palestinian Authority, and Jordan track migrating birds.[26] In 30 percent of Israeli high schools, students can matriculate in environmental studies. Students in special environmental programs carry out "ecotope" projects that involve fieldwork observations, surveys, and a paper on water scarcity or wetland conservation.[27] About 30,000 students carry out one of these projects annually. Teacher training centers have been set up to assist teachers in guiding students in these projects on the environment. In addition, a teaching center located on the banks of the Yarkon River has a fourth- to sixth-grade curriculum that deals with wetland rehabilitation and the pollution of Israel's rivers. Israel also participates in the Global Learning and Observations to Benefit the Environment project. Computer skills, water pollution, and conservation of flora and fauna are taught as part of the program in twenty-four schools throughout Israel. All Israeli universities offer courses on the "rational" use of wetlands and nature conservation.[28] Like hygiene lessons of the past, these types of projects teach the students the practice and awareness of new science and of their ecological surroundings. They uphold in practice the central place nature holds in the Zionist imagination. The strategy encourages public participation in planning and implementing "biological diversity

25. Ramsar Convention on Wetlands, National Report of Israel for COP7, 10.

26. Ortal and Gabbay, "Conservation of Wetlands in Israel," 38; Ramsar Convention on Wetlands, National Report of Israel for COP7, 25.

27. For more on the water scarcity debate in Israel/Palestine, see Samer Alatout, "From Water Abundance to Water Scarcity (1936–1959): A 'Fluid' History of Jewish Subjectivity in Historic Palestine and Israel," in *Reapproaching the Border: New Perspectives on the Study of Israel/Palestine*, ed. Mark LeVine and Sandy Sufian (Lanham: Rowman and Littlefield, 2007); Samer Alatout, "Towards a Bio-territorial Conception of Power: Territory, Population, and Environmental Narratives in Palestine and Israel," *Political Geography* 25, no. 6 (2006): 601–621.

28. National Report of Israel for COP7, 29.

policies" as an educational tool and strives to "achieve an optimal balance between regulatory action and education in order to promote responsible public behavior." In addition to these methods, the national plan claims to promote public awareness regarding the benefits of biodiversity, conservation, and sustainable development and encourages professional knowledge through international and regional conventions, ongoing research, and publications.[29]

A planned recreational water park in the Huleh includes boating, wind surfing, fishing, hot-air balloon rides, donkey rides, picnic areas, a safari park, and a hotel, all forging the bond between inhabitant and the land.[30] Referring to the Zionist image of barren, wasteland with no previous history, Shabtai Glass, a member of a kibbutz in the Huleh area, stated, "We could turn this place into an Eilat...After all, what was Eilat to begin with but a barren town Umm Rashrash at the end of the century. And now look at it."[31] Glass sees the tourist project in the Huleh as a way to keep the young Israeli generation from moving to the center of the country, to delay the relinquishment of the Zionist dream of working the land in order to become renewed. Glass's words point to a shift in Israeli society where the advocates of conquest of labor and physical work are no longer. Most Israelis today work in the service industry. Young Israelis and *kibbutz* members no longer discuss the great obstacles their predecessors faced in draining the swamps, but seek investors to help establish the tourist center. Ecotourism is supposed to supply the farmers with an additional source of income that would make up for the agricultural loss. According to the National Report of Israel for the Convention on Wetlands (no. 7), developing larger "restored" wetland areas into recreational sites in the coastal areas (En Afeq) are intended, in addition to recreation and education, to shift a portion of the human population from the small reserves in the coastal area. It is also meant to prevent urbanization of open-space areas and address economic needs there as well.

Yet plans for tourist facilities evoked protests in 1996 from the Society for the Protection of Nature in Israel.[32] Six thousand people came to

---

29. National Report of Israel for COP7. 5.

30. "Reflooding of Hula Could Restore Animal Life," *Hebrew University News,* 103 (Autumn/Winter 1993): 14.

31. Collins, "After the Flood," 14.

32. The Society for the Protection of Nature in Israel was actually founded to stop the draining of the Huleh in the 1950s. Collins, "After the Flood," 14.

oppose the plans for the water park.[33] The society argued that the tourist development plans would drive away the wildlife that had begun to return to the area. In addition, scientists of the Ministry of the Environment suspected that nighttime visitors to the hotel complex could be exposed to malaria-infested mosquitoes. Yet scientists like Moshe Gophen, the project's scientific coordinator for research, rejected this claim, stating that malaria has been controlled in Israel and that the mosquito species in Israel that could carry the malaria parasite are not limited to the Huleh region but are also in Tiberias, Tel Aviv, and the Dead Sea.[34] Even without a serious threat for a reemergence of malaria in the immediate future, the Society for the Protection of Nature in Israel argued that mosquitoes would cause an annoyance for visitors, requiring pesticides that could further damage the ecosystem. Resistance to pesticides could also occur, causing similar problems in Israel that prevail in the rest of the world. Gophen argued that DDT would not be used but rather a bacterium called BTI that attacks mosquito larvae in water. Fish would eat the larvae in the water.[35]

The Huleh Nature Reserve has retained some of its papyrus vegetation and waterfowl population, and it still serves as a stopping point for many migrating birds. Despite efforts to protect the Huleh Nature Reserve since 1971, however, the area has a low-quality water supply. Only four of the nineteen original fish species have survived. The Ministry of the Environment recently deemed it one of the most air-polluted areas in Israel. Local farmers allegedly burned tons of branches and the air lost in the valley was not readily replaced. A regional council in the Huleh encouraged these farmers to crush branches instead of burning them.[36]

Like Zionist swamp drainage projects of the past, the Huleh's reflooding project's long-term consequences are unknown. Mr. Yasur, a kibbutz member in the area, noted, "In some ways, it's just like forty years ago, when the pioneers drained the swamps and didn't know what the consequences would be . . . the farmers don't know what the results will be from reflooding the Huleh, but they'll come around in the end. Most of them

33. David Rudge, "Conservationists Protest Hula Valley Tourism Plans," *Jerusalem Post*, November 10, 1996, 12.

34. Collins, "After the Flood," 14.

35. Collins, "After the Flood," 14.

36. "Clearing Huleh Valley Once More," May 16, 2002. For a summary, see Arutz Sheva, http://www.britam.org/jerusalem/jerusalem31to60.html#tag8; National Report of Israel for COP7, 1999, 13.

FIGURE C.3. Birds gathering, Huleh Nature Reserve, December 2005. Photograph by the author.

realize that, just as it was for the first pioneers here, they don't have any alternative."

Whatever the consequences, the symbols of malaria and swamp drainage are charged in contemporary Israel. In addition to its reappearance in current environmental-tourist projects, malaria and swamp draining are frequent topics in contemporary Israeli culture.[37] Contemporary Israeli readers of Meir Shalev's book, *Roman Russi (The Blue Mountain)*, for instance, see his characters recall the *halutz* mythology where pioneers "got malarial fever 'until their body shook,'" where Jewish settlers cut the papyrus "into scythes until our shoulders were fossilized and our fingers hurt. And all this despite the harsh warnings of Dr. Yofe and despite the malaria that people caught."[38]

Just as swamps have reemerged as a nostalgic image or as a practical challenge in Israel, the topic of malaria and swamp draining has equally

37. Shafir, *Land, Labor and the Origins of the Israeli-Palestinian Conflict*, 173, 183.
38. Meir Shalev, *Roman Russi* (Tel Aviv: Am Oved, 1989), 290.

FIGURE 1.4. Close-up of reeds and flowers in the Huleh Nature Reserve, December 2005. Photograph by the author.

posed a challenge to the Zionist narrative within the Israeli academy and among the public. In 1984, an academic debate ensued about the extent of the swamps in the Valley of Jezreel; the gap between the "image and the reality."[39] The claim sparking the debate was that the swamp area in the Jezreel Valley in the 1920s was actually much smaller than remembered and the swamp myth is a complete exaggeration. Although I find this claim and the subsequent content of the academic debate (whether or not the myth actually exaggerates and how much swamp land actually existed) to be rather futile because myths by nature exaggerate, reactions

39. Bar-Gal and Shamai, "Swamps of the Jezreel Valley," 197.

to the argument are quite telling. My primary evidence shows that swamp drainage in the Jezreel Valley was quite extensive.[40]

Responses filled with emotion asserted that the claim of myth exaggeration was a political and historical affront to Zionism. The nature and strength of reactions in Israel was much like those to the New Historians of Zionism who are charged with deconstructing Zionist myths and producing scholarship that is informed by authors' positions in contemporary politics. Bar-Gal and Shamai were placed within the larger trend in post-Zionist historiography of deconstructing Zionist myths. In an attempt to hang on to the myth, to a glorious Zionist past, both academics and the media attacked Bar-Gal and Shamai, calling them traitors, anti-Zionists, anti-Labor and "PLO [Palestine Liberation Organization] supporters." They argued that these authors were trying to "rewrite the history of the landscape of Israel," and that they were "slaughtering a sacred cow."[41] Israeli academics, columnists, talk show hosts, students, and correspondents started a Save the Myth campaign, expressing their stake in a story that they felt proved their rights to a disputed land.[42] A radio show broadcasted in February 1984, invited founders of Jewish settlements in the Jezreel to counter the authors' claims. Other media distorted the original claim by incorrectly asserting that the original study said that no swamps existed at all in the Jezreel. Caricatures of the topic were also featured in *Davar* in February 1984.

This harsh reaction reflects the Israeli public's stake in the myth and also expresses the role of health and disease in Zionist mythology. The narrative of Zionist settlement is not just one of labor but also of illness, death, and an accompanying martyrdom. One rebuttal by Bar-Gal and Shamai to their colleagues was that these academics had extrapolated the wide area of malaria incidence to an image of a large swamp area, but that these two factors were not equal—malaria could be caused by any stagnant watered area, not only swamps; their initial claim centered around the *actual* area of swamps only.[43] In addition, Bar-Gal recognized in a separate rebuke that malaria (*kadachat*) and self-labor, not just swamps,

40. Bar-Gal and Shamai, "Swamps of the Jezreel Valley," 197–199; and Biger and Kartin, "Habitzot shel Emek Yizra'el-mitos or metziut?" 182.

41. Bar-Gal and Shamai, "Swamps of the Jezreel Valley," 198.

42. Bar-Gal and Shamai, "Swamps of the Jezreel Valley," 200; caricatures of the topic. *Davar*, February 29, 1984, 198–201.

43. M. Zentner, "Habitzot beEmek Yizra'el- metziut or mitos?" [The swamp in the Jezreel Valley—Myth or reality?], *Cathedra* 30 (1983): 177–178.

are central components of the myth.[44] So the debate about the *extent* of swampland was really less important than the deeper struggle it represented regarding the significance of the story of the swamps and malaria in the national history of Zionist settlement. Malaria and the swamps, and all that accompanies them, have signified a potential, promising future, and a powerful reminder of a pioneering past.

44. Bar-Gal, "Emek Yizra'el vebitzotav," 193–195.

Bibliography

Primary Sources

ARCHIVES AND LIBRARIES

Israel/Palestine

Central Zionist Archives, Jerusalem
Ein Kerem Medical Library, Hadassah Hospital, Jerusalem
Hebrew University Library at Mt. Scopus, Jerusalem
Histadrut Archives (Arkhiyon ha'avoda vehehalutz, makhon lavon leheker tnu'at ha'avoda), Tel Aviv
Islamic Center, Abu Dis
Israel State Archives, Jerusalem
Jabotinsky Archives, Tel Aviv
Keren Keyemet L'Israel Photo Archives, Jerusalem
National Library of Israel at Givat Ram, Jerusalem
Orient House Library, Jerusalem
Rav Kook Archives, Jerusalem
Tel Aviv Municipal Archives, Tel Aviv
Truman Institute for Peace Library, Jerusalem
Yad Ben Tzvi Archives and Library, Jerusalem

England

Public Records Office, Kew Gardens, London
Rhodes House, Oxford
St. Antony's Archives, Oxford

United States

College of Physicians of Philadelphia Library, Philadelphia, PA
Hadassah Archives, New York, NY

Hebrew Union College Library, New York, NY
Joint Distribution Archives, New York, NY
New York Academy of Medicine Library, New York, NY
New York Public Library, New York, NY
Rockefeller Foundation Archives, Tarrytown, NY
U.S. National Archives, Washington, DC

INTERVIEWS AND PERSONAL CORRESPONDENCE

Dr. Aftim Acra, September 26, 1998
Dr. Eli Davis, December 18, 1996
Muhammed Karakara, June 1996
Dr. Amin El-Khatib, July 18, 1996
Dr. Amin Majaj, June 18, 1996
Mr. Said Rabi, July 1, 1996
Dr. Srouj, September 26, 1998
Dr. John Tleel, DDS., June 7, 1996
Dr. Fritz Yekutiel, January 14, 1997 and February 4, 1997

PERIODICALS

Acta Medica Orientalia: The Palestine and Near East Medical Journal
Briut
Briut haʿAm
Briut haʿAmal
Briut haʿOved
Briut haTzibur
al-Difaa
Davar
Doar HaYom
Filastin
HaAretz
Hadassah Newsletter
HaGeh
HaRefuah
al-Kulliyeh
Luah haEm ve haYeled
al-Majala al-Tabiya al-ʿArabiya al-Filastiniya
Medical Leaves
Mishmar
New Palestine
Palestine
Palestine Post
Pioneers and Helpers
Shaʿarei Briut
Zichronot haDvarim

ARTICLES, BOOKS, AND PAMPHLETS

Aref al-Aref. *Bedouin Love, Law and Legend: Dealing Exclusively with the Bedu of Beersheba.* Jerusalem: Cosmos Publishing Co., 1944.

Bar-Adon, Dorothy. *A Trip through the Upper Galilee.* Tel Aviv: Lion the Printer, 1941–1942.

Berberian, D. A. "The Species of Anopheline Mosquitoes found in Syria and Lebanon: Their Habits, Distribution and Eradication." *Al-Majala al-Tabiya al-'Arabiya al-Filastiniya* [Journal of the Palestine Arab Medical Association] 1, no. 5 (July 1946): 120–146.

Berman, Dr. Tova. *Eich tishmore 'al briutcha beEretz Israel* [How to protect your health in Palestine]. Ed. Dr. Tova Berman. Tel Aviv: Kupat Holim, 1935.

Bihem, Arieh. *Hamilchama, habriut vehamachalot hamidbakot beEretz Israel* [War, health and infectious diseases in Palestine]. Jerusalem: Shirizli Press, 1914–1915.

Brachyahu, Dr. M. *Haderekh el habriut: hora'ot hahygiena bekitot hayesod shel beit sefer ha'amami* [The way to health: Hygiene instructions for elementary school classes of national schools]. Tel Aviv: Yavneh Press, 1947.

Canaan, Tawfiq. "Demons as an Aetiological Factor in Popular Medicine. Part 1." *Al-Kulliyeh* 3, no. 4 (February 1912): 150–154.

———. "Demons as an Aetiological Factor in Popular Medicine. Part 2." *Al-Kulliyeh* 3, no. 5 (March 1912): 183–189.

———. *Haunted Springs and Water Demons in Palestine.* Studies in Palestinian Customs and Folklore. Jerusalem: Palestine Oriental Society, 1922. Reprinted from *Journal of Palestine Oriental Society* 1, no. 1 (October 1920): 153–170.

———. *Mohammedan Saints and Sanctuaries in Palestine.* Jerusalem: Ariel Publishing House, 1927.

———. "Zionist Ambitions and the Palestine Crisis." *Palestine and Transjordan* 1, no. 3 (1936): 6–7.

———. *Qadia 'Arab Filastin* [The problem of the Arabs of Palestine]. Jerusalem, 1936.

ESCO Foundation for Palestine. *Palestine: A Study of Jewish, Arab and British Policies.* Volume 1. New Haven: Yale University Press, 1947.

Ettinger, Jacob. *Jewish Nuris: How the Keren haYesod Is Populating Historic Rural Districts in Palestine.* London, 1925.

Granovsky, Abraham. *Land Policy in Palestine.* Westport: Hyperion Press, 1937.

Hadassah Medical Organization. *Hishtadel l'hiot naki, bari, vechazak: hora'ot beshmirat habriut* [Try to be clean, healthy and strong: Instruction on protecting health]. Pamphlet no. 1. Jerusalem: Rafael Chaim Hacohen, 1921.

———. *Hamalaria: hitpashtuta, mecholeleya, darkei ha'avarata veemtzaei hahishtamrut mipneiya* [Malaria: Its spreading, its causative agent, means of infection and protective measures]. Pamphlet no. 2. Jerusalem, 1921.

———. *Sicha 'al hamalaria* [Conversation on malaria]. Pamphlet no. 3. Jerusalem, 1921.

———. *Hazvuvim* [Flies]. Pamphlet no. 4. Jerusalem, 1921.

———. *Sipur 'al HaGiveret Anopheles* [Story of Miss Anopheles]. Pamphlet no. 5. Jerusalem: Rafael Chaim Hacohen, 1922.

Hebrew Aid Society of American Zionists for Eretz Israel. *HaMalaria: Hitpashtuta, meholeleyha, darechai ha'avarata ve'Emtzai hahishramrut mepnaiha* [Malaria: Its spread, its generation, paths of transmission and the means of protecting against it]. Jerusalem, 1921.

Herzl, Theodor. *Altneuland: Old New Land.* Trans. Lotta Levensohn. New York: Markus Wiener Publishing, 1987.

Jewish Agency for Palestine. *The Area of Cultivable Land in Palestine.* Jerusalem, 1936.

———. *Jews and Arabs in Palestine: Some Current Questions.* Swann and Ibbott, 1936.

———. *Statistical Bases of Sir John Hope-Simpson's Report on Immigration, Land Settlement and the Development in Palestine.* London, 1931.

Kligler, Israel. *The Epidemiology and Control of Malaria in Palestine.* Chicago: University of Chicago Press, 1930.

———. "Malaria Control Demonstrations in Palestine." *American Journal of Tropical Medicine* 4 (1924): 139–174.

———. "The Control of Malaria in Palestine by Anti-anopheline measures." *Journal of Preventive Medicine* 1 (1927): 149–183.

———. "Flight of Anopheles Mosquitoes." *Trans. Royal Society of Tropical Medicine and Hygiene* 18 (1924): 199–206.

———. "The Epidemiology of Malaria in Palestine: A Contribution to the Epidemiology of Malaria." *American Journal of Hygiene* 6 (1926): 431–449.

Kligler, Israel, and R. Reitler. "Studies on Malaria in an Uncontrolled Hyperendemic Area (Hule, Palestine). I, The Seasonal Prevalence, Distribution and Intensity of Malaria in an Untreated Indigenous Population." *Journal of Preventive Medicine* 2 (1928): 415–432.

Kligler, Israel, Joseph Shapiro, and I. Weitzman. "Malaria in Rural Settlements in Palestine." *Journal of Hygiene* 23 (1924): 280–316.

Krikorian, K. S., and N. Bedrechi. *Atlas of the Anopheles Mosquitoes of Palestine.* Jerusalem: Department of Health, Palestine, 1940.

Kupat Holim. *Briut ha'oleh: Madrich le'oleh hahadash be'inyanei hygiena vebriut* [Health of the immigrant: A guide to the new immigration in matters of hygiene and health]. Jerusalem, 1940.

League of Nations Health Organization, Malaria Commission. *Principles and Methods of Antimalarial Measures in Europe.* Geneva, 1927.

———. *Reports on the Tour of Investigation in Palestine in 1925.* Geneva, 1925.

Levy, Dr. A. E. *Briut haTzibur: Drachim ve'emtzaei shmor neged hamachalot midbakot* [Health of the public: Paths and ways of protecting against infectious diseases]. Tel Aviv: Achiever Press, 1935.

Mandelberg, Dr. "Al haRechitza beHamei Tiberias." [On Bathing in the Tiberias Hot Springs]. *Briut Ha'Oved* 2 (1924): 3–9.

Masterman, E. W. G. *Hygiene and Disease in Palestine in Modern and in Biblical Times.* London: Palestine Exploration Fund, 1918.

Mer, Gideon. "Notes on the Bionomics of Anopheles Elutus." *Bulletin of Entomological Research* 22 (1931): 137–145.

————. "The Determination of the Age of Anopheles by Differences in the Size of the Common Oviduct." *Bulletin of Entomological Research* 23 (1932): 563–566.

Necheles, H. "A Contribution to the Physiology of Anopheles." *HaRefuah* 1, nos. 4–5 (April 1925): 2–3.

Nordau, Max. *Degeneration*. 1968. Lincoln: University of Nebraska Press, 1993.

————. "Yehadut hashririm" [Muscle Judaism]. *Max Nordau el'amo: ketavim mediniim*, 171–178. Tel Aviv: Medinit Press, 1936.

Norman, Dr. Y. *Darkei habriut: hygiena yom-yom* [Ways of health: Daily hygiene]. Tel Aviv: Shalom Am Library, 1926.

Palestine Arab Medical Association. *The Hygienic and Sanitary Conditions of the Arabs of Palestine*. Jerusalem, 1946.

Puchovsky, Dr. L. *Malaria: haofi shel hamachala vehamilchala negdo* [Malaria: The nature of the disease and the war against it]. Translated by Ezrachi-Krishovsky. Jaffa, Eitan and Shoshani, 1920.

Regeneration: A Reply to Max Nordau. Westminster: Archibald Constable and Co., 1895.

Rosenbaum, Dr. S. *Shmor'al briutecha beEretz Israel* [Protect your health in Palestine]. Tel Aviv: Olympia Press, n.d.

Ruppin, Arthur. *Three Decades of Palestine*. Jerusalem: Schocken Press, 1936.

————. *Abscrift Juedischen Nationalfonds*. The Hague, 1921.

Shapira, Joseph, and Zvi Saliternik. *HaMalaria beEretz-Israel: Sefer Shimushi* [Malaria in Eretz Israel: Practical book]. Jerusalem, 1930.

Tzernichovsky, Shaul. *Shirim*. Jerusalem: Schocken Publishing, 1943.

Vaad Leumi. *Palestine's Health in Figures: Report Submitted by the General Council (Vaad Leumi) of the Jewish Community of Palestine to the United Nations Special Committee*. Jerusalem, 1947.

Wilenska, Dr. "Harechitza beHamaei-Tiberias beshnat Tar"g" [Bathing in the Tiberias Hot Springs in 1923]. *Briut Ha'Oved* 1 (1924): 1.

Yofe, Emmanuel, and Sonya Kahana-Charag. *HaDerekh el HaBriut* [The way towards Health]. Tel Aviv: Yavneh, 1947.

Yofe, Hillel. *Dor Ma'apilim* [Generation of pioneers]. Tel Aviv: Tarmil Press, 1983.

————. "The Problem of Quinine Prophylaxis." *HaRefuah* 1, nos. 4–5 (April 1925): 1–2.

Zeidan, Emile. "Rise and Fall of Nations." *Al-Kulliyeh* 3, no. 4 (February 1912): 111–113.

Zondek, H. "Contemporary Medicine and Jewish Physicians." *Medical Leaves* 3, no. 1 (1940): 7–10.

Secondary Sources

Abu-Ghazaleh, Adnan. *Palestinian Arab Cultural Nationalism*. Brattleboro: Amana Books, 1991.

Abu El-Haj, Nadia. "Producing (Arti) Facts: Archaeology and Power during the British Mandate of Palestine." *Israel Studies* 7, no. 2 (2002): 33–61.

Ackerknecht, Erwin. *Malaria in the Upper Mississippi Valley, 1760–1900*. New York: Arno Press, 1977.

Adas, Michael. *Machines as the Measure of Men: Science, Technology, and Ideologies of Western Dominance*. Ithaca: Cornell University Press, 1989.

Alatout, Samer. "From Water Abundance to Water Scarcity (1936–1959): A 'Fluid' History of Jewish Subjectivity in Historic Palestine and Israel." In *Reapproaching the Border: New Perspectives on the Study of Israel/Palestine*. Edited by Mark LeVine and Sandy Sufian Lanham: Rowman and Littlefield, Forthcoming.

———. "Towards a Bio-territorial Conception of Power: Territory, Population, and Environmental Narratives in Palestine and Israel." *Political Geography* 25, no. 6 (2006): 601–621.

Alcalay, Reuben. *The Complete Hebrew-English Dictionary*. Ramat-Gan: Massada Publishing, n.d.

Almog, Shmuel. *Leumiut, Zionut, antishemiut: masot vemechkarim* [Nationalism, Zionism and Anti-Semitism: Essays and studies]. Jerusalem: Hassifria hatziyonit, 1992.

———. *Zionism and History: The Rise of a New Jewish Consciousness*. New York: St. Martin's Press, 1987.

Amadouny, Vartan. "The Campaign against Malaria in Transjordan, 1926–1946: Epidemiology, Geography and Politics." *Journal of the History of Medicine and Allied Sciences* 52, no. 4 (October 1997): 453–484.

Anderson, Benedict. *Imagined Communities: Reflections on the Origin and Spread of Nationalism*. London: Verso Press, 1983.

Anderson, Warwick. "Disease, Race and Empire." *Bulletin of the History of Medicine* 70, no. 1 (1996): 62–67.

———. "Where is the Postcolonial History of Medicine?" *Bulletin of the History of Medicine* 72, no. 3 (Fall 1998): 522–530.

———. "How's the Empire? An Essay Review." *Journal of the History of Medicine* 58 (October 2003): 459–465.

———. *The Cultivation of Whiteness: Science, Health, and Racial Destiny in Australia*. New York: Basic Books, 2003.

Arnold, David. *Colonizing the Body: State Medicine and Epidemic Disease in Nineteenth-Century India*. Berkeley: University of California Press, 1993.

———. *Imperial Medicine and Indigenous Societies*. Edited by D. Arnold. Manchester: Manchester University Press, 1988.

Arnold, David, ed. *Warm Climates and Western Medicine: The Emergence of Tropical Medicine, 1500–1900*. Amsterdam: Rodopi Press, 1996.

Arnold, David, and Ramachandra Guha, eds. *Nature, Culture, Imperialism: Essays on the Environmental history of South Asia*. Oxford: Oxford University Press, 1995.

Asad, Talal. "Anthropological Texts and Ideological Problem: An Analysis of Cohen on Arab Villages." *Review of Middle East Studies* 1 (1975): 1–40.

Ashbel, Rivka. *As Much As We Could Do: The Contributions Made by the Hebrew University of Jerusalem and Jewish Doctors and Scientists from Palestine during and after the Second World War*. Jerusalem: Magnes Press, 1989.

El-Asmar, Rouzi. "Portrayal of the Arab in Hebrew Children's Literature." *Journal of Palestine Studies* 16, no. 1 (1987): 81–94.

Atran, Scott. "The Surrogate Colonization of Palestine, 1917–1939." *American Ethnologist*, special section, "Tensions of Empire," 16, no. 4 (November 1989): 719–744.

Ayalon, Y. "Yibush habitzot Kabbara" [The drainage of the Kabbara swamps]. *Zev Vilnay Jubilee Volume* (1987): 233–239.

Bar-Gal, Yoram. "Emek Yizra'el vebitzotav-teshuva leteguvot" [The Jezreel Valley and its swamps—Reply to responses]. *Cathedra* 3 (1983): 185–195.

Bar-Gal, Yoram, and Shmuel Shamai. "Habitzot shel Emek Yizra'el-mitos or metziut?" [The Swamps of the Jezreel Valley—Myth or reality?]. *Cathedra* 27 (1983): 163–174.

———. "Swamps of the Jezreel Valley: Struggle over the Myth." *Sociologia Internationalis* 25, no. 2 (1987): 193–206.

Bar tal, Nira. "Hahachshara heteoretit vehama'asit shel ahayot yehudiyot beEretz Israel betkufat hamandat, 1918–1948 bi-re'i hitpathuto shel beit hasefer haahayot 'al shem Henrietta Szold, Hadassah Yerushalayim" [Theoretical and practical training of Jewish nurses in Mandatory Palestine, 1918–1948, as reflected by Hadassah's Henrietta Szold School, Jerusalem]. Ph.D. diss., Hebrew University of Jerusalem, 2000.

Bashford, Alison. *Imperial Hygiene: A Critical History of Colonialism, Nationalism and Public Health*. Houndsmills: Palgrave Macmillan, 2004.

Ben-Ari, Eyal, and Bilu, Yoram, eds. *Grasping Land: Space and Place in Contemporary Israeli Discourse and Experience*. Albany: State University of New York Press, 1997.

Ben-Arieh, Yehoshua. "Perceptions and Images of the Holy Land." In *The Land That Became Israel: Studies in Historical Geography*. Edited by Ruth Kark. New Haven: Yale University Press, 1989: 37–56.

Berkowitz, Michael. *Zionist Culture and West European Jewry before the First World War*. Chapel Hill: University of North Carolina Press, 1993.

———. *Western Jewry and the Zionist Project, 1914–1933*. Cambridge: Cambridge University Press, 1997.

Biale, David. *Eros and the Jews: From Biblical Israel to Contemporary America*. New York: Basic Books, 1992.

Biale, David, ed. *Cultures of the Jews: A New History*. New York: Schocken Books, 2002.

Biger, Gideon. *An Empire in the Holyland: Historical Geography of the British Administration in Palestine, 1917–1929*. New York: St. Martin's Press, 1994.

Biger, Gideon and A. Kartin. "Habitzot shel Emek Yizra'el-mitos or metziut?" [The Swamps of the Jezreel Valley—myth or reality?]. *Cathedra* 30 (1983): 179–182.

Birger, H. "Yibush hamitos" [Draining the myth]. *Cathedra* 3 (1983): 161–175.

Birn, Ann Emmanuelle. "Eradication, Control or Neither? Hookworm vs. Malaria Strategies and Rockefeller Public Health in Mexico." *Parassitologia* 40, nos. 1–2 (June 1998): 137–147.

Bolton Valencius, Conevery. *The Health of the Country: How American Settlers Understood Themselves and Their Land*. New York: Basic Books, 2002.

Bowden, T. "The Politics of Arab Rebellion in Palestine, 1936–1939." *Middle East Studies* 11, no. 2 (May 1975): 147–170.

Boyarin, Daniel. *Unheroic Conduct: The Rise of Heterosexuality and the Invention of the Jewish Man.* Berkeley: University of California Press, 1997.

Boyarin, Jonathan. *Palestine and Jewish History: Criticism at the Borders of Ethnography.* Minneapolis: University of Minnesota Press, 1996.

———. *Storm from Paradise: The Politics of Jewish Memory.* Minneapolis: University of Minnesota Press, 1992.

Bradley, D. J. "The Particular and the General: Issues of Specificity and Verticality in the History of Malaria Control." *Parassitologia* 40, nos. 1–2 (June 1998): 5–10.

Brandt, Allen. *No Magic Bullet: A Social History of Venereal Disease in the United States since 1880.* Oxford: Oxford University Press, 1987.

Brown, Michael. *The Israeli-American Connection: Its Roots in the Yishuv, 1914–1945.* Detroit: Wayne State University Press, 1996.

Brown, Peter. "Malaria, Miseria and Underpopulation in Sardinia: The 'Malaria Blocks Development' Cultural Model." Special issue on malaria and development, *Medical Anthropology* 17 (1997): 239–254.

Bruce-Chwatt, Leonard Jan, and Julien DeZulueta. *Rise and Fall of Malaria in Europe: A Historico-epidemiological Study.* Oxford: Oxford University Press, 1980.

Bynum, W. F. "'Reasons for Contentment': Malaria in India, 1900–1920." *Parassitologia* 40, nos. 1–2 (June 1998): 19–27.

Bunton, Martin. "Land Law and the 'Development' of Palestine: The Case of the Kabbara/Caesarea Concession." Unpublished paper. 1998.

———. "Demarcating the British Colonial State: Land Settlement in the Palestinian Jiftlik Villages of Sajad and Qazaza." In *New Perspectives on Property and Land in the Middle East.* Edited by Roger Owen, 121–160. Cambridge: Harvard Middle Eastern Monographs, 2000.

Calhoun, Craig, Edward LiPuma, and Moishe Postone. *Bourdieu: Critical Perspectives.* Chicago: University of Chicago Press, 1993.

Cambournac, F. J .C. "Contribution to the History of Malaria Epidemiology and Control in Portugal and Some Other Places." *Parassitologia* 36 (1994): 215–222.

Canguilhem, Georges. *The Normal and the Pathological.* New York: Zone Books, 1991.

Caplan, Neil. *Futile Diplomacy: Early Arab-Zionist Negotiation Attempts, 1913–1931.* Vol. 1. London: Frank Cass, 1983.

Carruthers, Jane. "Nationhood and National Parks: Comparative Examples from the Post-imperial Experience." In *Ecology and Empire: Environmental History of Settler Societies.* Edited by Tom Griffiths and Libby Robin, 125–138. Seattle: University of Washington Press, 1997.

Cohen, Mitchell. *Zion and State: Nation Class and the Shaping of Modern Israel.* Oxford: Basil Blackwell Inc., 1987.

Cohn, Bernard. *Colonialism and Its Forms of Knowledge: The British in India.* Princeton: Princeton University Press, 1996.

Collins, Liat. "After the Flood." *Jerusalem Post*, July 19, 1996, 14.

Comaroff, John, and Jean Comaroff. *Of Revelation and Revolution: The Dialectics of Modernity on a South African Frontier.* Volume 2. Chicago: University of Chicago Press, 1997.

Cooper, Fred, and Ann Stoler. "Between Metropole and Colony: Rethinking a Research Agenda." In *Technologies of Empire: Colonial Cultures in a Bourgeois World.* Edited by Fred Cooper and Ann Stoler, 1–56. Berkeley: University of Berkeley Press.

Corbellini, G. "Acquired Immunity against Malaria as a Tool for the Control of the Disease: The Strategy Proposed by the Malaria Commission of the League of Nations in 1933." *Parrasitologia* 40, nos. 1–2 (June 1998): 109–115.

Cooper, Fred, and Ann Stoler. "Introduction—Tensions of Empire: Colonial Control and Visions of Rule." *American Ethnologist* 16, no. 4 (November 1989): 609–621.

———. *Tensions of Empire: Colonial Cultures in a Bourgeois World.* Berkeley: University of California Press, 1997.

Cosgrove, Denis. *Social Formation and Symbolic Landscape.* Totowa: Barnes and Noble Books, 1984.

Cronon, W. *Changes in the Land: Indians, Colonists and the Ecology of New England.* New York: Hill and Wang, 1983.

Crosby, Alfred. *Ecological Imperialism: The Biological Expansion of Europe, 900–1900.* Cambridge: Cambridge University Press, 1986.

Crossette, Barbara. "UN and World Bank Unite to Wage War on Malaria." *New York Times,* October 31, 1998, A4.

Curtin, Philip. *Death by Migration: Europe's Encounter with the Tropical World in the Nineteenth Century.* Cambridge: Cambridge University Press, 1989.

Davidovitch, Nadav, and Rhona Scidelman. "Herzl's *Altneuland*: Zionist Utopia, Medical Science and Public Health." *Korot: The Israel Journal of the History of Medicine and Science* 17 (2004): 1–20.

Davidovitch, Nadav, and Shifra Shvarts. "Health and Zionist Ideology: Medical Selection of Jewish European Immigrants to Palestine." *Facing Illness in Troubled Times: Health in Europe in the Interwar Years, 1918–1939.* Edited by Iris Borowy and Wolf D. Gruner. Frankfurt: Peter Lang Publishing, 2005.

DeBevoise, Ken. *Agents of Apocalypse: Epidemic Disease in the Colonial Philippines.* Princeton: Princeton University Press, 1995.

Deichmann, Ute, and Anthony S. Travis. "A German Influence on Science in Mandate Palestine and Israel: Chemistry and Biochemistry." *Israel Studies* 9, no. 2 (Summer 2004): 34–70.

Dimentman, Ch., H. J. Bromley, and F. D. Por. *Lake Hula: Reconstruction of the Fauna and Hydrobiology of a Lost Lake.* Jerusalem: Israel Academy of Sciences and Humanities, 1992.

Dirks, Nicholas, ed. *Colonialism and Culture.* Ann Arbor: University of Michigan Press, 1992.

Dobson, Mary. *Contours of Death and Disease in Early Modern England.* Cambridge: Cambridge University Press, 1998.

Doleve-Gandelman, Tsili. "The Symbolic Inscription of Zionist Ideology in the Space of Eretz Israel: Why the Native Israeli is Called Tsabar." In *Judaism*

Viewed from Within and Without: Anthropological Studies. Edited by Harvey E. Goldberg, 257–284. Albany: State University of New York Press, 1987.

Doumani, Bishara. *Rediscovering Palestine: Merchants and Peasants in Jabal Nablus, 1700–1900*. Berkeley: University of California Press, 1995.

Efron, John. *Defenders of Race: Jewish Doctors and Race Science in Fin-de-siecle Europe*. New Haven: Yale University Press, 1994.

Ehrenreich John, ed. *The Cultural Crisis of Modern Medicine*. New York: Monthly Review Press, 1978.

El-Eini, Roza. "British Forestry Policy in Mandate Palestine 1929–48: Aims and Realities." *Middle Eastern Studies* 35, no. 3 (July 1999): 72–155.

———. "British Agricultural-educational Institutions in Mandate Palestine and Their Impress on the Rural Landscape." *Middle Eastern Studies (England)* (January 1999): 98–114.

———. *Mandated Landscape: British Imperial Rule in Palestine, 1929–1948*. New York: Routledge Press, 2006.

———. "Trade Agreements and the Continuation of Tariff Protection Policy in Palestine in the 1930s." *Middle Eastern Studies (England)* (January 1998): 164–191.

Even-Zohar, Itamar. *'Iyunim besifrut* [Studies in literature]. Jerusalem: Misrad hahinukh vehatarbut, 1965.

Eyles, John, and K. J. Woods. *The Social Geography of Medicine and Health*. New York: St. Martin's Press, 1983.

Falah, Ghazi. "Pre-State Jewish Colonization in Northern Palestine and its Impact on Local Bedouin Sedentarization, 1914–1948." *Journal of Historical Geography* 17, no. 3 (1991): 289–309.

———. *The Role of the British Administration in the Sedentarization of the Bedouin Tribes in Northern Palestine, 1918–1948*. Durham, England: Center for Middle Eastern and Islamic Studies, University of Durham, 1983.

Falk, Raphael. "Zionism and the Biology of the Jews." *Science in Context* 11, nos. 3–4 (1998): 587–607.

Fanon, Frantz. *Black Skin, White Masks*. New York: Grove Weidenfeld, 1967.

———. *A Dying Colonialism*. New York: Grove Press, 1967.

———. *Wretched of the Earth*. New York: Grove Press, 1965.

Fenner, F. "Malaria Control in Papua New Guinea in the Second World War: From Disaster to Successful Prophylaxis and the Dawn of DDT." *Parassitologia* 40, nos. 1–2 (June 1998): 55–64.

Firestone, Yaakov. "Crop-sharing Economics in Mandatory Palestine." Parts 1 and 2. *Middle East Studies* 11, nos. 1–2 (1975): 3–23, 175–194.

———. "Land Equalization and Factor Scarcities: Holding Size and the Burden of Impositions in Imperial Central Russia and the Late Ottoman Levant." *Journal of Economic History* 41, no. 4 (1981): 813–33.

———. "Production and Trade in an Islamic Context: Sharika Contracts in the Transitional Economy of Northern Samaria, 1983–1943." *International Journal of Middle East Studies* 6, no. 2 (April 1975): 185–209.

Fleck, Ludwik. *Genesis and Development of a Scientific Fact*. Chicago: University of Chicago Press, 1979.

Forman, Geremy, and Alexandre Kedar. "Colonialism, Colonization and Land Law in Mandate Palestine: The Zor al-Zarqa and Barrat Qisarya Lad Disputes in Historical Perspective." *Theoretical Inquiries in Law* 4, no. 2 (July 2003): 491–540.

Foucault, Michel. *The Archeology of Knowledge and the Discourse on Language.* New York: Pantheon Books, 1972.

———. *Birth of a Clinic: An Archeology of Medical Perception.* New York: Vintage Press, 1975.

———. *Discipline and Punish: The Birth of the Prison.* Vintage Books: New York, 1979.

———. *History of Sexuality: An Introduction. Volume 1.* New York: Vintage Books, 1988.

Frankenstein, Dr. C. *Child Care in Israel: A Guide to the Social Services for Children and Youth.* Jerusalem: Henrietta Szold Foundation for Child and Youth Welfare, 1950.

Gallagher, Nancy. *Egypt's Other Wars: Epidemics and the Politics of Public Health.* Syracuse: Syracuse University Press, 1990.

Gelber, Sylva. *No Balm in Gilead: A Personal Retrospective of Mandate Days in Palestine.* [Ottawa]: Carleton University Press, distributed by Oxford University Press, 1989.

Gerber, Haim. "The Ottoman Administration of the Sanjaq of Jerusalem, 1890–1908." *Asian and African Studies* 12, no. 1 (March 1978): 33–76.

Gertz, Nurith. *Sifrut ve-ideologyah be-Erets-Yisrael bi-shenot ha-sheloshim* [Literature and ideology in Eretz Israel in the 1930s]. Tel Aviv: Open University, 1988.

Gesler, Wilbert M. *The Cultural Geography of Health Care.* Pittsburgh: University of Pittsburgh Press, 1991.

Ghosh, Amitav. *Calcutta Chromosome.* Toronto: A. A. Knopf, 1996.

Giacaman, Rita. *Life and Health in Three Palestinian Villages.* Community Health Unit, Bir Zeit University. London: Ithaca Press, 1988.

Giblett, Rod. *Postmodern Wetlands: Culture, History, Ecology.* Edinburgh: Edinburgh University Press, 1996.

Giladi, Dan. "The Agronomic Development of the Old Colonies in Palestine (1882–1914)." In *Studies on Palestine during the Ottoman Period.* Edited by Moshe Ma'oz, 175–189. Jerusalem: Magnes Press, 1975.

Gilman, Sander. *Difference and Pathology: Stereotypes of Sexuality, Race and Madness.* Ithaca: Cornell University Press, 1985.

———. *The Jew's Body.* New York: Routledge, 1991.

———. *Picturing Health and Illness: Images of Identity and Difference.* Baltimore: Johns Hopkins University Press, 1995.

———. "By a Nose: On the Construction of 'Foreign Bodies.'" *Social Epistemology* 13, no. 1 (1999): 49–58.

Gilman, Sander. "The Jewish Nose: Are Jews white? or, the History of the Nose Job." In *The Other in Jewish Thought and History: Constructions of Jewish Culture and Identity.* Edited by Laurence J. Silberstein and Robert L. Cohn, 364–401. New York: New York University Press, 1994.

Gilmartin, David. "Models of the Hydraulic Environment: Colonial Irrigation, State Power and Community in the Indus Basin." In *Nature, Culture, Imperialism: Essays on the Environmental History of South Asia.* Edited by David Arnold and Ramachandra Guha, 210–236. Delhi: Oxford University Press, 1995.

Gilsenan, Michael. *Lords of the Lebanese Marches: Violence and Narrative in an Arab Society.* Berkeley: University of California Press, 1996.

———. *Recognizing Islam: Religion and Society in the Modern Middle East.* London: IB Taurus & Co., 1992.

Goldstein, Yaakov. "Tochnit zioni rishona leyibush bitzot haHuleh" [The first Zionist plan for draining the Huleh swamps]. *Cathedra* 45 (1985): 161–168.

Goubert, Jean-Pierre. *The Conquest of Water: The Advent of Health in the Industrial Age.* Princeton: Princeton University Press, 1986.

Greenblatt, C. L. "Historical Trends in the Anti-Malarial Campaign in Palestine." *Israel Journal of Medical Sciences* 14, no. 56 (May 1978): 508–517.

Griffiths, Tom. "Ecology and Empire: Towards an Australian history of the World." In *Ecology and Empire: Environmental History of Settler Societies.* Edited by Tom Griffiths and Libby Robin, 1–18. Seattle: University of Washington Press, 1997.

Grose, Peter. *Israel in the Mind of America.* New York: Knopf Press, 1983.

Grove, Richard. *Green Imperialism: Colonial Expansion, Tropical Island Edens and the Origins of Environmentalism, 1600–1860.* Cambridge: Cambridge University Press, 1995.

———. *Ecology, Climate and Empire: Colonialism and Global Environmental History (1400–1940).* Cambridge: White Horse Press, 1997.

———. "Conserving Eden: The (European) East India Companies and Their Environmental Policies on St. Helena, Mauritius and in Western India, 1660–1854." *Comparative Studies in Society and History* 35, no. 2 (April 1993): 318–351.

———. "Indigenous Knowledge and the Significance of South-West India for Portuguese and Dutch Constructions of Tropical Nature." *Modern Asian Studies* 30, no. 1 (February 1996): 121–143.

———. "Climatic Fears: Colonialism and the History of Environmentalism." *Harvard International Review* 23, no. 4 (Winter 2002): 50–55.

Grushka, Theodore. *The Health Services of Israel.* Jerusalem, 1952.

HaCohen, E. "He'arot 'al yibush habitzot vehazemer ha'ivri" [Comments on draining the swamps and the Hebrew song]. *Cathedra* 30 (1983): 183–184.

Halamish, Aviva. *Mediniut ha'aliya vehaklita shel Histadrut hazionit 1931–1937* [Policy-making of immigration and absorption of the Zionist Organization, 1931–1937]. Ph.D. diss., Tel Aviv University, 1995.

Halpern, Ben, and Jehuda Reinharz. *Zionism and the Creation of a New Society.* Hanover: Brandeis University Press, 2000.

Don Handelman and Lea Shamgar-Handelman, "Presence of Absence." In *Grasping Land: Space and Place in Contemporary Israeli Discourse and Experience* Edited by Eyal Ben-Ari and Yoram Bilu, 85–128. Albany: SUNY Press, 1997.

Hardiman, David. "Power in the Forest: The Dangs, 1820–1940." In *Subaltern Studies VIII: Essays in Honour of Ranajit Guha.* Edited by David Arnold and David Hardiman, 89–147. Delhi: Oxford University Press, 1994.

Harrison, Gordon. *Mosquitoes, Malaria and Man: A History of the Hostilities since 1880*. London: John Murray Press, 1978.

Harrison, Mark. "'Hot Beds of Disease': Malaria and Civilization in Nineteenth-Century British India." *Parrasitologia* 40, nos. 1–2 (June 1998): 11–18.

———. "'The Tender Frame of Man': Disease, Climate and Racial Difference in India and the West Indies, 1760–1860." *Bulletin of the History of Medicine* 70, no. 1 (1996): 68–93.

———. "Medicine and the Culture of command: the case of Malaria Control in the British Army during the Two World Wars." *Medical History* 40 (1996): 437–452.

———. *Climates and Constitutions: Health, Race, Environment and British Imperialism in India 1600–1850*. New Delhi: Oxford University Press, 1999.

Harshav, B. *Language in time of Revolution*. Berkeley: University of California Press, 1993.

Hart, Mitchell. "Picturing Jews: Iconography and Racial Science." In *Values, Interests and Identity: Jews and Politics in a Changing World. Studies in Contemporary Jewry an Annual XI*. Edited by Peter Y. Medding, 159–175. New York: Oxford University Press, 1995.

———. *Social Science and the Politics of Modern Jewish Identity*. Stanford: Stanford University Press, 2000.

Harvey, David. *The Condition of Postmodernity*. Oxford: Oxford University Press, 1989.

Headrick, Daniel. *The Tools of Empire: Technology and European Imperialism in the Nineteenth Century*. Oxford: Oxford University Press, 1981.

Hebdige, Dick. *Subculture: The meaning of style*. London: Routledge, 1979.

Heiberg, Marianne. *Palestinian Society in Gaza, West Bank and Arab Jerusalem: A Survey of Living Conditions*. Oslo: FAFO, 1993.

Hendricks, R. *A Model for National Health Care*. New Brunswick: Rutgers University Press, 1993.

Hertzberg, Arthur. *The Zionist Idea*. New York: Atheneum Press, 1959.

Hoffman, Frederick L. *A Plea and a Plan for the Eradication of Malaria throughout the Western Hemisphere*. Newark: Prudential Press, 1917.

Inbar, Moshe. "A Geomorphic and Environmental Evaluation of the Hula Drainage Project, Israel." *Australian Geographical Studies* 40, no. 2 (July 2002): 155–166.

Institute of Palestine Studies. *Survey of Palestine: Prepared in December 1945 and January 1946 for the Information of the Anglo-American Commission of Inquiry. Volume Two*. Washington D.C.: Institute of Palestine Studies, 1991.

Islamaglu, Huri. "Property as a Contested Domain: A Reevaluation of the Ottoman Land Code of 1858." In *New Perspectives on Property and Land in the Middle East*. Edited by Roger Owen, 3–62. Cambridge: Harvard Middle Eastern Monographs, 2000.

———. *State and Peasant in the Ottoman Empire: Agrarian Power Relations and Regional Economic Development in Ottoman Anatolia during the Sixteenth Century*. Leiden: E. J. Brill, 1994.

Jackson, John Brinckerhoff. *Discovering the Vernacular Landscape*. New Haven/London: Yale University Press, 1984.

Jankowski, J. "The Palestinian Revolt of 1936–1939." *Muslim World* 63 (1973): 220–233.

Jones, Margaret. "The Ceylon Malaria Epidemic of 1934–35: A Case Study in Colonial Medicine." *Social History of Medicine* 13, no. 1 (2000): 87–109.

Jurod, Dominique. *The Imperiled Red Cross and the Palestine-Eretz Israel Conflict 1945–1952*. London: Kegan Paul International, 1996.

Kalkas, B. "The Revolt of 1936: Chronicle of Events." In *The Transformation of Palestine: Essays on the Origins and Development of the Arab-Israeli Conflict*. Edited by Ibrahim Abu-Lughod, 237–274. Evanston: Northwestern University Press, 1971.

Kalpagam, U. "Colonising Power and Colonised Bodies." *Indian Journal of Social Science* 5, no. 1 (1992): 61–79.

Kamen, Charles. *Little Common Ground: Arab Agriculture and Jewish Settlement in Palestine, 1920–1948*. Pittsburgh: University of Pittsburgh Press, 1991.

Kanaaneh, Rhoda. "Desiring Modernity: Family Planning among Palestinians in Northern Israel." Ph.D. diss., Columbia University, 1998.

Karakara, Mohammed. *Ma'arechet habriut hamandatorit vehavoluntarit ve'aravi Eretz Israel: 1918–1948* [Development of public health services to the Palestinian under the British Mandate 1918–1948 (English title given)]. Master's thesis, University of Haifa, 1992.

Kark, Ruth. *American Consuls in the Holy Land, 1832–1914*. Jerusalem: Magnes Press, 1994.

———. *Geulat hakarka beEretz Israel: re'ayon vema'ase* [Redeeming the land in Eretz Israel: idea and practice]. Jerusalem: Yad Ben Zvi Press, 1990.

———. "Land-God-Man: Concepts of Land Ownership in Traditional Cultures in Eretz-Israel." In *Ideology and Landscape in Historical Perspective: Essays on the Meanings of Some Places in the Past*. Edited by Alan Baker and Gideon Biger, 1–14. Cambridge: Cambridge University Press, 1992.

Karlinsky, Nahum. *California Dreaming: Ideology, Society, and Technology in the Citrus Industry of Palestine, 1890–1939*. Albany: State University of New York Press, 2005.

Karmon, Yehuda. *Emek haHuleh hatzfoni* [The northern Huleh valley]. Jerusalem, 1956.

Kaufman, Menachem. *An Ambiguous Partnership: Non-Zionists and Zionists in America, 1939–1948*. Jerusalem: Magnes Press, 1991.

Kearns, Robin, and Wilbert Gesler, eds. *Putting Health in Place: Landscape, Identity and Well-being*. Syracuse: Syracuse University Press, 1998.

Khalaf, Issa. *Politics in Palestine: Arab Factionalism and Social Disintegration, 1939–1948*. Albany: State University of New York Press, 1991.

Khalidi, Aziza. "Indicators of Social Transformation and Infant Survival: A Conceptual Framework and an Application to the Populations of Palestine, 1927–1944." Ph.D. diss., Johns Hopkins University, 1996.

Khalidi, Rashid. *Palestinian Identity: The Construction of Modern National Consciousness*. New York: Columbia University Press, 1997.

Khalidi, Walid, ed. *All That Remains: The Palestinian Villages Occupied and Depopulated by Israel in 1948*. Washington, D.C.: Institute for Palestine Studies, 1992.

Khawalde, Sliman, and Dan Rabinowitz. "Race from the Bottom of the Tribe That Never was: Segmentary Narratives amongst the Ghawarna of Galilee." *Journal of Anthropological Research* 58, no. 2 (Summer 2002): 225–243.

Klein, Ira. "Development and Death: Reinterpreting Malaria, Economics and Ecology in British India." *Indian Economic and Social History Review* 38, no. 2 (2001): 147–179.

———. "Malaria, Economics and Ecology in British India." *Indian Economic and Social History Review* 38, no. 2 (2001):, 147–179.

Kimmerling, Baruch. *Zionism and Territory: The Socio-Territorial Dimensions of Zionist Politics.* Institute of International Studies. Berkeley: University of California, 1983.

Kimmerling, Baruch, and Joel Migdal. *Palestinians: The Making of a People.* New York: Free Press, 1993.

Kitron, Uriel. "Malaria, Agriculture and Development: Lessons from Past Campaigns." *International Journal of Health Sciences* 17, no. 2 (1987): 295–326.

Kohn, D., Z. Weiss, and E. Flatau. "The History of Malaria in the Jezreel Valley in the Years 1922–1928 in Ein Harod." *Korot* 8, nos. 3–4 (1982): 177–186.

Kottek, Samuel, and Manfred Wasserman, eds. *Health and Disease in the Holy Land: Studies in the History and Sociology of Medicine from Ancient Times to the Present.* Lewiston: Edwin Mellen Press, 1996.

Krieger, Nancy, and Ann-Emanuelle Birn. "A Vision of Social Justice as the Foundation of Public Health: Commemorating 150 Years of the Spirit of 1848." *American Journal of Public Health* 88, no. 11 (November 1998): 1603–1606.

Kuhkne, L. *Lives at Risk: Public Health in Nineteenth-Century Egypt.* New York: New World Press, 1970.

Kuhn, Thomas. *The Structure of Scientific Revolutions.* Third Edition. Chicago: University of Chicago Press, 1996.

Al-Labadi, Abd al-Aziz. *Al-Ihwal al-sahiya wa al- ijtima'ia lil-sha'ab al-Filastin, 1922–1972* [The health and social conditions of the Palestinian people, 1922–1972]. Amman: Dar-al Karmel, 1986.

Laquer, Walter, and Barry Rubin, eds. *Israel-Arab Reader: A Documentary History of the Middle East Conflict.* New York: Penguin Books, 1984.

Latour, Bruno. *Pasteurization of France.* Cambridge: Harvard University Press, 1988.

———. *We Have Never Been Modern.* Cambridge: Harvard University Press, 1993.

Lehn, Walter, and Uri Davis. *The Jewish National Fund.* London: Kegan Paul International, 1988.

Levin, Marlin. *It Takes a Dream: The Story of Hadassah.* Jerusalem: Gefen Press, 1997.

Levine, Mark. "The Discourse of Development in Mandate Palestine." *Arab Studies Quarterly* 17, nos. 1–2 (Winter–Spring 1995): 95–104.

Levy, Nissim. "Harofim Ha'ivriim beEretz Israel beme'oraot tarps"t" [Hebrew doctors in Palestine during the 1929 Riots]. *HaRefuah* 117, nos. 3–4 (August 1989): 92–95.

———. "Zichron Yaakov-mirkaz harefuah harishon betzafon Eretz Israel" [Zichon Yaakov: The First Medical Center in the North of Palestine]. *HaRefuah* 108, no. 2 (January 1985): 90–99.

Livneh, Micha. "Yibush haHuleh" [Draining the Huleh]. *Ecologia vesviva* 1, no. 4. (August 1994): 211–219.

Lockman, Zachary. *Comrades and Enemies: Arab and Jewish Workers in Palestine 1906–1948*. Berkeley: University of California Press, 1996.

Loustaunau, Martha, and Elisa Sobo. *The Cultural Context of Health, Illness and Medicine*. Westport: Bergin and Garvey, 1997.

Lowenthal, David. "Empires and Ecologies: Reflections on Environmental History." In *Ecology and Empire: Environmental History of Settler Societies*. Edited by Tom Griffiths and Libby Robin, 229–236. Seattle: University of Washington Press, 1997.

Lupton, Deborah. *Medicine as Culture: Illness, Disease and the Body in Western Societies*. London: Sage Publications, 1994.

MacKenzie, John M. "Empire and the Ecological Apocalypse: The Historiography of the Imperial Environment." In *Ecology and Empire: Environmental History of Settler Societies*. Edited by Tom Griffiths and Libby Robin, 215–228. Seattle: University of Washington Press, 1997.

MacLeod, Roy. "On Visiting the 'Moving Metropolis': Reflections on the Architecture of Imperial Science." *Historical Records of Australian Science* 5 (1982): 1–16.

MacLeod, Roy and Milton Lewis, eds. *Disease, Medicine and Empire: Perspectives on Western Medicine and the Experience of European Expansion*. London: Routledge, 1988.

Maegraith, Brian. "Tropical Medicine What It Is Not, What It Is." *Bulletin of the New York Academy of Medicine* 48, no. 10 (November 1972): 1210–1230.

Marcus, A. "Israel, Conceding Environmental Goof, Plans to Turn Valley back into Swamp." *Wall Street Journal*, July 16, 1993, 6A.

Marks, Shula. "What Is Colonial about Colonial Medicine? And What Has Happened to Imperialism and Health?" *Social History of Medicine* 10 (1997): 205–219.

Mayer, Michael. *Jewish Identity in the Modern World*. Seattle: University of Washington Press, 1990.

McCarthy, Justin. *The Population of Palestine: Population History and Statistics of the Late Ottoman Period and the Mandate*. New York: Columbia University Press, 1990.

McClintok, Anne. *Imperial Leather: Race, Gender and Sexuality in the Colonial Context*. New York: Routledge, 1995.

McNeill, William H. *Plagues and Peoples*. New York: Doubleday Press, 1977.

Meade, Melinda, John Florin, and Wilbert Gesler. *Medical Geography*. New York: Guilford Press, 1988.

Meinig, D.W. Introduction. *The Interpretation of Ordinary Landscapes: Geographical Essays*. Edited by D. W. Meining, 1–7. New York: Oxford University Press, 1979.

———. "Reading the Landscape: An Appreciation of W. G. Hoskins and J. B. Jackson." In *The Interpretation of Ordinary Landscapes: Geographical Essays*. Edited by D. W. Meining, 195–244. New York: Oxford University Press, 1979.

———. "Symbolic Landscapes: Some Idealizations of American Communities." In *The Interpretation of Ordinary Landscapes: Geographical Essays*. Edited by D. W. Meining, 165–192. New York: Oxford University Press, 1979.

Metzer, Jacob. *The Divided Economy of Mandatory Palestine.* Cambridge: Cambridge University Press, 1998.

Metzer, Jacob, and Oded Kaplan. "Jointly But Severally: Arab-Jewish Dualism and Economic Growth in Mandatory Palestine." *Journal of Economic History* 215, no. 2 (June 1985): 327–345.

Miron, Dan. *Im lo tihyeh Yerushalayim: masot'al ha-sifrut ha'Ivrit beheksher tarbuti-politi* [If Jerusalem will not be: Essays on Hebrew literature in cultural and political context]. Tel Aviv: Kibbutz HaMeuhad, 1987.

Miller, Donald. "A History of Hadassah, 1912–1935." Ph.D. diss., New York University, 1968.

Mitchell, Tim. *Colonizing Egypt.* Berkeley: University of California Press, 1988.

——. *Rule of Experts: Egypt, Techno-Politics, Modernity.* Berkeley: University of California Press, 2002

Morris, Benny. *Righteous Victims: A History of the Zionist-Arab Conflict, 1881–1999.* Knopf: New York, 1999.

Mosse, George. *Confronting the Nation: Jewish and Western Nationalism.* Hanover: Brandeis University Press, 1993.

——. *Image of Man: The Creation of Modern Masculinity.* Oxford: Oxford University Press, 1996.

——. *Nationalism and Sexuality: Middle Class Morality and Sexual Norms in Modern Europe.* Madison: University of Wisconsin Press, 1985.

Moulin, Anne. "Tropical without the Tropics: The Turning point of Pasteurian Medicine in North Africa." In *Warm Climates and Western Medicine: the Emergence of Tropical Medicine: 1500–1900.* Edited by David Arnold, 160–180. Amsterdam: Rodopi Press, 1996.

Muslih, Mohammed. *The Origins of Palestinian Nationalism.* New York: Columbia University Press, 1988.

Naff, Thomas, and Ruth Matson. *Water in the Middle East: Conflict or Cooperation?* MERI Special Studies, no. 2. Boulder: Westview Press, 1984.

Niederland, Doron. "Hashpa'a harefuim—ha'olim meGermania 'al hitpatchut harefuah beEretz Israel (1933–1948)" [The influence of the German immigrant doctors: On the development of medicine in Eretz Israel]. *Cathedra* 30 (1983): 111–160.

Nimrod, Yoram. "'Ma'asecha yekarvum': yachasei yehudim-'aravim bepeulo hatziburi shel Zvi Botkovsky" ['Your actions will bring them close': A new approach to Jewish/Arab relations; The economic initiative of Zvi Botkovsky]. *Tzion* 57, no. 4 (1992): 429–450.

Nye, Edwin, and Mary Gibson. *Ronald Ross: Malariologist and Polymath; A Biography.* New York: St. Martin's Press, 1997.

O'Connor, F. W. "Review: Epidemiology and Control of Malaria in Palestine by Israel J. Kligler." *Jewish Quarterly Review* n.s., 21, no. 3 (January 1931): 333–337.

Ohry, Abraham. "Haybetim refuim sheyekumiim betipul benechim betkufat hamandat beEretz Israel ubemilchemet hashichrur" [Medical aspects that were established for the treatment of the disabled/wounded during the Mandate period in Palestine and during the War of Independence]. *HaRefuah* 116, no. 10 (May 1989): 549–551.

Ortal, Reuven, and Shoshana Gabbay. "Conservation of Wetlands in Israel: Israel National Report on the Implementation of the Ramsar Convention, February 1999." Http://www.sviva.gov.il/Enviroment/Static/Binaries/index_pirsumim/p0379_1.pdf.

Ortner, Sherry. "Theory in Anthropology in the Sixties." *Comparative Studies in Society and History* 26 (1984): 144–160.

Osbourne, Michael. "Acclimatizing the World: A History of the Paradigmatic Colonial Science." *Osiris* 15 (2001): 135–151.

———. "Introduction: The Social History of Science, Technoscience and Imperialism." *Science, Technology and Society* 4, no. 2 (1999): 161–169.

Owen, Roger, ed. *Studies in the Economic and Social History of Palestine in the Nineteenth and Twentieth Centuries.* Carbondale: Southern Illinois University Press, 1982.

Owen, Roger. Introduction. *New Perspectives on Property and Land in the Middle East.* Edited by Roger Owen, ix–xxiv. Cambridge: Harvard Middle Eastern Monographs, 2000.

Packard, Randall. *White Plague, Black Labor: Tuberculosis and the Political Economy of Health and Disease in South Africa.* Berkeley: University of California Press, 1989.

———. "The Myth of the Malaria-Tolerant Native: Medical Knowledge and Agricultural Development in South Africa in the 1920s and 1930s." Paper presented at UCLA Programs in Medical Classics, Los Angeles, California, 19 October 2004.

———. "Visions of Postwar Health and Development and Their Impact on Public Health Interventions in the Developing World." In *International Development and the Social Sciences: Essays on the History and Politics of Knowledge.* Edited by Fred Cooper and Randall Packard, 93–115. Berkeley: University of California Press, 1997.

Packard, Randall, and Peter Brown. "Rethinking Health, Development and Malaria: Historicizing a Cultural Model in International Health." *Medical Anthropology* (special volume dedicated to malaria and development) 7, no. 3 (1997): 181–194.

Palladino, Paolo, and Michael Worboys. "Science and Imperialism." *Isis* 84 (1993): 91–102.

Pappe, Ilan. *A History of Modern Palestine: One Land, Two Peoples.* Cambridge: Cambridge University Press, 2004.

Parker, Andrew, et al. *Nationalisms and Sexualities.* London: Routledge, 1992.

Peard, Julyan G. *Race, Place and Medicine: The Idea of the Tropics in Nineteenth-Century Brazilian Medicine.* Durham: Duke University Press, 1999.

Penslar, Derek. *Zionism and Technocracy: The Engineering of Jewish Settlement in Palestine, 1870–1918.* Bloomington: Indiana University Press, 1991.

———. "Zionism, Colonialism and Technocracy: Otto Warburg and the Commission for the Exploration of Palestine 1903–07." *Journal of Contemporary History* 25 (1990): 143–160.

Petitjean, Patrick, Catherine Jami, and Anne Marie Moulin, eds. *Science and Empires: Historical Studies about Scientific Development and European Expansion.*

Boston Studies in the Philosophy of Science. 136. Dordrecht: Kluwer Academic Publishers, 1992.

Philip, Kavita. *Civilizing Natures: Race, Resources and Modernity in Colonial South India.* Rutgers University Press: New Brunswick, 2004.

———. "Global Botanical Networks, Environmentalist Discourses, and the Political Economy of Cinchona Transplantation to British India." *Revue Francaise d'Histoire d'Outre-Mer* 86 (1999): 119–142.

———. "Imperial Science Rescues a Tree: Global Botanical Networks, Local Knowledge and the Transcontinental Transplantation of Cinchona." *Environment and History* 1, no. 2 (1995): 173–200.

Pick, Daniel. *Faces of Degeneration: A European Disorder, 1848–1918.* Cambridge: Cambridge University Press, 1989.

Porath, Yehoshua. *The Emergence of the Palestinian Arab National Movement, 1918–1929.* London: Frank Cass, 1974.

———. *The Palestinian Arab National Movement: From Riots to Rebellion. Volume Two: 1929–1939.* London: Frank Cass, 1977.

Porter, Dorothy, ed. *Social Medicine and Medical Sociology in the Twentieth Century.* Atlanta: Rodopi Press, 1997.

Porter, Roy. *The Greatest Benefit to Mankind: A Medical History of Humanity.* New York: W. W. Norton & Company, 1997.

Poser, Charles M., and George W. Bruyn. *An Illustrated History of Malaria.* New York: Parthenon Press, 1999.

Pratt, Mary Louise. *Imperial Eyes: Travel Writing and Transculturation.* London: Routledge, 1992.

Presner, Todd Samuel. "'Clear Heads, Solid Stomachs, and Hard Muscles': Max Nordau and the Aesthetics of Jewish Regeneration." *Modernism/modernity* 10, no. 2 (2003): 269–296.

Pyenson, Lewis. *Civilizing Mission: Exact Sciences and French Overseas Expansion: 1830–1940.* Baltimore: Johns Hopkins University Press, 1993.

Rabinowitz, Dan. *Overlooking Nazareth: The Ethnography of Exclusion in Galilee.* Cambridge: Cambridge University Press, 1997.

Rabinowitz, Dan, and Sliman Khawalde. "Demilitarized, then Dispossessed: The Kirad Bedouins of the Hula Valley in the Context of Syrian-Israeli Relations." *International Journal of Middle East Studies* 32, no. 4 (Nov. 2000): 511–30.

Ram, Uri. "The Colonization Perspective in Israeli Sociology: Internal and External Comparisons." *Journal of Historical Sociology* 6, no. 3 (September 1993): 326–350.

———. "A Review of 'Comrades and Enemies: Arab and Jewish Workers in Palestine 1906–1948.'" *Israel Studies Bulletin* 14, no. 1 (Fall 1998): 17–19.

Ranger, Terence, and Paul Slack. *Epidemics and Ideas: Essays on the Historical Perception of Pestilence.* Cambridge: Cambridge University Press, 1992.

Raz-Krakotzkin, Amnon. "The Zionist Return to the West and the Mizrahi Jewish Perspective." *Orientalism and the Jews.* Eds. Ivan Kalmar and Derek Penslar. Waltham and Hanover: Brandeis University Press and University Press of New England, 2005.

Reinharz, Yehuda, and Anita Shapira, eds. *Essential Papers on Zionism.* New York: New York University Press, 1996.

Reiter, Yitzhak. *Islamic Endowments in Jerusalem under British Mandate.* London: Frank Cass, 1996.

Reiss, Nira. *Health Care of the Arabs in Israel.* Boulder: Westview Press, 1991.

"Reflooding of Hula Could Restore Animal Life." *Hebrew University News* 103 (Fall–Winter 1993): 14.

Robin, Libby. "Ecology: A Science of Empire?" In *Ecology and Empire: Environmental History of Settler Societies.* Edited by Tom Griffiths and Libby Robin, 63–75. Seattle: University of Washington Press, 1997.

Roll Back Malarian Web Site. "Roll Back Malaria Partnership Consensus Statement: Ensuring Access to Effective Malaria Case Management." March 2004. Http://www.rollbackmalaria.org/partnership/wg/wg_management/docs/RBMConsensusStatement_ACT.pdf

Romanucci-Ross, Lola, Daniel Moerman, and Laurence Tancredi, eds. *The Anthropology of Medicine.* New York: Bergin and Garvey, 1991.

Rosenberg, Charles. *Care of Strangers: The Rise of America's Hospital System.* New York: Basic Books, Inc, 1987.

———. *The Cholera Years: The United States in 1832, 1849 and 1866.* Chicago: University of Chicago Press, 1987.

———. *Explaining Epidemics and Other Studies in the History of Medicine.* Cambridge: Cambridge University Press, 1992.

Rubinstein, Shimon. *At the Height of Expectation: Land Policy of the Zionist Commission in 1918.* Jerusalem: 1992.

Rudge, David. "Conservationists Protest Hula Valley Tourism Plans." *Jerusalem Post*, November 10, 1996, 12.

Rupin, Tzipi. "Ha'avoda hachinuchit refuiut shel Hadassah beEretz Israel betkufat hamandat habriti" [The health education work of Hadassah in Eretz Israel during the British Mandate period]. Ph.D. diss., Ben Gurion University, 1999.

Rupke, Nicolaas A., ed. *Medical Geography in Historical Perspective.* London: Welcome Trust Centre for the History of Medicine at UCL, 2000.

Said, Edward. *Orientalism.* New York: Vintage Books, 1979.

Saliternik, Zvi. "Bitzot vekadachat beEmek Yizra'el" [Swamps and malaria in the Jezreel Valley]. *Cathedra* 32 (1984): 182–189.

———. "Hamalaria veshlavei hadbarata bearetz" [Malaria and the stages of its disinfestation in Eretz-Israel] *Briut HaTzibur* 2–3 (May 1964): 356–365.

———. "Reminiscences of the History of Malaria Eradication in Palestine and Israel." *Israel Journal of Medical Sciences* 14, no. 56 (May 1978): 518–520.

———. *A Guide to the Collection on the History of Malaria Control and Eradication in Palestine and in Israel.* Jerusalem: Israel Institute of the History of Medicine, 1980.

———. *Hamalaria vehadbarata beIsrael.* Jerusalem: Rafael Chaim HaCohen Press, n.d. [1971?].

———. *Korot Hamilchama bekadachat beEretz Irsael vehadbarata (kovetz prakim betoldot beYishuv].* Jerusalem: Israel Institute for the History of Medicine, 1979.

———. *Hakadachat vehadbarata beIsrael.* Jerusalem: Society for the Protection of Nature, 1984.

Schaebler, Birgit. "Practicing Musha': Common Lands and the Common Good in Southern Syria under the Ottomans and the French." In *New Perspectives on Property and Land in the Middle East.* Edited by Roger Owen, 241–312. Cambridge: Harvard Middle Eastern Monographs, 2000.

Schmelz, Usiel O. "Some Demographic Peculiarities of the Jews of Jerusalem in the Nineteeth Century." In *Studies on Palestine during the Ottoman Period.* Edited by Moshe Ma'oz, 119–141. Jerusalem: Magnes Press, 1975.

Scoones, Ian. "The Dynamics of Soil Fertility Change: Historical Perspectives on Environmental Transformation from Zimbabwe." *Geographical Journal* 163, no. 2 (July 1997): 161–169.

Shafir, Gershon. *Land, Labor and the Origins of the Israeli-Palestinian Conflict, 1882–1914.* Cambridge: Cambridge University Press, 1989.

Shalev, Meir. "Drying the Hula: The Dying Lake." In *Those Were the Years...* Edited by Nissim Mishal, 37. Tel Aviv: Miskal, 1998.

———. *Roman Rusi* [Russian novel]. Tel Aviv: Am Oved, 1989.

Shapira, Anita. *Land and Power: The Zionist Resort to Force, 1881–1948.* Oxford: Oxford University Press, 1992.

Shapiro, Haim. "Reclaimed Hula Now Being Recreated." *Jerusalem Post*, October 20, 1995, 11.

Shilony, Zvi. *Ideology and Settlement: The Jewish National Fund, 1897–1914.* Jerusalem: Magnes Press, 1998.

Shimoni, Gideon. *The Zionist Ideology.* Hanover: Brandeis University Press, 1995.

Shvarts, Shifra. *Kupat Holim Clalit: itzuva vehitpatchuta kegorem hamirkazi beshirutei habriut be Eretz Israel, 1911–1937* [Kupat Holim of the Histadrut: Its formation and development as the central actor in health services in Palestine, 1911–1937]. Beersheva: Ben Gurion University Press, 1997.

Shvarts, Shifra, and Ted Brown, "Kupat Holim, Dr. Isaac Max Rubinow and the American Zionist Medical Unit's Experiment to Establish Health Care Services in Palestine, 1918–1923." *Bulletin of History of Medicine* 72, no. 1 (Spring 1998): 28–46.

Shvarts, Shifra, Nadav Davidovitch, Rhona Seidelman, and Avishay Goldberg. "Medical Selection and the Debate over Mass Immigration in the New State of Israel (1948–1951)." *Canadian Bulletin of Medical History* 22, no. 1 (2005).

Silva, K. T. "'Public Health' for Whose Benefit? Multiple Discourses on Malaria in Sri Lanka." *Medical Anthropology* (special volume on malaria and development) 17, no. 3 (1997): 195–214.

Singer, Merrill, et al. "Why Does Juan Garcia Have a Drinking Problem? The Perspective of Critical Medical Anthropology." *Medical Anthropology* 14, no. 1 : 77–108.

Smith, Barbara. *The Roots of Separatism in Palestine: British Economic Policy 1920–1929.* Syracuse: Syracuse University Press, 1993.

Smith, C. Gordon. "The Geography and Natural Resources of Palestine as Seen by British Writers in the Nineteenth and Early Twentieth Century." In *Studies on Palestine during the Ottoman Period.* Edited by Moshe Ma'oz, 87–102. Jerusalem: Magnes Press, 1975.

Soder, Hans Peter. "Disease and Health as Contexts of Modernity: Max Nordau as a Critic of Fin-de-Siecle." *German Studies Review* 14, no. 3 (1991): 473–487.

Sonbol, A. *The Creation of a Medical Profession in Egypt, 1800–1922.* Syracuse: Syracuse University Press, 1991.

Sontag, Susan. *Illness as Metaphor and Aids and Its Metaphors.* New York: Doubleday Books, 1989.

Sopher, David E. "The Landscape of Home: Myth, Experience, Social Meaning."
In *The Interpretation of Ordinary Landscapes": Geographical Essays.* Edited
by D. W. Meinig, 129–149. Oxford: Oxford University Press, 1979.

Stein, Kenneth. *The Land Question in Palestine, 1917–1939.* Chapel Hill: University of North Carolina Press, 1984.

Stepan, Nancy. *The Idea of Race in Science: Great Britain, 1800–1960.* London:
MacMillan Press, 1982.

———. *Picturing Tropical Nature.* Ithaca: Cornell University Press, 2001.

Sternhell, Zeev. *The Founding Myths of Israel: Nationalism, Socialism and the
Making of the Jewish State.* Princeton: Princeton University Press, 1998.

Stockler, Rebecca. "Development of Public Health Nursing Practice as Related
to the Health Needs of the Jewish Population in Palestine, 1913–1948." Ph.D.
diss., Teachers College, Columbia University, 1975.

Stoler, Anne. *Race and the Education of Desire: Foucault's History of Sexuality
and the Colonial Order of Things.* Durham: Duke University Press, 1995.

Sufian, Sandy. "Anatomy of the 1936–1939 Revolt: Images of the Body in Political
Cartoons of Mandatory Palestine." *Journal of Palestine Studies.* Forthcoming.

———. "Colonial Malariology, Medical Borders and the Sharing of Scientific
Knowledge in Mandatory Palestine." *Science in Context* (Fall 2006): 381–400.

———. "Re-imagining Palestine: Scientific Knowledge and Malaria Control in
Mandatory Palestine." *Dynamis* (special volume: health in Palestine and the
Middle Eastern context, edited by Iris Borowy and Nadav Davidovitch) 25
(2005): 351–382.

———. "Defining National Medical Borders: Medical Terminology and the Making of Hebrew Medicine." In *Reapproaching the Border: New Perspectives on
the Study of Israel/Palestine.* Edited by Mark LeVine and Sandy Sufian. Lanham: Rowman and Littlefield Publishers, Inc., 2007.

Sufian, Sandy, and Shifra Shvarts. "Mission of Mercy: American Jewish Medical
Relief to Palestine during World War I." In *Proceedings of European Associ-
ation for Jewish Studies Conference, Toledo 1998: Jewish Studies at the Turn
of the Twentieth Century. Volume II: Judaism from the Renaissance to Modern
Times.* Edited by Judit Targarona Borras and Angel Saenz-Badillos, 389–398.
New York: E. J. Brill, 1999.

Swearingen, Will. *Moroccan Mirages: Agrarian Dreams and Deceptions, 1912–
1986.* Princeton: Princeton University Press, 1987.

Swedenburg, Ted. *Memories of Revolt: The 1936–1939 Rebellion and the Palestinian
National Past.* Minneapolis: University of Minnesota Press, 1995.

Swedenburg, Ted. "The Role of the Palestinian Peasantry in the Great Revolt,
1936–1939." In *Islam, Politics and Social Movements.* Edited by Edmund Burke
and Ira Lapidus, 169–203. Berkeley: University of California Press, 1988.

Tal, Alon. *Pollution in a Promised Land: An Environmental History of Israel.*
Berkeley: University of California Press, 2002.

Taubes, Gary. "A Mosquito Bites Back." *New York Times Magazine,* October 28,
1997, 40–46.

Tessler, Mark. *A History of the Israeli-Palestinian Conflict*. Bloomington: Indiana University Press, 1994.

Teveth, Shabtai. *Ben-Gurion and the Palestinian Arabs*. Oxford: Oxford University Press, 1985.

Tomes, Nancy. *Gospel of Germs: Men, Women and the Microbe in American Life*. Cambridge: Harvard University Press, 1998.

Troen, S. Ilan. *Imagining Zion: Dreams, Designs and Realities in a Century of Jewish Settlement*. New Haven: Yale University Press, 2003.

Tuan, Yi-Fu. "Thought and Landscape: The Eye and the Mind's Eye." In *The Interpretation of Ordinary Landscapes: Geographical Essays*. Edited by D. W. Meining, 89–102. New York: Oxford University Press, 1979.

Turner, Bryan. *Medical Power and Social Knowledge*. London: Sage Publications, 1987.

Turner, Craig. "UN Launches Joint Campaign to Fight Spread of Malaria." *Los Angeles Times*, October 31, 1998, A3.

Tyler, Warwick. "The Huleh Lands Issue in Mandatory Palestine, 1920–1934." *Middle East Studies* 27, no. 3 (July 1991): 343–373.

———. *State Lands and Rural Development in Mandatory Palestine*. Brighton: Sussex Academic Press, 2001.

Vaughn, Megan. *Curing Their Ills: Colonial Power and African Illness*. Stanford: Stanford University Press, 1991.

Watts, Sheldon. *Epidemics and History: Disease, Power and Imperialism*. New Haven: Yale University Press, 1997.

Weindling, Paul. "Eugenics and the Welfare State during the Weimar Republic." In *State, Social Policy and Social Change in Germany 1880–1994*. Edited by Robert Lee and Eve Rosenhaft, 134–163. Washington, D.C.: Berg, 1997.

Weindling, Paul. *Health, Race and German Politics between National Unification and Nazism, 1870–1945*. Cambridge: Cambridge University Press, 1989.

———. Introduction. *International Health Organizations and Movements, 1918–1939*. Edited by Paul Weindling. Cambridge: Cambridge University Press, 1995: 1–16.

———. "Social Medicine at the League of Nations." In *International Health Organizations and Movements, 1918–1939*. Edited by Paul Weindling. Cambridge: Cambridge University Press, 1995.

Weiss, Meira. *The Chosen Body: The Politics of the Body in Israeli Society*. Stanford: Stanford University Press, 2002.

Wenyon, C. M., A. G. Anderson, K. McLay, T. S. Hele, and J. Waterston. "Malaria in Macedonia, 1915–1919." *Journal of the Royal Army Medical Corps* 37, no. 2 (August 1921): 6–365.

Whitcombe, Elizabeth. "The Environmental Costs of Irrigation in British India: Waterlogging, Salinity and Malaria." In *Nature, Culture, Imperialism: Essays on the Environmental History of South Asia*. Edited by David Arnold and Ramachandra Guha, 237–259. Delhi: Oxford University Press, 1995.

Williams, Michael. "Ecology, Imperialism and Deforestation." In *Ecology and Empire: Environmental History of Settler Societies*. Edited by Tom Griffiths and Libby Robin, 169–184. Seattle: University of Washington Press, 1997.

Winter, S., and N. Levy, "Medicine in Palestine following the Flight of Jewish Phy-
    sicians from Nazi Germany." *Adler Museum Bulletin* 12, no. 19 (1986): 19–23.
Worboys, Michael. "Manson, Ross and Colonial Medical Policy: Tropical Medi-
    cine in London and Liverpool, 1899–1914." In *Disease, Medicine and Empire:
    Perspectives on Western Medicine and the Experience of European Expansion.*
    Edited by Roy MacLeod and Milton Lewis, 21–37. London: Routledge, 1988.
Yekutiel, Fritz. "Infective Diseases in Israel: Changing Patterns over Thirty
    Years." *Israel Journal of Medical Science* 15, no. 12 (December 1979): 976–982.
———. "Masterman, Muhlens and Malaria: Jerusalem 1913." Unpublished paper,
    1997.
Yona, Rachel. "Martivim et haHuleh" [Flooding the Huleh]. *Davar*, May 20, 1994,
    12.
Young, Allan. "Mode of Production of Medical Knowledge" *Medical Anthropol-
    ogy* 2, no. 2 (Spring 1978): 97–124.
Zentner, M. "Habitzot beEmek Yizra'el—metziut or mitos?" [The swamp in the
    Jezreel Valley—Myth or reality?] *Cathedra* 30 (1983): 176–178.
Zerubavel, Yael. *Recovered Roots: Collective Memory and the Making of Israeli
    National Tradition.* Chicago: University of Chicago Press, 1995.
———. "The Forest as a National Icon: Literature, Politics and the Archeology of
    Memory." *Israel Studies* 1, no. 1 (Spring 1996): 60–98.
———. "Revisiting the Pioneer Past: Continuity and Change in Hebrew Settle-
    ment Narratives." *Hebrew Studies* 41 (2000): 109–224.
Zureik, Elia. *The Palestinians in Israel: A Study in Internal Colonialism.* London:
    Routledge, 1979.

Index